Professional Development for Primary Teachers in Science and Technology

INTERNATIONAL TECHNOLOGY EDUCATION STUDIES

Volume 9

Scope
Technology Education has gone through a lot of changes in the past decades. It has developed from a craft oriented school subject to a learning area in which the meaning of technology as an important part of our contemporary culture is explored, both by the learning of theoretical concepts and through practical activities. This development has been accompanied by educational research. The output of research studies is published mostly as articles in scholarly Technology Education and Science Education journals. There is a need, however, for more than that. The field still lacks an international book series that is entirely dedicated to Technology Education. *The International Technology Education Studies* aim at providing the opportunity to publish more extensive texts than in journal articles, or to publish coherent collections of articles/chapters that focus on a certain theme. In this book series monographs and edited volumes will be published. The books will be peer reviewed in order to assure the quality of the texts.

Professional Development for Primary Teachers in Science and Technology

The Dutch VTB-Pro Project in an International Perspective

Edited by

Marc J. de Vries
Delft University of Technology, The Netherlands

Hanno van Keulen
Utrecht University & Fontys University of Applied Science, The Netherlands

Sylvia Peters
Kennisnet, public educational organization on ICT in Education, Research Department, The Netherlands

Juliette Walma van der Molen
University of Twente, The Netherlands

SENSE PUBLISHERS
ROTTERDAM/BOSTON/TAIPEI

A C.I.P. record for this book is available from the Library of Congress.

ISBN: 978-94-6091-710-3 (paperback)
ISBN: 978-94-6091-712-7 (hardback)
ISBN: 978-94-6091-713-4 (e-book)

Published by: Sense Publishers,
P.O. Box 21858,
3001 AW Rotterdam,
The Netherlands
www.sensepublishers.com

Printed on acid-free paper

TABLE OF CONTENTS

MICHEL ROCARD[i]

FOREWORD

In this beginning of the XXIst century, we seem to have many reasons to worry about the future: climate heating, financial disequilibrium, nuclear proliferation among others. . .

Let me call the reader's attention on another one, less visible and more forgotten, but more capable to produce some optimism when looked at, and treated: it can be remedied.

I am thinking here to this long term, slow and regular tendency which conducts in most developed countries boys and even more girls at school to choose scientific studies in a declining percentage. This evolution is threatening, for the future, for our competition position in front of emerging countries and especially China. The answer is to be found in deep changes and improvements in the way mathematics and sciences are taught.

In the Report I presented to the European Commission in 2007, Science Education Now: a Renewed Pedagogy for the Future of Europe, we stressed the fact that teachers are key players to renew science education, especially in primary schools. I was pleased to see the large impact of this Report, not only in the funding decisions taken by the Commission, but primarily in the numerous initiatives, experiments and creative projects which began to flourish in Europe. Besides, we observe an increasing involvement of the scientific community, which follows the pioneering path traced by Georges Charpak.

Among these many successful initiatives, I have been pleased to discover the VTB-Pro three-years project carried out in the Netherlands (Broadening technological education in primary school). Focusing on professional development of teachers and presenting first hand testimonies and research, the present book demonstrates how to deal with this issue, so critical for a renewed pedagogy. With proper methods, the knowledge of science, the interest in science and technology, the pedagogical skills can all be improved among teachers who often have no or little affection for science.

I congratulate the authors of this book and I hope that the new European strategy Europe 2020 will keep supporting such creative ventures, which are so important for our common future.

NOTES

[i] Michel Rocard is a former French Prime Minister and was also a member of the European Parliament. He chaired the High Level Group on Science Education that produced the report Science Education Now: a Renewed Pedagogy for the Future of Europe (European Commission, 2007).

MARC J. DE VRIES, HANNO VAN KEULEN, SYLVIA PETERS, AND JULIETTE H. WALMA VAN DER MOLEN

PREFACE

This book is the outcome of a major project on science and technology in primary education in the Netherlands that ran from May 2007 till December 2010. The project aimed at providing professional development to Dutch primary teachers in order to enable them to implement new activities in their curriculum that focus on science and technology. The name of the project was: VTB-Pro. VTB stands for Verbreding Techniek Basisonderwijs - Professionalisering, in English: Broadening Technological Education in Primary Education – Professional Development. A substantial part of this project was dedicated to educational research. This book contains a selection of research studies that have been conducted in the context of this project. As the themes that are dealt with in the research go beyond the specific situation in the Netherlands, this book is truly a publication that is of interest for an international readership. To emphasize this, we have asked two experts of international reputation to write the first chapter. Wynne Harlen (University of Bristol) and Pierre Léna (Université Paris Didérot) were prepared to do this.

The VTB-Pro project was related to the VTB project in which the introduction of science and technology activities in primary education was the main goal. This project was at school level. But it is well known that in order to make this introduction a success, teachers need to be well prepared for it. This is by no means obvious when it comes to science and technology in primary education. The large majority of primary teachers have no affection for science and technology. To the contrary, they often became primary teachers in the expectation that they would not need to be involved in that. Often, the mere thought of having to teach science and technology makes them feel quite uncomfortable. That is why the VTB-Pro project was initiated: as a response to this problem. The purpose of the VTB-Pro project was to create favourable conditions for primary science and technology education by helping primary teachers to acquire the necessary knowledge, skills, attitudes and pedagogy for teaching science and technology. The professional development activities and the research in this project were developed and conducted by consortiums of primary teacher training institutes and universities. These were organised in what was called 'Knowledge Centres', of which there were five in the country. A Project Management group was responsible for the organisation of the project; a Programme Council was installed to guard the scientific quality of the project. Two external assessment organisations were hired to monitor the project.

The VTB-Pro project was guided by a theoretical framework that described what primary teachers needed to know and be able to in order to implement science and technology in their classroom practice (Walma van der Molen, de Lange, & Kok, 2007). Three main elements were identified:

1. Science- and technology-related knowledge and skills
2. Favourable attitudes towards science and technology, and
3. Pedagogical skills for inquiry-based learning and learning-by-design.

These three elements formed the basis of the professional development activities, but also of the research part in the project. This is reflected in the structure of this book. Part I is about the first strand in the VTB-Pro research programme: knowledge and skills. In this part there are both studies on what knowledge and skills are desirable from a social and educational point of view and studies on what primary teachers already know and are able to. Part II is on attitudes. The studies in this chapter range from instrument development to identifying the actual attitude teachers have. Part III deals with concept learning and language development as the two main domains that have been studied in the context of the Pedagogical Content Knowledge for primary science and technology education. Studies in this part investigate to what extent the professional development activities have resulted in teachers acquiring this type of expertise. Part IV focuses on the nature of the professional development activities themselves: what makes such activities successful? Each of these Parts is further introduced in a separate chapter, one for each Part.

We want to thank all authors for their cooperation in this effort. In particular we want to thank Wynne Harlen and Pierre Léna for their input. We still have very positive remembrances of the meeting of us as the editors of this book with the two of you in Pierre's institute in Paris. We also want to thank Sense's Peter de Liefde for offering us the opportunity to present the outcomes of the VTB-Pro project in the International Technology Education Studies series. We hope this volume will prove to be a worthy addition to this successful series.

WYNNE HARLEN AND PIERRE LÉNA

1. INTRODUCTION TO THE THEME

INTRODUCTION

There is concern in many countries of the world and throughout Europe that there is a decline in young people's interest in studying science, technology and mathematics (European Commission 2007, OECD 2008a). The blame for this state of affairs is laid mainly at the door of teaching methods which have presented science as being a matter of facts and theories that seem to have little relevance to students' everyday lives. It is not surprising that alternative areas of study, in the arts and humanities for instance, appear more attractive to students living in a fast-moving, media-rich world.

However, the impact on the supply of future scientists and technologists is not the main issue; rather it is the importance of ensuring that all children derive from their education an understanding of scientific ideas, of how science works and of science as part of culture. In other words, what is required is a science education that engages and informs everyone; not just future scientists and technologists.

In today's world, science – understanding phenomena in nature – and technology – using this knowledge to design and make products for the use of mankind – overlap to such a degree that it seems difficult for education to consider only the former, despite differences in their epistemological status and possibly in their learning process. Both are therefore considered throughout this book, although this introductory chapter has a stronger focus on science.

Providing science education for all students is a considerable challenge; it forces attention to a range of questions about the content, the pedagogy, the role of assessment, how teachers are prepared for teaching science and where science education should start. These are questions with which we are concerned, but since the focus is on primary education it is important to have sound reasons for beginning science education at the very start of schooling.

WHY START SCIENCE IN THE PRIMARY SCHOOL?

In the 1960s, projects developing science in the primary school were begun in the United States (ESS, SCIS, SAPA) and in the United Kingdom (Nuffield Junior Science, Science 5/13) and spread to other countries, notably Africa (SEPA). The reasons given for these developments half a century ago are still relevant today. In

Marc J. de Vries, Hanno van Keulen, Sylvia Peters, and Juliette Walma van der Molen (Eds.),
Professional development for primary teachers in science and technology.
The Dutch VTB-Pro project in an international perspective.

addition to the need for more scientists, the main reasons were the need for everyone to:

- be able to relate to the rapid changes that science and technology make to the world around
- share in understanding and celebrating science as an important human achievement.
- know how to approach problems by seeking relevant information and basing decisions on evidence.

Since that time, further strong reasons have been added to the case through increasing attention to research into children's learning. Research into the ideas of young children, inspired by the work of Jean Piaget, was carried out in France by Guesne (1973) and Tiberghien and Delacôte (1978). Similar work by Driver (1973, 1983) and others, conducted with secondary school students, revealed that older students' ideas about scientific aspects of the natural world were at odds with the scientific view that secondary teachers assumed them to have.

From the early research, in which students were interviewed individually by strangers, there developed methods which enabled more systematic investigation and the use of quantitative as well as qualitative methods of analysis. Larger scale studies into primary children's ideas began with studies in New Zealand (Osborne and Freyberg 1985). In the UK the Science Processes and Concept Exploration (SPACE) project revealed a range of ideas about the scientific aspects of their surroundings that children had developed from their limited experience and ways of thinking (SPACE Research Reports 1990 – 1998). It was clear that these ideas could not be ignored; children believed them, had worked them out for themselves, and indeed it was clear that these pre-existing ideas had to be the starting point from which more scientific ideas could be developed.

Thus a further argument was added to the case for science in the primary school, that children's ideas about the natural world are developing throughout the primary years whether or not they are taught science. Without intervention to introduce a scientific approach in their exploration, many of the ideas they develop are non-scientific and if they persist may obstruct later learning.

Since then, other research has added to the importance of starting science early:

- Attitudes towards science develop in the pre-secondary years, earlier than attitudes to some other school subjects. This was first reported by Ormerod and Duckworth in 1975, but more recently research evidence reported by The Royal Society (2006, 2010) and by the French Académie des sciences (Charpak *et al* 2005) shows that most children develop interests and attitudes towards science well before the age of 14 and many before the age of 11.
- Studies made on renowned scientists or engineers have shown that their deep interest for science arose as early as age 6 or 7, and was often encouraged by parents or teachers (Guichard 2007).
- Gender differences in academic performance, which continue to be of concern in science education at higher levels, have not appeared at the primary stage (Haworth *et al* 2007, Royal Society 2010).

- At the primary level there is no correlation between attitudes to science and science achievement, so primary children can feel positive about science regardless of their level of achievement (Royal Society 2010).

Benefits of an early start to science education

How early could early be? Observations made on infants (Gopnik *et al* 1999, Dehaene-Lambertz *et al* 2008) indicate that making sense of their experience of the environment characterises the development of cognition from a very early age. With an appropriate pedagogy, science education can begin as early as at the age of 5 in preschool, as numerous experiments have demonstrated (Duschl *et al* 2007, Fleer 2007).

There are benefits of starting science in the primary school for the children themselves and for society. For children it helps them to understand aspects of the world around them, both satisfying and stimulating their curiosity. By learning to investigate and inquire they realise that they have the capability of answering some of their questions. Challenging each other, and being challenged, to say 'how do you know that...', as children frequently do, they begin to recognise the importance of having evidence to support claims. The benefits to society follow from young people developing understanding of key ideas that enable them to make informed choices both as children and later in life about, for instance, their diet, exercise, use of energy and care of the environment. Equally the development of scientific skills and attitudes supports a growing appreciation of the role of science in daily life.

The research clearly indicates that unless children's intuitive non-scientific ideas are addressed through appropriate primary science education, these benefits are less likely to be realised.

THE GOALS OF PRIMARY SCIENCE

Deciding what is 'appropriate primary science education' should start from considering the understanding, skills and attitudes we want primary school children to develop. Taking into account how children learn then leads to identifying the kinds of experiences that are likely to help that learning at various stages. To achieve the understandings that justify beginning science in the primary school it is apparent that we must be concerned not only to help them to gain knowledge of how things in the natural world behave, but also how this knowledge is achieved. This means developing both ideas *of* science and *about* science.

Science here means the natural sciences, and does not include mathematics. Although modern science would not exist without mathematics, it is fortunate that learning science at the primary level can be pursued without the need of formulae, using a great many qualitative observations with only a modest use of quantitative data. This does not mean that many of the inquiry principles developed in this book could not equally inspire mathematics education (Artigue 2010), nor that science and mathematics should be taught entirely independently.

Ideas of science

Considering first the ideas *of* science, primary school experiences have to serve both to enable children to understand and to enjoy finding out about the natural world around them today and to begin their understanding of the broad generalisations that will serve them in their later life; it has to benefit both the present and the future. We find it useful to describe the relationship between the ideas young children develop from exploration and observation of their immediate surroundings and the more abstract generalisations that enable understanding of a wide range of phenomena in terms of 'small' and 'big' ideas. For example small ideas are those that children might form through exploration of living and non-living things about the essential characteristics of organisms. These ideas form a basis for later understanding of how the functions of organisms can be explained in terms of their cellular composition. Similarly, finding that pushing and pulling things can make them move is a small idea that makes a contribution to a more general understanding of the relationship between movement of objects and the forces acting on them.

As children grow and expand their experiences they should be helped to link related small ideas together, gradually forming bigger ideas, constituting a progression in learning from the particular to the more general and abstract. This is not to limit the inquiries of young children to certain phenomena and events that lead to big ideas but rather to highlight the importance of constructing the broader understandings needed for scientific literacy by linking together smaller ideas. In other words, a big idea is not a collection of small ones but is built from them.

Ideas about science

Inclusion of ideas *about* science among the goals recognises that children encounter many facts, ideas and claims that purport to be scientifically based. It is important for them to develop the ability to evaluate the quality of this information, for otherwise they are powerless to resist claims based on false evidence or no evidence at all. Such evaluation requires an understanding of the ways of collecting, analysing and interpreting data to provide evidence and of the role of evidence in arriving at scientific explanations.

As with ideas of science, there are big and small ideas about science. For example, a big idea would be that 'science is a search for explanations which fit the evidence available at a particular time but may be changed if convincing conflicting evidence is found'. This level of abstraction is beyond primary children but, in trying to explain an observation, they can take a step towards this idea through becoming aware of the difference between, on the one hand, a guess at what causes a certain effect and, on the other, proposing a cause that is supported by evidence. In practice the best way to come to understand how science works is by participation, by children undertaking scientific inquiries of different kinds in which they have to decide what observations or measurements are needed to answer a question, they collect and use relevant data, they discuss possible explanations and then reflect critically on the processes they have carried out. In this way they develop understanding of the role of these skills in proposing explanations of events and phenomena.

Skills

Developing ideas about science requires knowledge of the skills involved in science inquiry. A further goal of science education is to complement this knowledge with the ability to use the skills in conducting investigations. This requires the ability to

- raise questions that can be answered by investigation
- develop hypotheses about how events and relationships can be explained
- make predictions based on the hypotheses
- use observation and measurement to gather data
- interpret data and draw valid conclusions from evidence
- communicate, report and reflect on procedures and conclusions.

Providing the subject matter is familiar, there is a discernible progression in the development of the skills. For example, children are likely to begin to 'interpret data and draw valid conclusions from evidence' by simply comparing what they find with what they expected or predicted. This matures into drawing conclusions consistent with all the evidence available and eventually to recognising that any conclusions are tentative and might be change by new evidence.

The last of the skills listed above connects with children's development of language, a corner stone of primary education underpinning learning across the curriculum. In the case of science, language is central to the development of understanding and at the same time its development benefits from the interaction with things and people that is the core of scientific activity. Naming objects enables them to be described and discussed in their absence; grouping and classifying according to criteria leads to development of concepts; expressing cause and effect requires the careful use of connecting words and of tenses. Some words used in science ('energy', 'work', 'animal' for instance) have particular meaning in science, more precise than their use in everyday speech. It is not possible to prevent the everyday usage, but children need to know when such words are being used strictly with their scientific meaning.

Chapters in Part 1 of this book take up the discussion of the ideas and abilities required for scientific literacy. But it would not be appropriate to leave the subject of goals without including reference development of attitudes.

Attitudes

Attitudes are generally taken to be 'potentially important determinants of behaviour, describing the state of being prepared or disposed to act in a certain way in relation to particular objects' (Royal Society 2010). It is useful to distinguish between attitudes that apply within scientific activity (scientific attitudes) and those that apply in relation to taking part in or having an affective response towards scientific activity (attitudes towards science). Attitudes of the former kind include openmindedness in collecting and interpreting data, being prepared to change or modify ideas in light of new evidence, and behaving responsibly in conducting investigations. Claims about attitudes of the second kind need to be treated with caution since they are generally derived from self-report of liking for the subject or for specific activities, rather than from observations of behaviour during scientific activities. Moreover, there is evidence that an affective response is not so much associated with the whole subject as with specific topics or activities, mediated by the self-concept as someone who is good or not good at science (Russell et al 1988, Martin 2010).

Attitudes of both kinds, towards the subject and within the subject, are not developed in the same way as scientific ideas and skills. They exist in the way people behave and are communicated largely through behaviour; they are 'caught' rather than 'taught', with implications for teachers to which we return later. Neither is there quite the same type of progression in developing attitude as in the case of ideas and skills. Indicative behaviours of attitudes are accumulated and depend more on experiences that foster them than on age or stage.

LEARNING EXPERIENCES IN PRIMARY SCIENCE

In order to identify the most fruitful learning experiences that will enable the goals of primary science to be achieved we must first consider what is known about how children learn. Information from research into children's learning leads to the following conclusions:

- Children are forming ideas about the world around them from birth and will use their own ideas in making sense of new events and phenomena they encounter
- Real understanding, rather than being received from others, is created by children in interaction with adults or other children
- Some of children's ideas are in conflict with the scientific views of things
- Language, particularly discussion and interaction with others, has an important role to play in forming children's ideas
- Direct physical action on objects is important for infants' learning, gradually giving way to reasoning which, at the primary level, is about real events and objects rather than abstractions. These experiences of real objects slowly lead to the construction of abstract notions, such as *velocity* or *energy*.

Neurosciences and learning

These conclusions from studies at the macroscopic level of behaviour are now confirmed by evidence at the microscopic level of brain activity. For example, studies of the activity in different parts of the brain when someone is engaged in various types of thought and action show that memory of events is aided when the original events are accompanied by talk, especially by conversation with adults who elaborate and evaluate the experience. Also, making notes or drawings helps in the solution of problems, particularly when attempting new types of problems involving several steps. It appears that such external representations can help offload some of the heavy demands upon working memory (Howard-Jones *et al* 2007).

Words are important because they represent objects or events, whilst being separate from the objects or events. This detachment of the symbol, the word, from what it represents enables the mental manipulation of experience, which is then no longer dependent on direct action. This representation is also essential to the development of metacognition – thinking about thinking – which is necessary for the development of control over mental processes, feelings and behaviour. According to Goswami and Bryant (2007) children can begin to gain awareness of their thinking and control behaviour in the later primary years. This enables them to improve their learning and memory by 'adopting effective cognitive strategies and by being aware of when they don't understand something' (p.14). The emergence of metacognition in the late primary years and continued increase in adolescence and adult life is consistent with findings about the nature and timing of the development of the brain (OECD 2007: 198).

Learning is influenced not only by the parts of the brain associated with cognition, but is also dependent on inner structures concerned with emotions. Excitement or anxiety causes stress and the release of chemicals in the brain which in turn release energy. Whilst excessive stress is damaging to cognitive functioning, as illustrated by recent research in China (Wei Yu 2010), at a level that it enables energy to be directed effectively in trying to understand new experiences or develop new skills, moderate stress may be positive, motivate learning and lead to the pleasure that comes from achieving a goal (Zull 2004). This supports the approach of finding what ideas and skills children have and using this information to give just the right amount of challenge for them to make progress. Engagement in learning depends on giving attention to certain stimuli, which depends on the brain's assessment of their importance to self identify, as made evident in attitudes to the subject.

A key message for science education from neurosciences is that the development of science concepts depends on the simultaneous activity in the visual, spatial, memory, deductive and kinaesthetic regions of the brain and in both hemispheres (Goswami and Bryant 2007). This indicates the need for a variety of different kinds of experience involving both physical and mental activity. It will involve being able to touch and manipulate objects, using language, linking to previous experience, reasoning, reflecting and interacting with others.

Implications for children's experiences

These considerations of learning lead to the conclusion that, in order to achieve the goals of primary science education, children should have experiences that:

- are a source of enjoyment and wonder, but at the same time enable them to develop their understanding of key ideas in science;
- concern real things in their experience that are seen by the children as relevant and appealing;

- build on their previous experience and pre-existing ideas, providing challenges within the reach of children so that they experience pleasure in learning;
- engage the emotions by making learning science exciting.

Over a period of time the activities should provide opportunities for

- developing skills of questioning, observing, measuring, hypothesising, predicting, planning controlled investigations, interpreting data, drawing conclusions, reporting findings, reflecting self-critically on procedures;
- talking to others, parents and the teachers about their ideas and activities;
- working collaboratively with others, considering others' ideas and sharing their own;
- expressing themselves, both orally and in writing, progressively using appropriate scientific terms and representations;
- applying their learning in real-life contexts.

Diversity among children

In view of globalisation and increased migration across cultural and national boundaries (OECD 2008b) it is important to ask: how universal are the requirements of children for learning science? In many developed or emerging countries school populations are highly heterogeneous, having different mother tongues, social groups, and economic, cultural and religious backgrounds. Such diversity broadens the category of children with special needs, a category often limited to the handicapped (OECD 2005). However, there are two factors which all children, no matter how diverse they are, have in common. First is their immediate perception of natural phenomena – the Sun and stars, water and air, falling stones, plants and animals. The laws governing these phenomena demonstrate the universality of science. Second is the universality of curiosity among children which, although less scientifically established, is empirically observed in classrooms throughout the world. Science education can build on these two factors to engage all children in a process of developing their scientific understanding, as has indeed been shown to be the case even for the most handicapped children (Centre Jean Lagarde 2006).

It is worth pointing out here that research in primary science education have shown little overall difference between boys' and girls' attitudes and performance, although there are differences in particular aspects, such as girls being better at writing plans for investigations and boys more willing to undertake practical investigations. (Russell *et al* 1988, Royal Society 2010). Obviously puberty will later introduce deeper differences, hence the importance of building up on a common ground before the age of 12-13.

It hardly needs to be pointed out that providing experiences listed above, however diverse the children are, is a task of considerable proportions; one which many primary teachers feel ill-prepared by their knowledge and skills to undertake effectively. The role of the teacher in this provision is so central that it deserves a detailed analysis, to which we turn in the next section.

A RENEWED PEDAGOGY

How will teachers choose activities of appropriate content and difficulty? Could it be by following a programme of activities that have been worked out by others and usually shown to 'work' for most children? Children are not all the same and following a pre-packaged course slavishly will inevitably mean that for some the experiences may be too distant to be understood or too familiar to challenge existing ideas.

However, finding activities to adopt or adapt is not the main problem in improving primary science education, particularly now that potentially useful activities are available in many countries for access from the internet (e.g. La main à la pâte 2010). Rather, what is required in many cases is a radical change in pedagogy. As the EC report cited at the start of this chapter suggests, what is needed is a greater use of inquiry-based methods. Although widely advocated across the world and in some cases for many years (NSF 1997, 1999; Michaels *et al* 2008; Harlen and Allende 2006, 2009) the spread is slow.

Ensuring that children have the kinds of opportunities we have argued are needed for real understanding requires a broad interpretation of inquiry-based science education. It is certainly more than children using skills for exploring and finding out; it is also more than providing first-hand experiences of materials and phenomena – even though these are important. It involves taking account of children's pre-existing ideas and promoting progression by adjusting challenge to match these starting ideas. In other words it shares elements of constructivist pedagogy and of formative assessment.

A broad meaning of inquiry-based science education

Inquiry-based teaching, as we would like it to be interpreted, shares with constructivism the importance of starting from children's ideas and sees the role of the teacher as providing children with the experiences, evidence and reasoning skills that will enable them to construct *scientific* ideas. Various strategies are available to teachers in this endeavour, such as extending experience, helping children to test ideas, linking ideas from one experience to a related one and involving children in seeking a range of ideas from various sources, importantly including discussion, dialogue and argumentation.

In recent thinking about learning, sharing and discussing ideas has been emphasised. There has been a perceptible shift away from the view that ideas are formed by individuals in isolation – that is, 'individual constructivism' – towards 'socio-cultural constructivism', which recognises the impact of others' ideas on the way learners make sense of things (Bransford *et al* 1999). This means a greater emphasis perhaps than before on communication through language, on the influences of cultural factors and on linking into a 'community of learners'.

Inquiry-based pedagogy shares with formative assessment the aim of developing understanding through learners taking charge of their learning. Formative assessment is a continuing cyclic process in which information about children's ideas and skills informs ongoing teaching and helps learners' active engagement in learning. It involves teachers gathering information about where children are in relation to the goals of their activities as part of teaching. This information is used in the identification of appropriate next steps and decisions about how to take them. It helps to ensure that there is progression by regulating teaching and learning to ensure the right degree of challenge to optimise progress.

Formative assessment, or assessment for learning, has been widely advocated in many countries since the review of research by Black and Wiliam (1998) showed that its use can raise levels of achievement. Among the key features is the active involvement of children in their own learning, which requires teachers to communicate to children the goals of their activities and the quality criteria to be applied so that children themselves can assess where they are in relation to the goals and, with their teacher, decide their next steps in learning. Teachers also provide feedback to children, not in the form of a judgment of how good their work is but of suggestions for how to improve it and to go further.

Although not all learning in science involves inquiry – there are some things, such as conventions, names and the basic skills of using equipment, that are more efficiently learned by direct instruction, as and when they are needed – it is important to ensure that inquiry is used where the aim is for real understanding that builds big ideas. On the other hand, teachers need to beware of pseudo-inquiry, where there is plenty of practical activity – observing, measuring and recording – but a lack of involvement of the children in making sense of phenomena or events in the natural world. This may be because the teacher is doing the interpretation for the children. It may also be that the content of the activities does not lead to the development of scientific models or explanations, a not uncommon occurrence at the primary level. Teachers who are unsure of their scientific understanding, tend to keep to rather trivial content, regarding science inquiry as a set of skills, rather than an opportunity to gather and interpret data and to reach conclusions based on available evidence (Harlen and Holroyd 1997).

In summary, interpreting inquiry-based teaching in this way means that teachers activities include:

- enabling children to reveal to themselves, their peers and the teacher, their pre-existing ideas and skills relevant to studying the phenomena or events involved in particular activities;
- probing children's ideas and skills by questioning, observing, and listening during the course of activities;
- communicating to children the purpose of their activities and how they can judge progress;
- ensuring children's access to a range of sources of information and ideas relating to their science activities;
- fostering written and oral expression in clear and correct language, while respecting free expression of children;
- providing feedback to children that reflects and communicates the criteria of good work, and helps them to see how to improve or move on;

- modelling scientific attitudes such as respect for evidence, openmindedness and care for living things and the environment;
- encouraging through appropriate questioning the use of inquiry skills in testing ideas;
- engaging children regularly in group and whole-class discussions where scientific ideas and ideas about science are shared and critically reviewed;
- using information about on-going progress to adjust the pace and challenge of activities;
- providing opportunities for children to reflect on their learning processes and outcome;
- identifying progress towards both short and longer-term goals of learning.

If we recall that pedagogy, in its broadest sense, means not only the act of teaching but also the theories, values and justifications that underpin it and the skills and creativity needed to provide effective learning activities and to engage children in them, then the size of the task of encouraging change becomes clear. This is the role of teacher education and continued professional development, to which we now turn.

TEACHER PROFESSIONAL DEVELOPMENT

Implementing inquiry-based pedagogy in primary schools in depth and on a large scale certainly is a challenging and lengthy process. But the effort required is justified by the need to prepare the new generations for the 21st century. There are several facets to the implementation: institutional, financial, cultural. The central one, however, is the professional development of teachers (European Commission 2007, OECD 2008a). Taking into account the variety of recruiting levels, practices, salaries and social status of teachers, size of classes – all factors which vary widely from place to place among countries and even in a given country – nevertheless some essentially common principles and difficulties can be discerned, relating to teachers' image of science, the gradual development of expertise in teaching and the role that formative evaluation can play in implementing change.

Teachers and science

While literature, arts, history or even mathematics are often familiar to primary school teachers, either through their past studies in secondary school, reinforced by their vocational training or reading, their understanding of natural sciences may be limited to a collection of facts about the phenomena of nature. This leads them to a view of science teaching as simply ensuring that students know and memorize these facts. In addition, reflecting the way science has often been taught to them, discoveries and explanations are presented without their historical perspective to reflect the flavour of science as a human adventure.

Today, the media present the great achievements of modern science using images of complex instruments and abstract concepts (black holes, genetic transmission). This can easily convey the impression of science as something quite outside the phenomena experienced in daily life (a shadow, a cloud, boiling water) and that study of these phenomena – precisely the ones available in a primary classroom – no longer qualifies as science. In addition, breaking science into specialized sub-disciplines (physics, chemistry, biology…) does not help understanding of the deep unity of science as a *process* of developing knowledge, aiming at unveiling the truth, without ever fully reaching it.

It is essential that professional development aims at progressively modifying teacher's vision of science (Murphy *et al.* 2007). Professional scientists have a unique role there, and may greatly contribute to this change, telling stories, sharing their own practices, coaching teachers. As Lord Kelvin (1824-1907) said: *Blow a soap bubble and observe it for a whole life, you will discover the whole physics in it.* Using the environment, connecting science to other subjects of knowledge (especially history), making activities relevant to children's lives and, whenever appropriate, blending scientific knowledge with indigenous knowledge, are ways of breaking the vision of science as an ivory tower out of reach for anyone but specialists.

Teachers and inquiry in the classroom: from novice to expert

Prior to being an active pedagogy, inquiry is a mental attitude which has to be developed. Pilot projects of the implementation of inquiry-based science programmes in various countries show that it may well take five years of practice for a *novice* teacher to become first *competent*, then eventually *expert* in inquiry-based pedagogy (Bransford *et al.* 1999), assuming the help of professional development. Once expert, the teacher can contribute to the dissemination of inquiry practice among colleagues. This stage can only be reached with a full and active participation of the teachers in learning communities (Peer Learning Activity 2006, Sarmant *et al* 2010) involving:

- first-hand experience of inquiry in observing daily-life phenomena, making experiments, proposing hypotheses, writing conclusions in relation to investigations similar to those undertaken by children; then conducting them with children and reflecting on their own practices. In this process, contact with reality should be always preferred to ICT simulations;
- developing their own pedagogical resources, including material, events or phenomena for investigation;
- learning to accept awkward or unexpected questions from students, recognising some of these questions make deep sense, and knowing how to deal with them;
- understanding the subtleties of the science learning process in the primary school, where the role of the teacher is to be a guide for stimulating and satisfying curiosity rather than a reservoir of knowledge; being prepared to say *I do not know!*
- meeting required curriculum demands – usually mandatory and not necessarily built around inquiry – through lessons organised for inquiry-based learning.

This progressive acquisition of expertise requires great patience from the teachers: therefore it is not surprising that, in all pilot projects, coaching by scientists and teachers trainers has proven necessary for the process to continue and succeed. It is crucial to break down the isolation of the single teacher and tackle lack of confidence. This will be achieved through various means: collective work of teachers, interactive resources through Internet, community involvement (e.g. the EU funded *Pollen* project, Pollen 2009) and scientists' support. Experience shows that distance training with ICT tools, if efficient in terms of cost and coverage, cannot fully replace human contact organized at local or national levels.

Formative evaluation of teachers' practice

Inquiry-based pedagogy in science departs so deeply from traditional pedagogy that teachers need measurement tools to appreciate their own progress (James and Pedder 2006: 30). Tools can be developed to collect data by: observing teachers' classrooms practice in a systematic, comparative way; observing student's activities in detail; studying of students' notebooks and teachers' plans, analysing the science content of activities; noting how material is used in a classroom sequence, etc. These data enable the teachers and others to recognise progress in inquiry-based practice, identifying strong and weak points to inform further action (Saltiel and Duclaux 2010).

A WORLDWIDE MOVEMENT IN RENEWED SCIENCE EDUCATION`

Stressing the value of the principles involved in inquiry pedagogy for science education is certainly not a new story. In the Western culture, Socrates, Comenius, Michael Faraday, Maria Montessori, Henri Bergson, Célestin Freinet, Jean Perrin, Frank Oppenheimer and many others have for centuries and decades shown the value of an active, questioning pedagogy in place of learning facts by heart. What is really new within the last decade is, on one hand, a better scientific understanding of the process of cognitive learning as discussed earlier and, on the other hand, the global concern that school systems, especially the science education they provide, appear unsuitable for the challenges of the times, for developed countries as well as for emerging ones (Berthélémy 2007, OECD 2008c). There is a remarkable consensus concerning the value of an inquiry-based pedagogy, as we define it here in depth. It has inspired a great number of pilot projects across the world, irrespective of the state of the development of a country or the resources of its education system (OECD 2008a). Certainly, it is particularly noteworthy and interesting as a key to the future that the concern for science education is not limited to scientists but seems to be shared by politicians and economists.

What is striking is the unforeseen and decisive role that the scientific community is playing in proposing, implementing and supporting these projects. It is certainly not the simplicity of the scientific concepts taught in primary school which deserves such attention, nor is it only concern about the lack of interest of the younger generation in scientific careers in developed countries. Rather it is the clear perception of the part that a renewed science education has in the development in all people of 'the capacity to use science knowledge, to identify questions and to draw evidence-based conclusions in order to understand and make decisions about the natural world and changes made to it through human activity' (OECD 2003: 33). The role of the science community is manifest in the involvement of Science Academies across the world, organized within the InterAcademy Panel (Harlen and Allende 2006, 2009, Allende 2008, Alberts 2009, Léna 2009). First-hand practitioners of science are needed to make the fundamental change that is required in the image of scientific process among teachers. At the same time the magnitude of the change from traditional teaching methods cannot be achieved by the education institutions alone and greatly benefits from international cooperation, exchange of good practice and resources, and collaboration in research.

CONCLUSION

Two common themes stand out in this worldwide movement to reform science education: the response to evidence of the effectiveness of inquiry-based learning and teaching; and the importance of beginning science education in the early years of schooling and continuing it through primary education. Whilst there are undeniable reasons for making the changes required in these themes there are equally severe problems to overcome in doing so effectively. It is not a matter of creating new curricula, textbooks or materials for children to use. Important though good classroom materials are, real change is ultimately dependent on the teacher. Massive evidence and sound arguments relating to how learning takes place, supported by findings of neuroscience, create the case that children learn best when they are actively engaged in making sense of their experiences, when they are talking about and explaining their ideas and when, particularly in their early years, they have direct physical contact with the objects they are studying. In order to provide these conditions for learning, for many teachers, radical changes are required in their view of their role – from one as controller of learning and source of information, to one that acknowledges that children do the learning with the support of the teacher – as well as in their attitudes towards science and their personal understanding of science and of scientific activity.

Radical changes of these kinds require professional development of teachers on a large scale and a permanent basis, involving inputs from many different sources. The cost of this provision is not negligible but should be seen as an indispensible investment and one which can benefit from the existing contribution of the scientific community. Science education is crucial to our society and genuinely continuing professional development is central to the necessary renewal of science education.

Wynne Harlen
University of Bristol, UK
wynne@torphin.freeserve.co.uk

Pierre Léna
Université Paris Diderot, France
pierre.lena@obspm.fr

PART I

KNOWLEDGE AND SKILLS IN SCIENCE AND TECHNOLOGY PROFESSIONAL DEVELOPMENT

HANNO VAN KEULEN

2. INTRODUCTION TO PART I

Knowledge and skills in science and technology professional development

On Friday, March 11 2011, an earthquake with magnitude 9.0 and a devastating tsunami hit Japan. Undoubtedly, the next Monday, countless children all over the world will have asked their teachers about these events. "What is a tsunami?" "Where do earthquakes come from?" "How is it possible that so many buildings still stand after such an earthquake?" Probably, the next few days, children will have asked about nuclear energy and its risks and benefits. Everyday, children come to school poised with questions about the natural and material world they live in. "Why do leaves turn red in October?" "How does bubble gum work?" Some questions may be easy to answer and depend upon common sense knowledge, but in general, this will not be the case. Do you know exactly what happens in leaves that turn red (Hanson, 2007)? How do we expect teachers to react to all these questions? Or, rather, how do we prefer them to react? What knowledge do they need for adequate reactions and how would they acquire such knowledge?

It is hard to imagine a domain that is as knowledge intensive and expanding as science and technology. Pliny's *Historia Naturalis*, which was compiled in the first century AD and can be considered to be one of the first attempts to map the area, describes some 20.000 facts. The *Encyclopaedia Britannica* from 1768 contains more than 100.000 articles, whereas nowadays Wikipedia has more than 2 million articles in the English language alone. Some areas of science produce over 40.000 academic articles each month (Börner, 2010). Everyday, new specimens are discovered in tropical forests. New sub-microscopic particles are created in cyclotrons. New materials result from research in nanotechnology. New galaxies are spotted in remote parts of the universe. And who knows what discoveries will be made when, for example, the successor of the Very Large Telescope will be in place at the European Southern Observatory at Cerro Armazones in Chile? This successor will either be the Extremely Large Telescope with a mirror of 42 meter in diameter or maybe even the Overwhelmingly Large Telescope, if engineers succeed in finding ways to construct and stabilise its proposed 100-meter diameter mirror.

The sense of humour with which these ambitious projects are baptized may be wasted on primary school teachers, however. 'Overwhelmed' indeed is an

Marc J. de Vries, Hanno van Keulen, Sylvia Peters, and Juliette Walma van der Molen (Eds.), Professional development for primary teachers in science and technology. The Dutch VTB-Pro project in an international perspective.

adequate description of their reaction to state of the art science and technology. For a large majority of teachers, their ambition is to work with children, not to be a scientist or engineer. Many of them even qualify for the stigma of being a 'non tech' (Motivation and Young Works, 2010). Their knowledge is reported to be shallow, as is their self-efficacy with respect to teaching science and technology (Murphy, 2007; Traianou, 2007). Yet, the preparation of the next generation of scientists, engineers and technicians, as is the general scientific and technological literacy of all citizens, is in their hands. And it is vital not to wait too long because children start to make career choices at an early age.

For children, an understanding of the material world already starts during infancy. Babies banging objects together develop the law-like concept that solid things cannot penetrate each other. Consequently, as toddlers, they know better than try to walk through a closed door (cf. Duschl, Schweinsgruber and Shouse, 2007; Siegal, 2008; Goswami, 2008; Wolfe, Kluender and Levi, 2009). During pre-school and kindergarten, children elaborate upon this intuitive, tacit knowledge and expand their understanding of such concepts as force, motion, equilibrium or change. Playing the seesaw can be the first step towards the lever rule; the experience of swinging helps to prepare for an understanding of pendulum motion. What children do with sand and water at the playground is not so radically different from the way geoscientists study delta formation in large estuaries or even on Mars (Kleinhans, Bierkens and Van der Perk, 2010). The rise and fall of a cake that wasn't properly battered intrigues children and food technologists alike (McGee, 2004). So many chances are waiting to be exploited by sharp and anticipating teachers!

But does this mean that secondary school and teacher training college will have to supply primary teachers with knowledge of the lever rule, Huygens' formula for determining pendulum motion, the basics of hydrodynamics, and differences in solubility of air bubbles in margarine versus butter? In it self, each piece of knowledge from the wonderful world of science and technology is worth to be known, but each fact is also in a way trivial and non-essential. There is simply too much to know to cramp it in your head pre-service.

To make things even more complicated, primary schools are preparing children for a future nobody is able to forecast with sufficient level of precision. Educators all over the world failed to anticipate the computer or the Internet. Which new technologies that will change our lives will emerge during the next decades? Will it be ways to extend and remake our bodies using biotechnology and robotics? Lightweight infrastructure using new materials and distributed intelligence? Embedded systems that can sense, understand and act upon their environment? Computer simulations of complex social problems that will assist citizens making better choices in their daily lives (IFTF, 2006; 2010)? It seems obvious that our planet faces serious sustainability problems. Dealing with our climate, improving energy efficiency, supplying water and food for all, and preparing for pandemics are a few of the challenges our children will have to take on (Van Santen, Khoe and Vermeer, 2010). We'd better prepare them at an early age, then.

In this part of the book, questions pertaining the knowledge and skills in science and technology professional development are elaborated from four different perspectives.

In chapter 3, Baartman and Gravemeijer focus on the key competencies of the future workforce in the context of social and technological changes. They make an effort to identify scientific and technological literacy for the 21st century and focus on the skills that will be in demand in the near future. They suggest to emphasize categorising, thinking in terms of variables, understanding cause-and-effect, means-end, and function-realisation relationships, visualising, schematising and modelling as core technological and scientific thinking skills.

In chapter 4, Rohaan and Van Keulen investigate the so-called Canon of the Sciences, an attempt (perhaps one in many) to pinpoint 'what everyone should know about the sciences'. The fifty items in the Canon qualify for the kind of cultural scientific literacy that enables children to participate in debate and decision-making. The chapter focuses on what primary school teachers and students and teachers in teacher training know of these topics.

In chapter 5, Van den Berg and Van Keulen investigate the possibilities for (mandatory) knowledge bases for science and technology teaching. They also point to the problem that, in order to combine the conflicting demands of being comprehensive and concise at the same time, there is a tendency to escape to lists of items that are conceptual rather than factual. Words like 'gravity' and 'sustainability' certainly indicate important areas of knowledge, but you cannot 'know' gravity like you can know Newton's formula for gravitation.

This part of the book closes with chapter 6, by Steenbeek, Van Geert and their co-authors. They shift the attention from teacher knowledge to teacher capacity to jump to the occasion: to recognise talents of children and interact productively. "Having eyes, giving eyes, receiving eyes". This may be the clue.

J. (Hanno) van Keulen
Utrecht University and Fontys University of Applied Science, The Netherlands
j.vankeulen@uu.nl

LIESBETH BAARTMAN AND KOENO GRAVEMEIJER

3. SCIENCE AND TECHNOLOGY EDUCATION FOR THE FUTURE

INTRODUCTION

Our current society is deeply influenced and shaped by artefacts, ideas and values of science and technology, for example in health care, energy, transportation and communication. Also, issues such as pollution and nuclear energy become objects of public debate. In their jobs, professionals are confronted with an increased use of information and communication technologies and the need for flexibility and life-long learning. 'Non-science-jobs', such as nursing, increasingly require an understanding of science and technology. It is thus not only important to educate people in science and technology for science-related jobs, but for work in general (Rodrigues et al., 2007). Science and technology education should enable future citizens to live and work in this society with reasonable confidence and comfort (Forman & Steen, 1994; Osborne, 2007). In the Netherlands, a start is being made with innovating science and technology education in primary school with the Dutch VTB-Pro project that aims at promoting and improving science and technology education. A sustainable innovation has to anticipate the demands of the society in which the students will come to live in. Changes in the curricula of primary education and professional development of primary teachers are long-term endeavours. Therefore, this chapter looks at the content of science and technology education from the perspective of the needs of employees of the future.

Levy and Murnane (2005) present an economist perspective on current labour market developments, which is increasingly shaped by computers and globalization. For example, computers can substitute for human workers when tasks can be expressed in series of rules. This implies that routine cognitive and manual tasks are likely to be taken over by computers, leading to the loss of this type of jobs. On the other hand, computers can complement or help professionals in other types of jobs, in which computers for example visualize complex processes by means of graphs or models (Gravemeijer, 2009). This requires an understanding of science and technology, as using a model without understanding leaves one vulnerable to mistakes. Goos and Manning (2007) confirm the Levy and Murnane study. They studied labour market developments in the UK in the last decades, and found that jobs in upper wages and the lowest wages (i.e., non-routine cognitive and non-routine manual) indeed increase, whereas jobs in the middle

Marc J. de Vries, Hanno van Keulen, Sylvia Peters, and Juliette Walma van der Molen (Eds.),
Professional development for primary teachers in science and technology.
The Dutch VTB-Pro project in an international perspective.

region (i.e., routine jobs) disappear. Related to these labour market developments, Binkley et al. (2010) pose that success in working life does not lie in content knowledge, but in the ability to communicate, share and use information to solve complex problems, and to adapt and innovate in response to new demands and changing circumstances. The fact that non-routine jobs seem to prevail also indicates that at all levels of an organization, flexibility, creativity and innovation are necessary to stand up to global competition (Forman & Steen, 1994). Telling in this respect is that in Singapore a discussion arose about the existing labour force being non-critical and obedient to their superiors, leading to a lack of creativity, risk-taking and continuous learning in organizations (Gopinathan, 1999).

Altogether, various studies have argued for the need to redefine the scientific and technological knowledge and skills taught at school. However, these studies tend to focus on very general skills such as communication and problem solving (e.g., Holbrook & Rannikmae, 2007), and do not provide any specification of what should actually be taught. Other studies provide long and detailed lists of scientific and technological content to be taught in schools (e.g., AAAS, 1993). These lists run the risk of quickly becoming outdated as the amount of new technical information is doubling every two years (Binkley et al. 2010). That is, the knowledge and skills we now teach our children will have become obsolete by the time they enter the workplace. This begs the question whether we can identify knowledge and skills that are general in the sense that we may expect them to stay valuable for a long time, but at the same time are not too general to offer directions for a science and technology curriculum. Research questions are: What knowledge and skills in the domains of science and technology do they need to adequately function in their jobs? In answering this question we hope to provide a basis for the discussion about the contents of science and technology education.

DEFINING SCIENCE AND TECHNOLOGY

There is no clear consensus even among scientists on the definition of 'science' or 'technology', or on the relationship between science and technology. In most English literature, science refers to the natural sciences, that is, chemistry, physics and biology. The OECD and PISA (2006), for example, divide the knowledge of science into the domains of physical systems, living systems, earth and space systems and technical systems, categories that go back to the Science Standards defined in the US (1996). Technology is sometimes referred to as the range of man-made materials and processes developed in society to address people's needs, including 'simple' tools like shopping bags and nail clippers (Holbrook & Rannikmae, 2007). It includes not only the artefacts themselves, but also their analysis, design, evaluation and the procedures to organise and use them (Benenson, 2001). Unlike scientific knowledge, technological knowledge comprises normative judgments: a function is well or badly fulfilled (De Vries, 2005). Barnett (1995) argues that non-expert users have a tunnel vision on technology which is restricted to its context outside the 'black box', that is, the ability to use devices. Experts also have a tunnel vision, but centred on the content or the inside of the black box. Employees and citizens of the future cannot know all that experts have ever known, but they might be expected to know something about the internal working of the black boxes they use.

Disagreement exists as to whether science and technology are two distinct domains with their own knowledge base, though most authors agree that science and technology are intrinsically related. People often do not distinguish between science and technology, as technological implications are very closely related to science (Osborne, 2007). A single problem often has both scientific and technological aspects (Benenson, 2001) and most Dutch curricula shape technology and science as two mutually constitutive practices (Van Eijck & Claxton, 2008). Also, technology-related activities contribute to learning science if they focus on the design and testing of artefacts and critical analysis and explanation (Roth, 2001). The other way around, science often depends on the appropriate technology in the instruments it uses. Van Eijck and Claxton argue that technology has its own epistemology, and is bound up with many more human practices other than science. It also needs knowledge that is much more specific and practical. Scientists are in the pursuit of truth, while industrial companies developing technologies are not (Osborne, 2007). Rocard (2007) also defines the aim of science as the modelling of the objective world, whereas technology aims to adapt reality to the needs of different people. Benenson (2001) describes a similar difference: the goal of science is to produce knowledge, whereas the goal of technology is to solve practical problems.

A similar argument can be made about the relations with mathematics. Mathematics is historically intimately tied to science and technology. On the one hand, mathematics emerges from scientific and technological activity; on the other hand, mathematics is applied in the same activity. Almost all scientific and technological models are mathematical in nature both in the way they are generated and in the way they are expressed. Even though mathematics is a discipline of its own, mathematics, science and technology are tightly intertwined in applications, which advocates for a similar integration in education. This especially comes to the fore in activities that involve measurement, geometry, statistics, formulas and graphs.

This research views science, technology and mathematics as intrinsically related domains with their own bodies of knowledge, which are context-bound and shaped in the social and collaborative construction process (Latour, 1986). As our purpose is to define what science and technology is needed in different jobs, this characterisation best suits the purpose of this research.

THINKING SKILLS FOR SCIENCE AND TECHNOLOGY

To avoid too general or too specific content specifications, we focus on what we call ‘thinking skills’. In his research, Costa (n.d.; 2000) distinguishes between content, thinking skills, cognitive operations, and habits of mind, which are represented in four concentric circles. The circles represent a hierarchical relationship in which the outer circles encompass the smaller elements in the inner circles. The centre circle represents the content or subject area and the required content knowledge: rules, skills, concepts and principles. The second circle represents the thinking skills. Adequately functioning in work and society depends upon cognitive functions such as recalling, comparing, classifying, generalising and evaluating. The third circle represents how these thinking skills are employed within larger and more complex cognitive operations such as problem solving and decision-making. Finally, habits of mind or dispositions specify how people must be willing and alert to apply their thinking skills in appropriate situations. A similar model is the IOWA Integrated Thinking Model (Burklund, Garvin, Lawrence, & Yoder, 1989), which specifies thinking skills and cognitive operations in more detail. It describes how cognitive operations like problem solving and designing require more specific thinking skills, which are further worked out in components such as analysing, evaluating, and synthesising. Also, the model specifies how cognitive strategies require content knowledge: rules, skills, concepts and principles from the subject area.

Based on these ideas, we make a distinction between content knowledge, attitudes, operational skills, cognitive skills, and complex (cognitive) operations. Our scientific and technological attitudes are comparable to what Costa calls habit of mind. The different elements are described as follows:

- *Complex cognitive operations:* like in Costa's model (2000), this refers to more encompassing cognitive operations such as problem solving and decision-making. Also, the cycle of empirical research in science and the cycle of designing, making and evaluating in technology, are seen as part of the cognitive operations. These complex cognitive operations require the use of content knowledge, attitudes, and operational and thinking skills.
- *Content knowledge:* in our case knowledge in the domain of science and technology, including physical systems, living systems, earth and space systems and technological systems. This also includes the relationships between these different systems, and between science and technology, and mathematics.
- *Attitudes:* we characterise attitudes as described in social psychology (e.g., Fazio, 2009): object-evaluation associations, which can be the result affective reactions, belief-based judgments, past experiences, etc. Baartman and De Bruijn (submitted) apply these attitude definitions to vocational competence and the integration of knowledge, skills and attitudes. This description is narrower than Costa's idea of habits of mind, which he describes as 'what human beings do when they behave intelligently' and include values, inclination, sensitivity, capability and commitment (Costa, 2000).
- *Operational skills:* Burklund et al. (1989) further work out the thinking skills as defined by Costa (2000). Comparable to Burklund et al., we make a distinction between operational skills and thinking skills. Operational skills are more basic, practical and concrete skills like the use of instruments and software.
- *Thinking skills:* as in Costa's model (2000), thinking skills are described as the cognitive functions that are necessary to carry out the broader cognitive operations. They are comparable to higher cognitive functions or processes (Krathwohl, 2002). As such, thinking skills are more abstract than content

knowledge or operational skills, but more specific than cognitive operations such as problem solving.

The research presented in this chapter focuses on the thinking skills as described above, which are more specific than the very general skills like problem solving often mentioned in previous research. On the other hand, they are less specific and different from the long lists of content knowledge and therefore do not run the risk of quickly becoming outdated. Two methods were used to distinguish thinking skills that are important for science and technology, to define them and to generate examples of these thinking skills in practice: a literature study and interviews with experts in the domains of mathematics, science and technology.

LITERATURE STUDY

Two domains of study could provide input to the question what science and technology are needed for future employees: (1) studies on 21st century skills, and (2) studies into the actual use of science and technology at the workplace.

21st century skills. The need for 21st century skills is mostly attributed to changes in society: the rapid development of technology, and changes in the labour market caused by globalisation and internationalisation. Also, some studies mention the need for individuals to flexibly adapt to a changing society and the need for sustainable democratic development (OECD, 2009). However, the need for 21st century skills is mostly uttered by private or business initiatives, while educational leaders, practitioners and the educational community do not actively participate in the debate (Voogt & Pareja Roblin, 2010). In relation to this, Lesgold (2009) observes how curriculum developments in the US are slowed down because the traditional curriculum is backed up by experiences of those who succeeded in previous generations through success in that curriculum, including teachers.

Binkley et al. (2010) analyzed various curriculum and assessment frameworks that have been developed around the world. They see creativity as the key component of 21st century skills, for example in web technologies where users produce and share content in new ways (e.g., in open source software). Ten competencies are identified and categorised:

- Ways of thinking, including (1) creativity and innovation, (2) critical thinking, problem solving and decision making, (3) learning, metacognition.
- Ways of working, including (4) communication, (5) collaboration
- Tools for working, including (6) information literacy, (7) ICT literacy
- Living in the world, including (8) citizenship, (9) life and career, and (10) personal and social responsibility.

Lesgold (2009) reports curriculum developments for the 21st century in the US. Here, five Applied Learning Standards have been developed: problem solving, communication tools and techniques, information tools and techniques, learning and self-management, and working with others. Problem solving for example includes designing new 'things' to meet identified needs making existing things work more effectively. Information tools and techniques include information gathering, the use of research techniques and information technology, for example word-processing and spreadsheet software. Lesgold also mentions modelling as an important skill: because the scale of society and information space far exceeds our ability to manage it, we need to model important processes and large amounts of information.

In their UK report about science education for the future, Millar and Osborne (1998) describe several 'ideas of science', that is, what pupils at different key stages need to know and understand. At lower key stages (primary education and the start of secondary education), this includes: the measurement of quantities, careful observation and measurement, comparisons between objects by studying patterns and regularities in events, an understanding of uncertainty, interactions between variables, and modelling. Later, pupils need to understand scientific methods, appreciate the strengths and weaknesses of scientific evidence and make a sensible assessment of risks and moral and ethical implications.

In the Netherlands, Savelsbergh (2007) wrote a report about the required curriculum changes in the domain of mathematics and the natural sciences. He also mentions the importance of modelling: the development of mathematical and scientific models and applying them in concrete situations. The steps of modelling described by Savelsbergh are very similar to the empirical cycle in scientific research. Also, content topics are distinguished in physics, chemistry, biology and mathematics in which modelling plays an important role.

A review of five reports about 21st century skills was carried out by Voogt and Pareja Roblin (2010). The reports included are: Partnership for 21st century skills, a national organisation with the sponsorship of the US government and several organisations from the private sector; EnGauge, developed by the Metiri Group and the Learning Point Associates; Assessment and Teaching of 21st Century Skills (ATCS), sponsored by Cisco, Intel and Microsoft, National Educational Technology Standards (NET5), and Technological Literacy for the 2012 National Assessment of Educational Progress (NAEP). Dede (2009) compares similar frameworks in his review of 21st century skills and takes the Partnership for 21st century skills as a starting point. The frameworks appear to strongly agree on the need for skills in the areas of communication, collaboration, ICT literacy, and social/cultural awareness. Most frameworks also regard creativity, critical thinking, problems solving and the capacity to develop relevant and high quality products as important skills. Problematic, however, is the fact that explicit references to educational levels are missing. Also, the skills are defined in very general terms. They need to be further specified in terms of knowledge, skills, attitudes, values and ethics to enable a comparison with current curricula, integration into these curricula and assessment.

Altogether, studies about 21st century skills seem to provide quite general descriptions of required skills, for example communication, collaboration, problem solving and ICT literacy. Other skills such as creativity, critical thinking and the assessment of risks and moral and ethical implications could fall in the category of what we called attitudes. Finally, some skills are more specific and seem to come closer to what we termed thinking skills: comparing between objects and studying patterns and regularities, understanding uncertainty, interactions between variables, and modelling. However, we agree with Voogt and Pareja Roblin that most skills identified in 21st century skills literature need to be further worked out to be applicable in education.

Workplace studies. The number of studies on the use of science and technology at the workplace is very limited, especially when compared to the amount of studies on the use of mathematics. Many studies have been conducted into how people use mathematics in their work (e.g., Bakker et al., 2006; Pozzi, Noss, &Hoyles, 1998; Roth, 2005). These are usually in-depth studies involving a single company and a few employees. They provide insight in how mathematics gets meaning in work contexts and how this meaning may differ from what and how is learned in school. Some of the studies described below focus on mathematics, but are useful with regard to science and technology as well because we are looking for more general thinking skills. Further research into the use of science and technology specifically seems warranted, though.

Anglo-Saxon studies (e.g., Bakker et al., 2006; Pozzi, et al., 1998; Roth, 2005) on the use of mathematics at the workplace in general show the importance of: (1) knowing what processes are 'hidden' in computers and machines to be able to react in case of break downs, and (2) analyzing relationships between variables, based on quantitative data in a specific context.

Due to technological developments, the first point becomes increasingly important, as routine work is more and more taken over by machines. Therefore, it seems less useful to teach science and mathematics, as they are built into these machines. For example, Noss et al., (2007) showed that in a factory producing transparent packaging film, the production process is only visible through the mediation of measuring instruments, process control systems and graphical and numerical representations. Bakker et al. (2010) found that in laboratory work lower-level analysts do not know what happens inside machines. This did not immediately lead to problems, but the analysts did find it important to understand what they were doing and blindly following procedures could lead to waste of time and materials. On the other hand, Wynne (1991) found that employees did not always feel the need to understand the processes inside the machines they were working with. It requires a lot of effort and motivation to translate general knowledge to one's own practical situation and employees preferred 'uncritically trusting the institution'. In the workplace, people mainly need simple mathematics to solve complex problems, whereas people practice complex mathematics on simple problems at school (Forman and Steen, 2000). This transfer problem does not seem to be specific to mathematics (e.g., Resnick, 1987). Prevailing science education also generates problems of transfer. Scientific concepts such as gravity and force are often taught as general laws or principles in school, causing problems in the application in practical problems in reality (Wynne, 1991). Workplace knowledge is very idiosyncratic and bound to specific contexts and tools (e.g., Roth, 2005; Noss et al., 2007). Because science and mathematics are not easily recognizable in workplace situations, modelling is necessary. In a work context, models have a highly situated nature. They are developed to understand practical problems in a particular situations, and modelling here is different from 'standard' modelling in which a problem is simplified, mathematised and mapped back to the real world (Bakker et al., 2006; Pozzi et al., 1998).

The second point is related to the first one. Working with machines and computers has changed many jobs. For example, bank employees have got more complex tasks since word processors and ATMs have taken over the more simple tasks of these jobs (Gravemeijer, 2010). More and more employees, also at lower levels, are faced with handling complex data and computer output. The need to engage with technical information especially increased for intermediate level employees, whose educational background is often basic and who have achieved promotion through effort and self-improvement (Bakker et al., 2009). Whereas machines do the calculations, employees need to recognize (ir)regularities, structures and patterns (Van der Kooij, 2002). Many more data are generated, which need to be handled, interpreted and analysed. In a US study, Forman and Steen (2000) found that to better serve the needs of future employees, education needs to focus on problem solving and the use of complex tools such as systems analysis and statistical quality control, besides more basic skills such as reading graphs and tables. Bakker et al. (2009) identified the following techno-mathematical literacies in a food factory: defining problems, seeing the need to quantify, identifying and measuring key variables, systematic measurement and sampling, and representing and interpreting data. Important skills are modelling, visualising, schematising: reducing reality by means of symbols, diagrams, graphics and other means (Latour, 1986; Gravemeijer, 2009).

Altogether, these studies show that employees seem to need different skills than often taught at schools. Important skills that were mentioned in studies on the use of science, technology and mathematics at the workplace are: recognizing (ir)regularities, structures and patterns, handling, interpreting and analysing data, problem solving, using complex tools such as systems analysis, reading graphs and tables, modelling, visualising, schematising, identifying and measuring key variables, systematic measurement and sampling, and representing and interpreting data. In the model presented at the end of this chapter, these skills will be further worked out and the relationships between the different skills will be explained. We will first turn to the interview study.

INTERVIEW STUDY

Next to the literature study, interviews were conducted with 10 university researchers working in the domain of science, technology and mathematics education. The researchers were selected from our own network, based on their work on the innovation of education and/or their focus on the use of science, technology and mathematics at the workplace. These experts were asked:

- Whether they think future employees need new and/or additional knowledge and skills to function in their jobs and society;
- To what extent they think employees need insight into the black boxes they work with;

- Examples of science and technology in jobs and society;
- Whether they think current curricula prepare learners for the future.

Besides this, they were shown a short preliminary list of thinking skills found in the literature, for example: recognition of problems, modelling, observing, measuring, thinking in terms of variables and method-goal relationships. They were asked to generate more thinking strategies and to specify possible relationships between thinking strategies. Also, they were asked which thinking strategies they deemed especially important.

Below, we present a summary of the answers of the experts on the needs of future employees, the required insight in black boxes and their reactions on the preliminary list of thinking skills.

Regarding the needs of future employees, the experts stated that employees need different knowledge and skills because part of their work is taken over by computer and machines. For example, many part-processes are carried out by machines, while the human worker has to oversee the bigger picture: thinking in terms of variables that stand in relation to each other, having insight in what causes what, interpreting output in terms of tables and graphs generated by computers. In society in general, driving lessons have changed because people learn how to navigate using a gps, which partly replaced the skill to read a map. Also, a critical attitude was deemed important: employees have to keep up with constant innovations, and not be afraid to act in novel situations. Finally, the relationship between technology and society has become more important: employees have to translate consumer wishes to technological specifications, taking into account ethical and environmental issues.

With regard to insight in the black boxes people work with, the experts seemed to agree on the fact that the required insight differs depending on people's working level. Employees working at lower levels of an organisation do not need to exactly know the scientific and mathematical principles machines work on. They often work in standard situations. They need to know *that* machines and computers work according to certain algorithms and these might be quite complex processes including uncertainties. Some knowledge of science and technology seems valuable, also for employees in non-science jobs. A hairdresser who knows that the paint she uses is toxic will be more inclined to wear gloves. A shop assistant in a paint shop needs to know that paint based on linseed oil – often used in old houses – cannot be repainted with water-based paint. On the other hand, these employees do not need to know the exact chemical processes involved. At higher levels, employees often work in non-standard situations in which they need more abstract knowledge and skills and specific insight into the workings of machines to be able to solve unexpected problems and to be creative. One researcher made a distinction between users and designers of computers and machines. Users only need a basic understanding of the workings of machines, for example which buttons to push and what consequences this has (causality). Designers need a deeper understanding, for example someone who has to select new machines to be bought by a factory. On the other hand, some researchers mentioned that people often feel the need to know 'what is going on inside machines'. It gives them a feeling of safety and control; it helps to learn the workings of new machines and reduces the risk of mistakes. Also, the need to be critical requires some insight into the black box: you need a sense of the problems that could occur inside machines to be able to be critical and creative.

With regard to the preliminary list of thinking skills, most experts agreed that the terms differed: some are more specific (e.g., measuring), others are more general (e.g., problem solving). This corroborates the difference between thinking skills and more general cognitive operations described above. The thinking skills need to be further worked out, as they differ for different occupations. However, the experts warned against a too detailed description, as examples can easily become guiding for curricula: only the examples are taught and the relationship between different thinking skills gets lost. A thinking skill that was deemed important by most experts is modelling, which they described as 'translating or transferring' between practical situations and science and technology. Recognizing problems in practical situations was deemed important and prerequisite for modelling. In school, the problem is mostly given, so this skill often is not trained very well. Other skills that are prerequisite for modelling are: the ability to quantify, decontextualising, and distinguishing major and minor issues. Critical thinking was deemed important as well. This is necessary for continuous innovation, to see where things can be improved and on its turn requires collecting evidence to show improvements indeed work and setting up of experiments. Finally, due to technological developments, people need insight in (automatic) feedback systems, input-process-output systems, data collection, and graphs and tables. They need to work with large amounts of data, compress these data and find and interpret trends. As a citizen, one needs insight into the means and goals of different technological artefacts to be able to decide whether they want to buy and use new artefacts.

Altogether, the expert interviews partly yielded the same thinking skills that were found in the literature study. Also, the experts added other skills. Summarizing, the thinking skills that became apparent in the interview are: thinking in terms of variables, insight in causality, interpreting tables and graphs, a critical attitude, not being afraid in new situations, taking into account ethical and environmental aspects, creativity, modelling, recognizing problems, being able to quantify, decontextualise and distinguish between major and minor issues, having insight in (automatic) feedback systems, visualising, and working with large amounts of data.

SCIENTIFIC AND TECHNOLOGICAL THINKING SKILLS: A MODEL

The literature study and the interviews with the experts generated a number of concepts that were put together in a model, distinguishing between more encompassing cognitive operations and underlying thinking skills, operational skills, attitudes, and content knowledge. This model is presented in Figure 1. As the focus of this research is on thinking skills, they are further elaborated below. Note that the thinking skills described here are not mutually exclusive.

Relationships and overlaps exist between them, which will be described when necessary. We think, however, that in an educational context it is necessary to include all of them, in order to make sure that all essential aspects of these thinking skills will be addressed.

Categorising/classifying: This relates to structuring reality, which can ultimately lead to abstraction (and for example modelling). While machines carry out the simple work, employees have to recognise (ir)regularities, structures and patterns (Van der Kooij, 2002; Levy & Murnane, 2006). To be able to do this, pupils in schools have to learn to make comparisons between object, materials and events, study different patterns and possible explanations for them (Millar & Osborne, 1998). For example, children can go on a scavenger hunt, collect examples of artefacts, classify them and sort them into categories they name themselves (Benenson, 2001)

Thinking in terms of variables: The increased use of technology has led to increasing role of quantitative information. Therefore, students need to be familiar with a quantitative approach to reality, which requires them to discern properties and to quantify them. This means, thinking in terms of variables: understanding that what is measured is not an object, but a property or attribute of an object (Gravemeijer, 2010). Children appear to find it difficult to appreciate different properties as variables (Hancock, Kaput, & Goldsmith, 1992)

Understanding relationships between variables: Quantitative data are often understood dynamically, including if-then relations and dependencies between variables (e.g., if the room temperature rises above 20 degree, the thermostat will turn off). In technology, the practical design of artefacts requires an understanding of means-end relationships between variables and the relationship between the functions of an artefact and it physical realisation (Van Keulen & Van der Molen, 2009). In school, pupils can analyse existing artefacts and answer questions like 'what problems was it designed to solve, what are the elements of the design that contribute to its functioning, and how well does it accomplish this?' (Benenson, 2001). Slangen et al (2010) present an example of how primary school children learn the principles of system thinking, design and form-function by working with robots. This way, they learn to distinguish the different variables that matter in the design of artefacts and their functioning. Also, they can design simple artefacts to meet identified needs or make existing artefacts work more effectively (Lesgold, 2009).

Visualising: Visualising is related to modelling, as both imply a simplification of reality. Visualising, however, concerns a specific element, that of making and interpreting what Latour (1986) calls 'inscriptions': marks on paper or on a computer screen. This may concern a variety of visual representations, such as maps, symbols, pictures, drawings and diagrams. All inscriptions pertain a reduction of reality to simplify complex phenomena to be able to manipulate and understand them (Gravemeijer, 2009). In technology, visualising also implies that the designer has an image of what an artefact will look like and how it will work (Ferguson, 1977; De Vries, 2005).

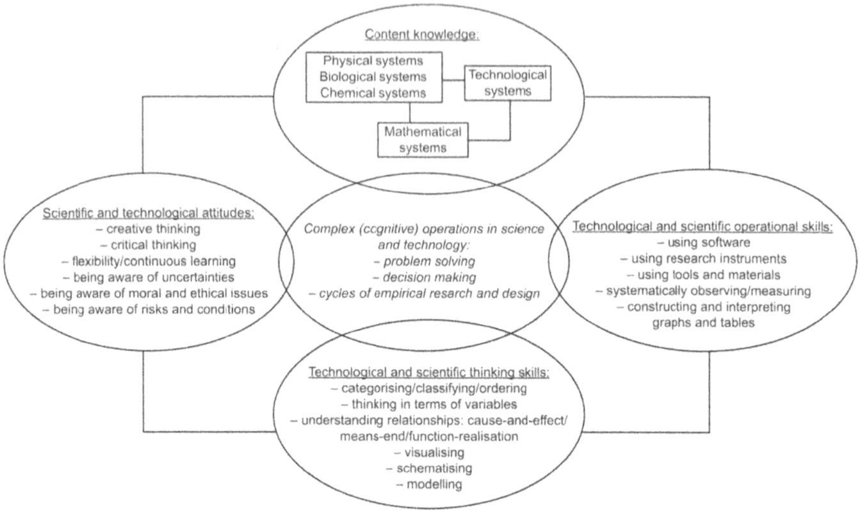

Figure 1. Scientific and technological thinking skills model

Schematising: In addition to proclaiming visualising as a central activity in science and technology, Latour (1986) points to the trend to develop simpler and simpler inscriptions that mobilize larger and larger numbers of events in one spot. We therefore make a distinction between visualizing and schematizing. We reserve visualising for rather straightforward transpositions, and we will use the term schematising for activities that involve the transition from more elaborate to more condense inscriptions that have a more general meaning or application.

Modelling: The thinking skill 'modelling' was often mentioned in the literature and by the experts we interviewed in this research. Different descriptions were given of the process of modelling. In some cases, modelling was seen as quite similar to the empirical cycle in scientific research, including the analysis of a problem, the choice of variables and the use of mathematical and scientific knowledge and skills to solve the problem (Savelsbergh, 2007). Models, however, may also have a more heuristic function and may concern metaphors or metonymies. We may further note that in scientific research, the focus is on the development of models, while in practice people often work with existing models. Often, modelling is described as the development of a scientific model as a manageable representation of reality. (Millar & Osborne, 1998; Savelsbergh)

SCIENTIFIC AND PRACTICAL IMPLICATIONS

Science and technology appear to be perfect topics for fostering complex cognitive operations such as problem solving and critical thinking. We want to argue however, that in order to operate on a higher level, students need operational skills and thinking skills. Assuming that operational skills will be taken care of in regular science and technology lessons, we focus on thinking skills. Since thinking skills transcend specific topics, we believe that it should not be too difficult to integrate thinking skills in a science and technology curriculum. We want to stress, however, that this will take a conscious effort. In the average school, the thinking skills distinguished in this research are no part of the curriculum, and teachers will have to learn what thinking skills are and how to teach them. We may note in this respect that thinking skills ask for a problem-centred approach to instruction, which in turn requires a corresponding set of beliefs of students and teachers about their roles. Moreover, in order to be able to teach thinking skills, the teachers themselves should dispose of the corresponding pedagogical content knowledge. Instructional activities, instructional materials, and theories about how students acquire thinking skills and how such learning processes can be supported, will have to be developed. It seems reasonable to expect that information technology can successfully be employed here. In this respect we may point to Kaput and Schorr's (2007) observation that information technology allows for dynamic representations, which can be used in helping students to come to grips with dynamic phenomena such as change and co-variation. Exploratory design experiments in Dutch schools are very encouraging in this respect, for example concerning the use of graphs to reason about variables and their relations (Gravemeijer, 2010).

Concluding, we think it is very important to keep looking at the future to prepare our children for the world they come to live and work in. As such, this research contributes to making innovations in science and technology education, like the Dutch VTB-Pro project, long-lasting an enduring.

Liesbeth K.J. Baartman
Eindhoven University of Technology, The Netherlands
l.baartman@tue.nl

Koeno Gravemeijer
Eindhoven School of Education, The Netherlands
koeno.gravemeijer@esoe.nl

ELLEN ROHAAN AND HANNO VAN KEULEN

4. WHAT EVERYONE SHOULD KNOW ABOUT SCIENCE AND TECHNOLOGY

A study on the applicability of The Canon of Science in primary education

INTRODUCTION

For many teachers in primary schools, teaching science and technology is a challenge. Recent surveys in the Netherlands (Walma van der Molen, 2009) indicate that teachers think science and technology education is important to their pupils. Besides, teachers express that pupils in general enjoy science and technology lessons. Yet, it is known that primary school teachers have little knowledge of science and technology and their self-efficacy with regard to teaching in this domain is rather low (e.g., Akerson, Morrison & McDuffie, 2006; Appleton, 2008; Traianou, 2007; Yilmaz-Tuzun, 2008). Hence, teachers might welcome a concise handbook or framework that tells them what is really important to know and teach about science and technology.

In 2008, prominent Dutch scientists collaborated in the making of a book called 'The Canon of Science' (Dijkgraaf, Fresco, Gualthérie van Weezel & Van Calmthout, 2008). The book contains 50 short articles of three to four pages on topics that, according to the somewhat provocative sub title, constitute 'what everyone should know about the natural sciences'. The articles themselves are written by 50 young scientists at the start of their scientific career, which emphasizes the importance of these topics for current research. Table 1 provides an overview of the book's content. All articles provide examples and outline the impact on society but formulae and complicated definitions are avoided. There also is a website from which all articles can be downloaded[i].

The editors emphasize the importance of the topics in the book to pupils in primary schools by writing in the preface that "(…) this Canon should find its way to the public at large, starting with education, and preferably with primary education. If there is one group of people whose lives will be affected by these topics, it is the group of young children" (Dijkgraaf et al, 2008, p. 19). This phrase could be dismissed as unrealistic and wishful thinking by scientists who long for more public recognition of their efforts. But this kind of reaction would do injustice to the genuine concern that many scientists feel regarding the ignorance of science and technology that seems to prevail in society and in schools.

---

[i] http://www.volkskrant.nl/vk/nl/2864/Betacanon/index.dhtml

Marc J. de Vries, Hanno van Keulen, Sylvia Peters, and Juliette Walma van der Molen (Eds.), Professional development for primary teachers in science and technology. The Dutch VTB-Pro project in an international perspective.

Table 1. The 50 topics in The Canon of Science

No	*Topic*	*No*	*Topic*	*No*	*Topic*	*No*	*Topic*	*No*	*Topic*
1	Zero	11	Quantum	21	Periodic Table	31	Water Infrastructure	41	Chaos
2	Plate Tectonics	12	Photo Synthesis	22	Pavlovian Reflex	32	Entropy	42	Robots
3	Hygiene	13	Enzymes	23	Micro Organisms	33	DNA	43	Computers
4	Transistor	14	Symbols and Formulae	24	Isaac Newton	34	Agriculture	44	Solar System
5	Energy	15	Climate	25	Life Span	35	Sex	45	Telephone
6	Evolution	16	Big Bang	26	GPS	36	Ocean flows	46	Time
7	Algorithms	17	Ecosystem	27	Money	37	Electro-magnetism	47	Language
8	Memory	18	Error	28	Albert Einstein	38	Homo Sapiens	48	City
9	Nuclear Bomb	19	Standard Model	29	Catastrophes	39	Food	49	Mobility
10	Cognition	20	Plastics	30	Normal Distribution	40	Avogadro Constant	50	Nano Technology

In the Netherlands, the need for strengthening science and technology in education is deeply felt nowadays, because the percentage of students with interest in science and technology is significantly lower (about 16% in 2007) than the European average (about 25%) and compared to many other comparable countries (Rocard et al., 2007). Shortages occur in many areas depending on a workforce that is well educated in science or technology (ROA, 2008).

It is estimated that students who enter teacher training colleges for primary education are in majority 'non techs' (Motivaction & YoungWorks, 2010). That is, they perceived STEM (science, technology, engineering & mathematics) subjects during secondary school as difficult, uninteresting and irrelevant to their personal life. They don't feel confident in science and technology subjects, sometimes even actively ignoring them. Large governmental programmes were initiated in 2004 to raise interest and improve performance in science and technology in primary education. Recent surveys show that these programmes have reached primary teachers quite well (Research Ned, 2010). Also, the percentage of secondary school students opting for STEM subjects and choosing a STEM track in higher education is rising. However, so far little effect has been achieved regarding students entering teacher-training colleges. This cohort is still recruited from the 'non techs'.

In this research project the possibilities for a science and technology curriculum derived from the 'big ideas' presented in the Canon of Science are investigated. We do not discuss its content. The choice of topics that 'everyone should know' will always remain contestable. Others (Angier, 2007; Fischer, 2001; Ganten, Deichmann & Spahl, 2005) have published similar books with almost identical titles and claims, but not with identical content. More important, a science and technology curriculum cannot just focus on big ideas and the kind of cultural scientific literacy that enables children to participate in political debate or read the newspaper. Other aims, such as developing skills for reasoning and manipulating, and knowledge of more mundane topics (e.g., ball bearings or farm animals) is arguably more important for successful participation in modern-day society. For the sake of this investigation, we take the choices of the Canon for granted and focus on resulting questions pertaining teachers and schools. One of the important aims for education is to provide citizens with enough scientific and technological literacy to participate in debate and decision making (Bybee, 2006; Garmire & Pearson, 2006; National Science Board, 2010; Jenkins, 1990; Laugksch, 2000). The Canon certainly suggests an agenda for this element of literacy. It may be the kind of resource that a 'non tech' student appreciates.

In the next sections we explore the presence of Canon-like ideas in current Dutch curriculum and other countries. It is investigated what pre-service teachers, in-service teachers, and teacher trainers know about science, defined by the content of The Canon of Science, what they think about the book as a guiding framework, and how they think the topics could be taught in school.

SCIENCE AND TECHNOLOGY EDUCATION IN DUTCH PRIMARY SCHOOLS

Primary education in the Netherlands starts at age 4 and continues for 8 years (grade 1 to 8). The Netherlands do not have a governmentally approved national curriculum, nor detailed guidelines on what should be taught or known. A short document stipulates 58 core objectives, covering all subjects and grouped into domains called Language, Maths/Arithmetic, Personal and World Orientation, Art Education, and Physical Education (Ministry of Education, Culture, and the Sciences, 2006). Science and technology (in the document called 'Nature and Technology') is part of Personal and World Orientation and has seven learning objectives:

- Pupils learn to recognize and name plants and animals that are common in their own environment, and they learn how these plants and animals function.
- Pupils learn about the constitution of plants, animals and humans and the form and function of their parts.
- Pupils learn to do research on materials and physics phenomena, including light, sound, electricity, force, magnetism, and temperature.
- Pupils learn to describe weather and climate in terms of temperature, precipitation, and wind.
- Pupils learn to relate function, form, and use of materials in products from their own environment.
- Pupils learn to design, carry out, and evaluate solutions for technological problems.
- Pupils learn that the position of the earth in relation to the sun causes the seasons and the rhythm of day and night.

Two other related core objectives are grouped in the sub-domains 'Man and Society' and 'Space':

- Pupils learn to handle the environment with care.
- Pupils learn about the measures that are taken in the Netherlands to enable living in areas threatened by water.

These core objectives should be reached by the end of the final (eighth) year of primary education, approximately at age 12. Core objectives in the Netherlands may be considered as general indicators of common educational contents. They are supposed to guide the programme that schools should offer their pupils. Schools are free to develop and choose their own specific goals, activities, pedagogical approach, textbooks, and educational materials. There is no national assessment that is obligatory for all, however, many schools do have their pupils take a similar national test at the end of the eighth year. This test mainly focuses on knowledge and skills in mother tongue (Dutch) and mathematics. Although it does contain a few items on science and technology, these items do not contribute to the over all test result.

The core objectives themselves are clearly not phrased in terms of major topics as in The Canon of Science. Although there is some overlap (e.g., with Climate and Solar System), most topics are not explicitly mentioned and there is no core objective simply stating that pupils should have conceptual knowledge of major topics in science and technology.

Given the large amount of freedom for schools and teachers to arrange their programmes, the Canon topics may nevertheless be part of what is taught. An analysis of four Dutch science and technology textbooks revealed that the contents could easily be linked to one (or more) of the 50 topics in The Canon of Science. E.g., various textbooks for primary schools contain chapters on Robots and Water Infrastructure. Only a few topics were not found in any textbook (i.c., Plate Tectonics, Enzymes, Entropy, Standard Model, and Nanotechnology). Textbooks that were published during the last three years include more topics on physics and technology than older books, which usually have a focus on biological topics (animals and plants).

In summary, there are no clear incentives for primary school teachers in the Netherlands to focus on topics that can be found in The Canon of Science, but there is also nothing that prevents them doing so.

AN INTERNATIONAL COMPARISON OF CURRICULA

Each country has its own traditions regarding science and technology education in primary schools. A striking difference is how countries teach ‘science’ and ‘technology’. England and the USA, for example, have a tradition of keeping science and design & technology (or engineering) separate as two different school subjects and there is opposition against combining these (Benson, 2009). Certainly, science education cannot be reduced to technology education and vice versa (Lewis, 2006). Nevertheless, in other countries, like the Netherlands, France and Belgium, it is more or less tradition that science and technology belong to the same disciplinary area and that similarities are more important than differences. The Canon of Science is an obvious example of the combination paradigm, including fundamental scientific topics, and technological innovations. The analysis of curricular documents of five other countries (England, France, USA, Germany, and Belgium) showed that the objectives and contents of science and technology education were not so much different in character from the Netherlands. Typically, in the investigated curricula the objectives and content were described in more detail.

For example, primary ('elementary') objectives for science education in the USA are organized into clusters called inquiry, physical science, life science, earth and space science, science and technology, science in personal and social perspectives, and history and nature of science. In grades K-4, children should develop abilities necessary to do scientific inquiry; of technological design; and to distinguish between natural objects and objects made by human. They should also develop an understanding of or about scientific inquiry; properties of objects and materials; position and motion of objects; light, heat, electricity, and magnetism; the characteristics of organisms; life cycles of organisms; organisms and environments; properties of earth materials; objects in the sky; changes in earth and sky; science and technology; personal health; characteristics and changes in populations; types of resources; changes in environments; science and technology in local challenges; and science as a human endeavour. In grades 5-8 this is supplemented with motions and forces; transfer of energy; properties and changes of properties in matter; structure and function in living systems; reproduction and heredity; regulation and behaviour; populations and ecosystems; diversity and adaptations of organisms; structure of the earth system; earth's history; earth in the solar system; abilities of technological design; populations, resources, and environments; natural hazards; risks and benefits; science and technology in society; nature of science; and history of science (AAAS,1993; NSES, 1996; NSTA, 2003).

Another example is Germany, a federal union of states, which has no federal curriculum for primary education ('Grundschule'). The approaches on the level of states (‘Länder’) resemble each other, however. Typically, in science, children should be able to describe specific properties of common materials and explain their use; investigate change in chemical reactions; investigate physical change of materials; explain the structure of matter using a simple corpuscular model; apply the concepts of force and energy change to examples from nature and technology; give examples of interacting systems; explain relations between structure and function in nature and technology; clarify differences and similarities among animals and plants; give examples of the accommodation of organisms to their surroundings; describe organs of living beings and their functions; give examples of healthy and unhealthy behaviour; describe the plurality of human sexuality. In technology, it is deemed important that children can work safely and correctly with tools, machines and materials; know simple construction parts; are able to solve simple problems; analyse technical constellations; plan activities; expand knowledge through experiments; keep the area of work tidy; remain dedicated while working; accept mistakes and learn from these; cooperate with others; specify and evaluate criteria for products and processes; appreciate the effects of technology for humans, society, economics and ecology; and stick to safety procedures (cf. Ministerium Brandenburg, 2008; Ministerium Schleswig-Holstein, 2009).

All investigated curricula contain topics that can easily be aligned with topics from The Canon of Science. England and USA have documents referring to standards for teachers, which also contain references to fundamental concepts in science and technology. However, the emphasis seems to be on rather broad concepts that cover as many other things as possible (e.g., Systems, Models, and Forces) in contrast with The Canon of Science in which the end point entities are emphasized.

PRE-SERVICE AND IN-SERVICE TEACHERS' AND TEACHER TRAINERS' KNOWLEDGE OF SCIENCE AND TECHNOLOGY

To examine knowledge of the topics in The Canon of Science, a new instrument with 50 multiple choice ($\alpha = 0.76$) and 10 open-ended items was constructed and administered online to a sample of pre-service teachers ($n = 140$), in-service teachers ($n = 81$), teacher trainers ($n = 29$), and science and technology programme coordinators ($n = 14$). All instruments are available on request from the authors.

The 50 multiple-choice items, one for each topic, had four answer alternatives (one correct and three incorrect) each. A test score between 0 (lowest) and 10 (highest) was computed by counting the number of correct answers and deviding by 5. An example of a question is given below:

What was the likely cause of the extinction of dinosaurs?

A A large volcanic eruption
B The impact of a huge meteorite
C A period of global warming
D A brief ice age

The 10 open-ended items were introduced with a cartoon about one of the topics from the Canon and then asked for a description in your own words of the central concept in the cartoon (i.c., Natural Sciences, Plate Tectonics, Energy, Evolution, Big Bang, Standard Model, Pavlovian Reflex, E=mc^2, Avogadro Constant, and Chaos). See Figure 1 for an example. The cartoons come from 'The Canon of Science according to Fokke and Sukke' (Reid, Geleinse & Van Tol, 2008). In the introduction to this compilation of cartoons made by well-known Dutch cartoonists the editor-in-chief of The Canon of Science, Robbert Dijkgraaf writes: "When people can laugh about these cartoons, we have achieved our goal".

Figure 1. Example of a cartoon on a topic in the Canon of Science: "Fokke and Sukke do not believe in evolution: 'The fittest never participate in a survival'"

The main criterion to select these ten cartoons was that conceptual knowledge was needed to fully understand the cartoon. The answers were qualitatively analysed by comparing the answers with expert answers and definitions on the online encyclopaedia Wikipedia. Codes were given on a 5-point scale: from no congruence (0) with the expert answer and Wikipedia to almost complete congruence (4). One rater coded all the answers and a second rater about 10% of the answers per item. The weighted inter-rater-reliability (ICC) ranged from 0.71 (good) to 0.95 (excellent) (Shrout & Fleiss, 1979).

In table 2 the mean scores on the multiple choice items for the different sub samples are presented. Correlation values (Pearson's product-moment correlation coefficient) between test score and interest in science are also included in this table. A t-test showed that men (M=6.53) scored significantly higher than women (M=5.50). Furthermore, the correlation between the test score and interest in science is relatively strong ($r=0.473$; $p<0.01$). In other words, respondents with an interest in science had, on average, a higher score on the test. The respondents who said to have no or little interest in science (n=33), correctly answered only 50% of the items (M=5.04; SD=0.99). Additionally, differences between groups with a different educational background were found. Respondents who were educated in the natural sciences (n=45) correctly answered 69% of the items, while respondents with a background in humanities (n=25) correctly answered 63% and respondents with a social and behavioural science background (n=44) 60% of the items.

Table 2. Mean scores multiple choice items (scale: 0-10)

Sample	*N*	*Mean*	*SD*	*Correlation with interest in science*
Pre-service teachers	140	5,40	1,226	0,443**
male	27	5,96*		
female	113	5,27*		
Teacher trainers	29	7,08	1,186	0,661**
male	17	7,52*		
female	12	6,45*		
In-service teachers	81	6,08	1,116	0,464**
male	43	6,28		
female	38	5,85		
Coordinators	14	6,79	1,420	0,677**
male	11	7,11		
female	3	5,60		
Total	276	5,88	1,323	0,473**
male	104	6,53**		
female	172	5,50**		

*p < 0,05; **p < 0,01

It is not uncommon in such tests to arbitrarily set the pass/fail limit at 50%. When correcting for guessing (one out of four), one could speak of a 'pass' when more than 62.5% of the items are answered correctly. In that case, pre and in-service teachers failed, but not blatantly. Relatively speaking, the level of knowledge of pre and in-service teachers of the natural sciences as it appears in the Canon is not that low compared to the sub samples of respondents who have been educated in the natural sciences (69% correct), coordinators (68%) or the teacher trainers (71%).

Table 3 presents the mean scores on the open-ended items for the different sub samples. The highest score (0.95) was obtained by the teacher trainers, closely followed by the coordinators (0.94). The teacher trainers knew best how to describe the Pavlovian Reflex, while the coordinators and in-service teachers scored highest on Energy. Pre-service and in-service teachers almost obtained similar mean scores (respectively, 0.56 en 0.57), but pre-service teachers scored highest on Plate Tectonics. It can be concluded that the examined sub samples have very limited knowledge about the ten key concepts. Especially pre and in-service teachers obtained a very low mean score. Many incorrect answers were given and many correct answers only contained a single element of the experts' answer. Hence, there is a high degree of unfamiliarity with the concepts from the natural sciences.

Table 3. Mean scores open-ended items (scale: 0-4)

Topic	*Pre-service teachers (n≈100)*	*Teacher trainers (n≈20)*	*In-service teachers (n≈50)*	*Coordinators (n≈10)*
Natural sciences	0,68	1,00	0,76	1,00
Plate tectonics	1,02	0,90	0,82	1,11
Energy	0,92	1,10	1,04	1,56
Evolution	0,61	1,37	0,67	1,00
Big Bang	0,31	0,84	0,45	0,78
Standard Model	0,08	0,21	0,06	1,00
Pavlov Reflex	0,89	1,50	0,96	1,22
E=mc2	0,50	1,00	0,60	0,78
Avogadro Constant	0,25	0,58	0,12	0,56
Chaos	0,47	1,00	0,32	0,38
Mean	*0,57*	*0,95*	*0,58*	*0,94*

PRE-SERVICE AND IN-SERVICE TEACHERS' AND TEACHER TRAINERS' OPINIONS ABOUT THE CANON OF SCIENCE AS A GUIDING FRAMEWORK

To investigate opinions, an online questionnaire (part A) of 5 items with a five 5-point Likert-scale (from 'no, not at all' to 'yes, certainly') and an alternative 'do not know', was constructed and administered to pre-service teachers, in-service teachers, teacher trainers, and science and technology programme coordinators. The items and percentages of answers in each category are shown in table 4.

It appears that the respondents broadly support a definition of the domain of science and technology for education through a book like The Canon of Science. They think it is important that everyone has sufficient knowledge to understand newspaper articles on science and technology, feel the need for a clearly defined framework for science and technology, and would (advice to) use a Canon of Science handbook for teachers, even though a considerable part of the respondents is unfamiliar with The Canon of Science. Understandably, they do not know whether the book provides a correct picture of the natural sciences or if it is useful as a guiding framework for science and technology education.

Table 4. Items and answers questionnaire part A (N = 230)

Items	*1 No, not at all*	*2*	*3*	*4*	*5 Yes, certainly*	*Don't know*
1 Do you think it is important that everyone has sufficient knowledge to understand newspaper articles on science and technology?	1,7%	12,6%	37,8%	38,7%	7,4%	1,7%
2 Does The Canon of Science give a correct picture of the natural sciences according to you?	1,7%	5,7%	17,8%	21,3%	1,7%	51,7%
3 Do you feel the need for a well-defined description of science and technology in primary schools?	4,8%	12,2%	15,7%	38,7%	24,8%	3,9%
4 To what extent do you think The Canon of Science is suitable as a guiding framework for science and technology in primary schools?	3,9%	10,0%	20,0%	15,7%	3,9%	46,5%
5 If there would be written a Canon of Science for primary education (a handbook for teachers), would you (advice to) use it?	2,2%	3,9%	15,2%	34,8%	27,4%	16,5%

THE APPLICABILITY OF THE CANON OF SCIENCE IN PRIMARY SCHOOLS

The second part of the questionnaire (part B) concerned the applicability of 20 selected topics from The Canon of Science for lower and upper grades, a self-assessment of knowledge about the topics, and self-efficacy rating in teaching the topics. A selection of 20 topics was made, because the questionnaire would otherwise be too lengthy. The selected topics were: Plate Tectonics, Hygiene, Energy, Evolution, Cognition, Photosynthesis, Climate, Big Bang, Ecosystem, Isaac Newton, GPS, Albert Einstein, Water Infrastructure, DNA, Electro-magnetism, Robots, the Computer, Solar System, Mobility, and Nanotechnology. For each topic four items were formulated: 1) I think this topic is suitable for lower grades, 2) I think this topic is suitable for upper grades, 3) I know much about this topic, and 4) I am good at teaching this topic. The answers were given on a 5-point Likert scale.

Based on the mean answers ranking lists of topics were made. Table 5 shows the top 5 of topics that are thought to be the most suitable for lower and upper grades.

Hygiene scored highest for both lower as upper grades within all sub samples, excluding the coordinators, who indicated Energy as the most suitable topic for the upper grades. Climate also scored high for the upper grades, while Mobility acquired a high score for the lower grades. Regarding the lower grades, only Hygiene had a mean score that was higher than 3.5, which implies that most topics are not thought to be suitable for pupils in the lower grades. All topics were found (highly) suitable for pupils in the upper grades.

Table 6 shows the 10 highest ranked topics on knowledge and self-efficacy (for the total sample). The mean scores on self-assessment of own knowledge ('I know much about this subject') ranged from moderate to considerable. Many topics were assessed with a mean score higher than 3 on a scale from 1 to 5. Three out of 20 topics (Hygiene, Climate, and Ecosystem) obtained a score higher than 3.5. Two topics (Nanotechnology and Albert Einstein) scored lower than 2.5.

Table 5.Top 5 of most suitable topics for lower and upper grades (scale: 1-5)

	Pre-service teachers		*Teacher trainers*		*In-service teachers*		*Coordinators*	
Lower grades								
1	Hygiene	4,43	Hygiene	4,41	Hygiene	4,50	Hygiene	3,82
2	Climate	3,25	Water Infra	3,14	Energy	3,35	Mobility	3,18
3	Mobility	3,22	Mobility	3,10	Computers	3,27	Energy	3,18
4	Computers	3,16	Ecosystem	3,09	Mobility	3,21	Solar system	3,00
5	Water Infra	2,87	Solar system	3,09	Water Infra	3,17	Climate	3,00
Upper grades								
1	Hygiene	4,65	Hygiene	4,73	Hygiene	4,60	Energy	4,73
2	Climate	4,56	Climate	4,68	Energy	4,51	Climate	4,55
3	Computers	4,45	Water Infra	4,59	Climate	4,49	Water Infra	4,55
4	Water Infra	4,43	Energy	4,45	Solar system	4,43	Solar system	4,45
5	Energy	4,36	Computers	4,45	Water Infra	4,40	Ecosystem	4,36

The self-efficacy ('I am good at teaching this topic') scores were also rather high. The assessment of the respondents' own teaching was slightly more positive than their own knowledge. Five topics (Hygiene, Climate, Ecosystem, Water Infrastructure, and Solar System) had a score that was higher than 3.5. Again, Nanotechnology and Albert Einstein scored lower than 2.5. Hygiene was the highest scoring topic on both knowledge and self-efficacy. In other words, the respondents judged their own knowledge and efficacy in teaching best for this topic. As expected, a clear correlation between (self-assessed) knowledge and self-efficacy could be noticed.

CONCLUSION

The data allow for the following conclusions to be drawn. The analysis of curricula reveals that neither the Canon's topics nor the Canon's objective to achieve a broad and up-to-date cultural scientific literacy are explicitly used as a framework for primary science and technology education. Participants, however, indicate that many topics are suitable to primary education, especially in the upper grades.

Table 6. Self-assessment knowledge and self-efficacy mean scores (scale 1-5)

Knowledge (self-assessment)				*Self-efficacy*			
Topic	*N*	*M*	*SD*	*Topic*	*N*	*M*	*SD*
Hygiene	223	3,91	,815	Hygiene	216	3,92	,800
Climate	224	3,76	,810	Climate	218	3,80	,893
Ecosystem	223	3,61	,962	Ecosystem	215	3,54	1,013
Water Infra	223	3,49	1,022	Water Infra	216	3,52	1,039
Solar system	223	3,46	,971	Solar System	217	3,51	1,023
Evolution	223	3,46	1,008	Computers	211	3,43	1,064
Computers	222	3,43	1,034	Mobility	198	3,39	,965
Mobility	208	3,37	,938	Evolution	214	3,34	1,092
Photosynthesis	220	3,36	,995	Energy	218	3,32	,968
Energy	224	3,27	,908	Photosynthesis	210	3,31	1,023

Participants, however, indicate that many topics are suitable to primary education, especially in the upper grades. Participants' self-efficacy with respect to teaching these topics is considerable. Knowledge of these topics as measured with the multiple-choice test is rather low in the most important cohort (pre-service and in-service teachers) but not negligible. Men do better than women; teacher trainers do better than teachers; in-service teachers do better than pre-service teachers. The open questions, however, reveal that the ability to describe topics in a conceptually meaningful way is quite poor. Participants indicate that they do not know much about the Canon but that they welcome materials or handbooks that would support them teaching science and technology through the Canon. Clearly, there is a need to 'translate' the Canon for educational purposes. As a matter of fact, a 'junior' version of the Canon on audio CD has appeared recently (Dijkgraaf, Fresco, Haring & Groothof, 2009) which has been supplemented with suggestions for teaching in this research project (Rohaan & Van Keulen, 2010).

DISCUSSION

This study on the applicability of The Canon of Science gave rise to several questions and points of discussion.

An important question pertains the theoretical and practical possibilities to establish 'what everyone should know about the sciences'. Arguably, there is no one, not even a Nobel Prize winner, who can answer all questions in this domain. There is no benchmark to be used for comparison. Consensus on which 50 (or 10 or 1000) topics define the domain of the natural sciences will probably never be reached. In our study, we had to work out an operational definition for the domain, in order to derive tests. We reduced the domain of science and technology to the Canon, and defined knowledge of science and technology as knowledge of what is written in the Canon. E.g., since the Canon writes about causes for the extinction of dinosaurs, this qualifies for questions. Most participants in our study had not read the Canon; they certainly had learned the texts by heart. Hence, our instruments revealed what was known by way of prior education, continuous professional development, reading newspapers, et cetera. When the Canon would be used as a standard or reference, teaching to the test in order to raise scores would be very tempting. More importantly, the Canon focuses on cultural scientific and technological literacy but ignores other important aspects of science and technology education, such as thinking skills, practical skills and functional literacy. Isn't it also important to attain general competencies and higher order cognitive or executive skills such as curiosity, asking questions, reasoning, modelling, doing research, solving problems and designing solutions? This may be much more valid to pursue, from the point of view that science and engineering are not primarily bodies of factual knowledge but human practices that derive their meaning from human needs and interests in ever changing contexts.

Hence, the use of the Canon of Science as an educational tool among others seems to be a more feasible and realistic goal. It is appealing that in the Canon the domain of the natural sciences is not constructed from the so-called 'basics' (formulae, laws, taxonomies, definitions) to concepts, but conversely: it starts with the current state of research and development. The Canon of Sciences addresses a range of important scientific and technological topics in a way that is accessible for and even appealing to 'non techs'. Another positive element of the Canon is that the format ignores the classical disciplinary structures (physics, chemistry, biology, mathematics, astronomy, geology, engineering, et cetera), but emphasizes inter disciplinarity.

Whether it is about transmitting factual knowledge or stimulating the development of process skills, it would be practically impossible and undesirable to teach all topics from 1 to 50. The selection of topics should surely depend on the pupils' age and developmental level, i.e., sufficiently concrete and related to prior experiences. Several topics (e.g., Entropy and Quantum) are merely too abstract and do not easily relate stages or prior experiences of most pupils. But what if the learning objective is not to fully understand the concept of Entropy, but rather to get acquainted with vocabulary that is often used in this context, for example? It seems that different kind of learning objectives could be attached to different topics.

The findings of this study are not sufficient to provide answers and solutions to all fundamental questions and issues. Therefore, further research, including field consultations and expert discussions, is necessary.

Ellen Rohaan
Eindhoven University of Technology & Fontys University of Applied Science, The Netherlands
e.rohaan@fontys.nl

Hanno van Keulen
Utrecht University & Fontys University of Applied Science, The Netherlands
j.vankeulen@uu.nl

ED VAN DEN BERG AND HANNO VAN KEULEN

5. A SCIENCE AND TECHNOLOGY KNOWLEDGE BASE FOR PRIMARY TEACHERS

In this study, we reflect on questions pertaining to the knowledge and skills that are needed to teach science and technology in primary education. What should primary teachers know of science and technology? As non-specialists in a huge and ever-expanding domain, primary teachers can use some advice on the priorities. A second incentive for pursuing this question concerns the doubts expressed in various parts of society as to the generic level of knowledge of primary teachers. Trust in teacher competence, so it is said, can be regained when teachers themselves pass rigorous tests. In this study, we investigate what the content of a science and technology knowledge base could be, and which challenges will have to be resolved for its use in practice.

INTRODUCTION

Undoubtedly, a strong foundation in subject matter knowledge and skills with regard to science and technology is an excellent starting point for a teaching career in primary education. However, this is also true for mathematics, language, history, music, child development and all the other issues that make teaching in primary education so many-sided and challenging. What knowledge and skills can we realistically expect of students who enter teacher-training colleges and what is possible to achieve pre-service? Can a knowledge base serve as a basis to develop the science and technology program for teacher training? Is it possible to define and test a knowledge base that could be used as a prerequisite for entering the job?

In the Netherlands, as in most countries, the primary teacher is a jack-of-all-trades and teaches all subjects. Subject specialization does occur in some schools but is rare. As elsewhere, Language and Math/Arithmetic are the prominent subjects. Science and technology have been included in the curriculum ever since 1857 but rank low in the priorities and lower yet in the preferences of teachers. A pupil might experience an estimated mere 250 hours of teaching related in one way or another to nature, science, design and/or technology between pre-school and the end of 8th grade, in many schools without clear learning objectives or cohesion. Teacher preparation does not pay much attention to science and technology either. Van Graft (2003) estimated that of a total of 6000 study hours (including self-study and internships) in the 4-year teacher education bachelor program about 200 are devoted to science, technology and/or nature. That is a meager 3%. Apparently, teacher-training colleges expect their students to have acquired sufficient

Marc J. de Vries, Hanno van Keulen, Sylvia Peters, and Juliette Walma van der Molen (Eds.), Professional development for primary teachers in science and technology. The Dutch VTB-Pro project in an international perspective.

knowledge during secondary education. Yet, as in many other countries (Harlen, 1997; Appleton, 2003), Dutch primary teachers lack both background and confidence to teach science and technology. Most pre-service teachers are 'non-techs' (Motivaction, 2010) and took their last science course when they were about 15 years old themselves.

In the Netherlands, from 2004 onwards, the VTB and VTB-Pro projects tried to strengthen the position of science and technology in primary education and stressed the importance for society of primary science and technology and of learning by inquiry and design. 'Technology' was to rise above handicraft; 'science' should comprise more than a few encounters with animals and plants (cf. Walma van der Molen, de Lange, & Kok, 2007). This raised awareness of the lack of background in science and technology of both pre-service and in-service teachers, drawing attention to the question of a knowledge base.

In the mean time, on the political level, serious doubts were expressed concerning the subject matter knowledge per se of students graduating from teacher-training colleges. This provoked the Government and the Council of Universities of Applied Science to establish Committees for knowledge bases for all subjects taught in teacher-training colleges. The first knowledge bases, on language and arithmetic/mathematics, were published in 2009. Other Committees are expected to publish their work in the course of 2011. Preliminary results have already been communicated in order to incite debate with professionals.

This study emerged from both strands. The first author headed the Committee on Nature, Science and Technology (cf. Van den Berg, Louman & Marell, 2010). The second author was commissioned by the National Platform Science and Technology to provide a contemporary description of the domain of primary science and technology to be used as a frame of reference for teachers in primary schools (Van Keulen, 2010)[ii].

Clearly, a knowledge base for teachers should have strong relations to the current state of affairs in science and technology. The danger of such a knowledge base, however, is that it grows into a long list of knowledge and skills to be acquired and that teachers, teacher educators and especially policy makers and politicians feel obliged to mandate, teach, and assess everything in the knowledge base. Knowledge bases and curricula should be used with wisdom. As Nobel Prize winner Viktor Weisskopf said: *"It is better to uncover a little than to cover a lot"*. Otherwise, one gets an overstuffed and undernourished curriculum. Other fallacies are the naive notion that all content knowledge required for teaching in the primary school can be acquired during pre-service teacher education, and that content knowledge per se is a sufficient prerequisite for teaching (Appleton, 2006).

[ii] This text can be downloaded from http://www.vtbprogramma.nl//docs/Beleidsdocumenten/wt-ijkpunten.pdf.

KEY ELEMENTS OF A PRIMARY SCIENCE AND TECHNOLOGY CURRICULUM

The aim of science and technology in secondary education is not primarily to prepare students for teaching in primary schools. In order to reflect on a knowledge base for primary teachers it is therefore better to start with what pupils in primary education should learn of science and technology. We use a publication of the National Research Council of the US Academies (Duschl, Schweingruber & Shouse, 2007, p334) as a starting point. Here, four fundamental strands of learning for science[iii] from kindergarten to grade 8 are identified.

Pupils who understand science:

1. *Know, use, and interpret scientific explanations of the natural world.*
2. *Generate and evaluate scientific evidence and explanations.*
3. *Understand the nature and development of scientific knowledge.*
4. *Participate productively in scientific practices and discourse.*

- Strand 1 is on content and concepts with an emphasis on using and interpreting scientific explanations in action.
- Strand 2 concerns the link between theory, hypotheses, models, predictions and experiments with data. Skills in this area develop with age but development is significantly enhanced by prior knowledge, experience, and instruction and recent studies indicate that children are far more competent than suspected (Duschl et al., 2007, p159).
- Strand 3 involves the nature of science and follows a tentative series of progressions in this area from grade 1 – 6 suggested by Smith, Maclin, Houghton, and Hennessey (2000).
- Strand 4 concerns doing science which can be observing or experimenting but should also involve discourse. Note that the phrasing does not follow the often used demarcation between content and process. In the past, the focus was often on process skills (e.g., *Science A Process Approach* of the American Academy of Science). Current insight is that content and process are intertwined (Millar & Driver, 1987; Duschl el al., 2007). When *reasoning with concepts and reasoning with evidence* becomes a major goal, pupils have to be able to use - and not just to know - concepts to obtain evidence. The authors also emphasize the importance of well-chosen core concepts of science which should be carefully developed following learning progressions based on what is known on how children learn (Duschl et al. p. 340). Concepts like 'matter' could gradually be developed from kindergarten to grade 8 and at some time reach the level of models that use particles like atoms and molecules.

[iii] We note that the Duschl report, as many other reports, are on *science* and hence follow Anglo-Saxon traditions separating science education from technology education. In our study, we focus on science *and* technology.

The England and Wales 2011 curriculum (APP, 2010) for key stages 1 and 2 (children of age 4 – 11) emphasises the link between concepts and inquiry skills. Worldwide learning of science by inquiry has been promoted in science education at least since Dewey (1910; 1916). Recent policy documents across the globe emphasize learning by inquiry and design (NSES, 1998; Duschl et al., 2007; Rocard et al., 2007; QCA, 1999; 2010). Inquiry learning can be considered as a *means* to learn science knowledge and concepts or as a *purpose* of science education: learning to investigate, to collect and reason with evidence and learning about the nature of scientific knowledge. As a means, inquiry can enhance motivation but can also lead to frustration. It can generate children's ownership of knowledge and open up new perspectives, but this requires sophisticated teaching and scaffolding. The process of inquiry requires lots of time and is an inefficient way to acquire knowledge that already exists. However, elements of 'doing science', if chosen well, help to see facts and results in the proper perspective. Teaching by inquiry and design is essential as it is the only way to expose the nature of scientific knowledge (Van Keulen, 2009).

The National Science Teacher Association in the USA proposed standards for science teachers, which are currently under revision (NSTA, 2003; 2010). An important standard is on knowledge and states: *"Effective teachers of science understand and demonstrate the knowledge and practices of contemporary science. They interrelate and interpret important concepts, ideas, and applications in their fields of licensure"*. Other standards focus on how pupils learn, on developing and delivering lessons and on assessing learning results.

Summarizing: the knowledge, skills and attitudes required for teaching primary science and technology can be organized in the following categories, keeping in mind that teachers should be able to use these categories in an integrated manner:

1. Knowledge of important concepts and theories.
2. Knowledge of the nature of science and technology.
3. Knowledge and skills concerning inquiry and design.
4. Scientific attitudes (curiosity, respect for evidence, creativity, perseverance, critical and open mind).
5. Knowledge and skills with regard to teaching and learning science and technology (Pedagogical Content Knowledge).

KNOWLEDGE OF CONCEPTS

Consequently, one of the first questions for a knowledge base to be answered is what the important concepts and theories are. This question is easily put forward but not so easy to answer. A bewildering number of documents exist on the content of science and technology, from inside and outside of education. An understanding of the core concepts, theories or big ideas of science and technology should be helpful in giving meaning to a wide range of experiences and pieces of information that pupils (and, later on, citizens) encounter across contexts. As a source for a knowledge base core concepts or big ideas should also have links to meaningful activities with children.

Sources from within education are of course the curricula and standards for primary science and technology in the Netherlands and elsewhere (e.g., OCW, 2006; NSES, 1996; NSTA, 2003; QCA, 1999; 2004; Australian Academy of Science, 2005; La main à la pâte, 2008; Vlaamse Overheid, 2010). Other sources are the many resources and books for primary teachers (e.g., TULE, 2010; Harlen & Qualter, 2004; De Vaan & Marell, 2006; Kersbergen & Haarhuis, 2010, Gillespie & Gillespie, 2007; Farrow, 2006; Cross & Bowden, 2009; Allen, 2010; Heywood & Parker, 2009; Van Keulen, 2010). In this paragraph we elaborate on a few examples.

In the USA, Project 2061 was charged by the American Association for the Advancement of Science (AAAS) to rethink the whole science curriculum from the bottom up. This led to *Science for All Americans (*Rutherford, 1990). The National Research Council of the Academy of Sciences then started a national project that involved many scientists and educators and resulted in the National Science Education Standards (NSES, 1996). NSES identified five big ideas or unifying concepts in science and used these and other considerations to choose science content standards for age 5 – 18 and standards for teaching as well. In 2007, the same NRC produced a major policy document *Taking Science to School - Learning and Teaching Science in Grades K-8* (Duschl et al, 2007). This book draws on recent research evidence on science learning and teaching of children and is widely quoted.

NSES	Systems, order and organization; change, constancy, and measurement; evolution and equilibrium; form and function; evidence, models, and explanation.

The College Board (2009) in the USA produces widely used tests for college admission and placement. They produced standards for grades 6 – 8 and 9 – 12 for all sciences.

College Board	Scale; evolution; equilibrium; matter and energy; form and function; models as explanations, evidence, and representations; interaction.

Katehi, Pearson and Feder (2009) produced a well-researched and extensive policy report for K-12 Engineering Education for the US National Academy and identified core concepts and skills.

Katehi	System; modeling; designing; social interaction; optimization.

Hacker, De Vries and Rossouw (2009) developed a set of core technology concepts and contexts with expert panels in a Delphi study.

Hacker	*Concepts*: Design (as a verb and as a noun), system, modeling, social interaction, optimization, innovation, specification, sustainability, energy, materials, resource, trade-offs, technology assessment, invention, function. *Contexts*: Energy in society, biotechnology, sustainable technology, transportation, medical technologies, nanotechnology, food, industrial production, water resource management, construction, two-way communication, global warming, domestic technologies.

In documents of the International Technology and Engineering Educators Association (ITEEA, 2007), the knowledge part of technological literacy is defined.

ITEEA	*Knowledge:* Recognizes the pervasiveness of technology in everyday life; Understands basic engineering concepts and terms, such as systems, constraints, and trade-offs; Is familiar with the nature and limitations of the engineering design process; Knows some of the ways technology has shaped human history and how people have shaped technology; Knows that all technologies entail risk, only some of which can be anticipated; Appreciates that the development and use of technology involve trade-offs and a balance of costs and benefits; Understands that technology reflects the values and culture of society.

Harlen (2010) with a group of international experts produced a set of ten big ideas *in* science and four important ideas *about* science for primary and secondary education.

Harlen	1) particle nature of matter, 2) objects can influence each other at a distance, 3) changing movement requires a net force, 4) energy conservation and energy transformation whenever "something happens", 5) composition of earth and atmosphere and processes in them shape surface and climate, 6) the solar system is a tiny part of the millions of galaxies in the Universe, 7) cellular basis of organisms, 8) organisms require energy/materials for which they compete, 9) genetic information is passed down generations, 10) diversity of organisms is result of evolution, 11) causality, 12) scientific models are those that best fit the facts known at a particular time (as opposed to striving for "eternal truth") 13) knowledge produced by science can be used in technology and serve mankind, 14) applications of science often have ethical, social, economic, and political implications.

Another approach was taken by a group of Dutch scientists led by the current president of the Royal Academy of Sciences. They produced a 'Canon' of fifty topics that capture the essence of science and technology. Knowledge is important to be a scientifically and technologically literate citizen (Dijkgraaf, Fresco, Gualthérie van Wezel & Van Calmthoud, 2008).

Dijkgraaf	Zero; Quantum; Periodic Table; Water Infrastructure; Chaos; Plate Tectonics; Photo Synthesis; Pavlov Reflex; Entropy; Robots; Hygiene; Enzymes; Micro Organisms; DNA; Computers; Transistor; Symbols and Formulae; Isaac Newton; Agriculture; Solar System; Energy; Climate; Life Span; Sex; Telephone; Evolution; Big Bang; GPS; Ocean flows; Time; Algorithms; Ecosystem; Money; Electromagnetism; Language; Memory; Error; Albert Einstein; Homo Sapiens; City; Nuclear Bomb; Standard Model; Catastrophes; Food; Mobility; Cognition; Plastics; Normal Distribution; Avogadro Constant; Nanotechnology.

In the Netherlands, Boersma and others (2007) produced detailed learning progressions for age 4 to18 for biology.

Boersma	*Organisation levels* (Molecule; Cell; Organ system; Organism; Population; Ecosystem; Biosphere); *Biological unities* (DNA; Cell; Organ; Plant; Animal; Human; Fungi; Bacteria; Virus; Species; Population; Ecosystem; Biosphere); *Self-regulation, self-organization* (Metabolism; Growth and maintenance; Respiration; Digestion; Blood circulatory system; Excretion; Transport; Defense; Movement; Homeostasis; Photosynthesis; Nutrition; Life cycle; Health; Food chain; Cycle; Equilibrium; Sustainable development); *Interaction* (Senses; Nervous system; Hormonal system; Behavior; Interaction with (a-)biotic factors); *Reproduction* (Cell cycle; Propagation; Inheritance); *Evolution* (Fossil; Genetic variation; Natural selection; Biodiversity).

In the PISA Programme for International Student Assessment (OECD, 2009), contexts have been defined that are important for children to understand science and technology.

PISA	*Health* (maintenance of health, accidents, nutrition, control of disease, social, transmission, food choices, community health, epidemics, spread of infectious, diseases). *Natural resources* (personal consumption of materials and energy, maintenance of human populations, quality of life, security, production and distribution of food, energy supply, renewable and non-renewable energy sources, natural systems, population growth, sustainable use of species). *Environment* (environmentally friendly behaviour, use and disposal of materials, population distribution, disposal of waste, environmental impact, local weather, biodiversity, ecological sustainability, control of pollution, production and loss of soil). *Hazard* (natural and human-induced, decisions about housing, rapid changes (earthquakes, severe weather), slow and progressive changes (coastal erosion, sedimentation), risk assessment, climate change, impact of modern warfare). *Frontiers of science and technology* (interest in science's explanations of natural phenomena, science-based hobbies, sport and leisure, music and personal technology, new materials, devices and processes, genetic modification, weapons technology, transport, extinction of species, exploration of space, origin and structure of the universe).

Van Keulen (2010), in the context of the Dutch VTB, VTB-Pro and Master Plan projects, and without suggesting that this is also what teachers are obliged to know, elaborated on the concepts used in PISA to make a domain description of the science and technology relevant for primary education, using a context-concept approach (cf. Walma van der Molen, de Lange, & Kok, 2009). He describes a day in the life as seen through the eyes of children. Between rising and going to sleep children encounter all kinds of science and technology related phenomena, processes, systems and products, and this can result in questions on dreaming, breathing, sound, sneezing, the sun, glasses, needles, soap, water, mirror, microwave, bicycle, etc. Each of these experiences is linked to one or more concepts that are organized in five systems.

Van Keulen	*Non-living systems* (structure, qualities and change of matter; motion and force; energy and energy transformations; interactions: light, warmth, sound, radio waves, X-rays, seismic waves; electricity and magnetism); *Living systems* (cells, organs and organisms; build of humans, plants and animals; respiration, circulation and digestion; life cycle, growth and reproduction; species, diversity, evolution and extinction; ecosystems, food chain, agriculture; proteins, sugars, lipids, enzymes; hormones, DNA, receptors); *Earth and space systems* (structure of the earth (lithosphere, hydrosphere and atmosphere); soil, rock, mountains, erosion, plate tectonics; water: fresh and salt, streams, evaporation, tides; air: atmosphere, stratosphere; seasons, climate and weather; history of the earth, influence of humans; solar system, galaxy, universe, stars, big bang, gravity); *Technology systems* (design: function; specifications; constraints, innovation, invention, modeling, systems, interface, control; construction: shaping, tools, materials, design, optimizing; categories: biotechnology, nanotechnology, transportation technology, medical technology, household technology, communication technology, process technology, food technology, water management; effects: quality of life, trade-offs, sustainability; climate change); *Mathematics systems* (quantity: numerical phenomena, quantitative relations and patterns, sense of number, logical operations; form and space: orientation, navigation, representation, forms and figures; change: relations, diagrams, tables; uncertainty: data, chance.

Van den Berg and co-workers followed the instructions to the Dutch Knowledge Base Committees to start with a big idea approach and derive second and third order concepts from a limited set of core concepts. This approach led to a set of eight terms. For example, with *unity in diversity* as a first order concept, one can generate species, descriptive anatomical concepts and biodiversity as derived concepts (Van den Berg et al., 2010).

Van den Berg	Unity in diversity; systems and organization; change; cause and effect; matter and energy; evidence, models and explanations; purpose and means; sustainable development.

Clearly, there is overlap between these approaches but they are not identical. Notable differences are with the topics and with the level of abstraction. To combine comprehensiveness with brevity, rich conceptual words like ‘system’, ‘change’, ‘energy’ or ‘diversity’ are used. Such concepts do not signify clear-cut facts or pieces of information. This may work out all right when teachers use a knowledge base to define their teaching priorities and which concepts they consequently should elaborate on themselves. When a knowledge base is used for testing pre-service teachers, this may not work out since conceptual terms may evoke very different levels of complexity and detail. When this is left open, test makers will define what the knowledge base really refers to.

KNOWLEDGE OF THE NATURE OF SCIENCE AND TECHNOLOGY AND OF INQUIRY AND DESIGN

Science and technology is knowledge intensive, but this does not mean that the ability to reproduce facts is sufficient or even quintessential. Whether or not a (prospective) teacher really knows or understands a concept depends upon his or her ability to use it to interpret the natural world, to generate and evaluate scientific evidence and explanations and to participate productively in scientific practices and discourse (cf. Duschl et al., 2007). Hence, whether the concept to be mastered is *working with evidence, models and explanations*, or *the particle nature of matter*, it requires an understanding of the nature of science and technology and reasoning, design and inquiry skills. This process is often pictured as a series of linear steps: the scientific method. In reality, inquiry is a messy iterative process in which concepts and skills are intertwined (Millar & Driver, 1987). Worth, Duque and Saltiel (2009) pictured this nicely in Figure 1. Various other representations and cycles have been proposed. De Vaan & Marell (2006) use a 5-step model. Bybee (2010) and others (Hackling, 2005; Coe, 2008) use the 5E model ('engage, explore, explain, elaborate, evaluate'). Van Graft and Kemmers (2007) developed a 7-step model. The differences between these approaches are not essential. The start and the end of the process have clear steps while the in-between steps can be repeated iteratively in different cycles and with retracing of steps. The design process is quite similar to the inquiry process although the focus is more on solving (unique) problems rather than generating (generic) models and explanations (cf. Katehi et al., 2009).

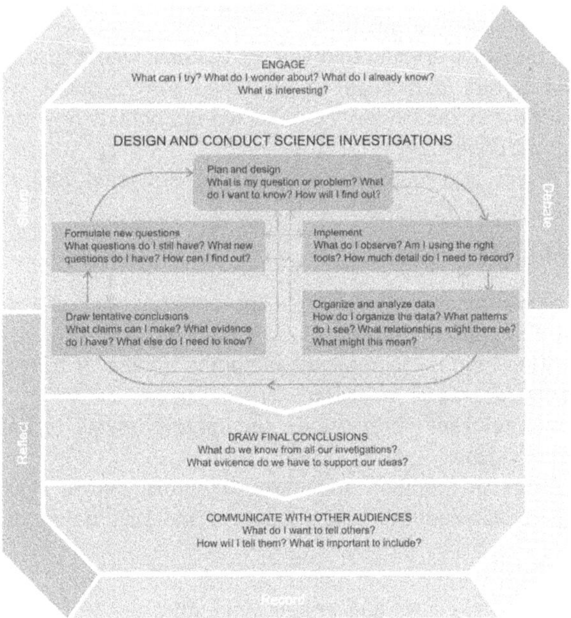

Figure 1: The inquiry process

With regard to the nature of science and technology and the knowledge and skills for inquiry based teaching we follow the 1999 National Curriculum for England and Wales (QCA, 1999) as it adequately emphasizes the connection between data, concepts, and evidence:

Ideas and evidence in science

Pupils should be taught:

- about the interplay between empirical questions, evidence and scientific explanations using historical and contemporary examples
- that it is important to test explanations by using them to make predictions and by seeing if evidence matches the predictions.
- about the ways in which scientists work today and how they worked in the past, including the roles of experimentation, evidence and creative thought in the development of scientific ideas.

Investigative skills

Pupils should be taught to:

Planning

- a use scientific knowledge and understanding to turn ideas into a form that can be investigated, and to decide on an appropriate approach
- decide whether to use evidence from first-hand experience or secondary sources
- carry out preliminary work and to make predictions, where appropriate
- consider key factors that need to be taken into account when collecting evidence, and how evidence may be collected in contexts in which the variables cannot readily be controlled
- decide the extent and range of data to be collected and the techniques, equipment and materials to use

Obtaining and presenting evidence

- use a range of equipment and materials appropriately and take action to control risks to themselves and to others
- make observations and measurements, including the use of ICT for data logging to an appropriate degree of precision
- make sufficient relevant observations and measurements to reduce error and obtain reliable evidence
- use a wide range of methods, including diagrams, tables, charts, graphs and ICT, to represent and communicate qualitative and quantitative data

Considering evidence

- use diagrams, tables, charts and graphs, including lines of best fit, to identify and describe patterns or relationships in data
- use observations, measurements and other data to draw conclusions

- decide to what extent these conclusions support a prediction or enable further predictions to be made
- use their scientific knowledge and understanding to explain and interpret observations, measurements or other data, and conclusions

Evaluating

- consider anomalies in observations or measurements and try to explain them
- consider whether the evidence is sufficient to support any conclusions or interpretations made
- suggest improvements to the methods used, where appropriate.

Students may have had experiences with these skills during secondary education. Again, the question is to what extent and level of complexity. Hence, it will be important to pay due attention to this during in-service teacher training. The inclusion of skills in a knowledge base seems a bit of a paradox

ATTITUDES AND PEDAGOGICAL CONTENT KNOWLEDGE

Prospective teachers of science and technology preferably have curious minds themselves and have a positive inclination towards the teaching of science and technology (Walma van der Molen, 2009). Many primary education students, however, arrive not only with a weak science and technology background but also with memories of frustrating school experiences, resulting in low confidence (Harlen, 1997; Appleton, 2003; Murphy, Neil & Beggs, 2007). Attitudes, moreover, are difficult to determine reliably since students may provide desirable answers or simulate desirable behavior when they know they are being evaluated.

Key issues in teaching science and technology are how to generate scientific curiosity, how to develop knowledge and understanding for a given topic in a given context, how to deal with prior knowledge and alternative conceptions, and how to develop student skills in reasoning with evidence and with concepts. This amalgam is often referred to as Pedagogical Content Knowledge (Shulman, 1987; Abell, Rogers, Hanuscin, Lee, & Gagnon, 2009). It develops over time and with experience, is closely linked to situational parameters and is difficult to assess with standardized means (Rohaan, 2009). Subject matter knowledge will be a prerequisite for pedagogical content knowledge, though.

REFLECTIONS AND DISCUSSION

The idea of a knowledge base for primary science and technology has its complications. There is a rich set of literature on scientific and technological concepts, topics and theories that are important to primary education. A conceptual knowledge base for the purpose of inspiring and informing teachers, to help them choose and prioritize in the immense domain of science and technology, and to guide their preparation, certainly is a possibility. We suggest that teachers who are acquainted with Harlen's (2010) set of fourteen big ideas in and about science, with the topics of Dijkgraaf's (2008) Canon, with the priorities for technology propsed by Hacker (2009), and with the skills for inquiry and design as stated by the Engeland and Wales curricula (QCA, 2009), are well-prepared for primary science and technology education.

Problems arise when 'knowledge' is taken more literally for the sake of testing and selecting. In that case, reference to broad concepts does not suffice. Those who have to take a high-stakes test will demand to know which texts, facts, details, definitions or formulae they need to learn. When the focus is too much on factual knowledge such that knowing appears to be privileged over understanding, skills, and attitudes and pedagogical content knowledge, we deny the true nature of science and technology as open-ended design and inquiry activities. We should value understanding over knowledge as a worthwhile and valued goal in education because of its enabling properties with regard to satisfying curiosity, promoting feelings of confidence and competence and facilitating further learning (Parker, 2004).

We also have to deal with the fact that many students nowadays enter teacher-training colleges with little previous knowledge of science and technology, and that only a subset of what is needed can realistically be developed in pre-service programs, even when the amount of credits and contact hours devoted to science and technology is increased. Many educators therefore prefer to teach by example (cf. Wagenschein, 1989; Heywood & Parker, 2009). In their book *Elementary Science Teacher Education,* Olson and Appleton (2006, p134) suggest to choose a limited number of topics, to integrate the acquisition of subject matter knowledge with the acquisition of teaching skills and pedagogical knowledge, and to make clear to students that their earlier failure in science was a failure of the system and not a personal failure. We certainly agree with the wisdom in this, but note that this approach does not guarantee a more comprehensive knowledge base.

In order to arrive at a knowledge base that does justice to the comprehensiveness of the domain, to the limit possibilities of expanding knowledge within the teacher-training colleges and with respect to the starting position of students, it may be well advised to focus on solutions that raise the pre-knowledge of students. Finland, for example, requires that all teachers have a Masters degree, which ensures that students have sufficient cognitive abilities. USA programs require students to take science courses outside Schools of Education, such as completion of the General Education science requirements where students typically take 7 US credits of non-major science courses in a liberal arts college setting. UK programs require a C grade or better on the GCSE exam, which is administered at age 16.

When students who want to become primary teachers have to take more science courses, whether they are in secondary or in higher education, it will be important to turn these experiences into a success. Many prospective teachers felt alienated and frustrated by the abstract, theoretical approaches with its emphasis on formulae and calculations (Murphy, Neil & Begg, 2007; Traianou, 2007). A qualitative and conceptually oriented approach (Hewitt, 2010) matches the interests of this group much better and could raise attitudes and self-efficacy.

In the current situation, much of a teacher's knowledge base, both with respect to subject matter knowledge but also with regard to skills, understanding, and pedagogical content knowledge, will have to develop through the teaching experience of active and reflective teachers. Support by high quality and life long professional development programs will be needed.

Ed van den Berg
Amsterdam University of Applied Science & University of Amsterdam, The Netherlands
e.vandenberg@uva.nl

J. (Hanno) van Keulen
Fontys University of Applied Science & Utrecht University, The Netherlands
j.vankeulen@uu.nl

HENDERIEN STEENBEEK, PAUL VAN GEERT, NENA HAGEMAN, RUTGER MEISSNER, SABINE VAN VONDEL, KEES BROEKHOF, AND HANS COHEN DE LARA

6. "HAVING EYES, GIVING EYES, RECEIVING EYES"

INTRODUCTION

In order to maintain its economic growth and the current level of wealth, the Netherlands need people with a scientific education who can ensure continuous technical capability of the highest standards in all fields of expertise. Unfortunately, a shortage of people with such qualifications is predicted (Van Benthem, Dijkgraaf & De Lange, 2005). Several explanations are given for this prediction. One of them is given by Creighton (2008), who states that teachers have a negative attitude towards science education and this has a negative influence on their thinking and teaching of this subject (Van Aalderen-Smeets, Walma van der Molen, & Asma, 2011). Further, teachers also seem to have low self-efficacy with regard to teaching science and have other curricular priorities (Schwartz, Lederman, Norman & Abd – El - Khalick, 2000). There is evidence that teachers do not sufficiently promote science education in elementary school (Bransford, Brown & Cocking, 2004) and that children loose their interest in science in the course of their elementary school career (Osborne, Simon and Collins, 2003). The result of these findings is that children do not get the education necessary to let their science and technology talents prosper.

In the Netherlands the program VTB-pro has been initiated with the aim of enabling teachers and prospective teachers to deepen their knowledge and become proficient in the domain of science and technology in primary schools by providing a training program (www.vtbpro.nl). The outputs of an associated program, namely the Curious Minds project, will be important for achieving this aim. Curious Minds investigates beta talents of young children in the preschool age. The project emphasizes that it is important to stimulate the science talents of children when they are still young, in the age between 3 – 6 years. Young children show talent in science and technology in the form of natural inquisitiveness, enthusiasm, asking questions and actively exploring their world. The concept of talent of the Curious Minds project entails that every child is talented, in the sense of having a particular developmental potential. Talent is a transaction between a talented child and a talent-stimulating environment. A child is seen as talented if the child is intrinsically motivated, cognitively active and exhibits a level of performance leading to further learning (Van Geert & Steenbeek, 2007; Steenbeek, Van Geert and Van Dijk, 2011). In this respect, talent also includes the capacity to elicit high quality information and instruction of the material- and social environment (Van Geert & Steenbeek,

Marc J. de Vries, Hanno van Keulen, Sylvia Peters, and Juliette Walma van der Molen (Eds.),
Professional development for primary teachers in science and technology.
The Dutch VTB-Pro project in an international perspective.

2007; Steenbeek, Van Geert and Van Dijk, 2011). In short, the talent of children in science manifests itself and develops in the dynamic relationship between three components, namely the child, the adult and the task (the Curious Minds talent triangle). The child and the environment mutually stimulate each other, in such a way that a positive developmental talent-spiral can develop over time (Steenbeek & Van Geert, 2009; Steenbeek, Van Geert and Van Dijk, 2011). The educational philosophy of the Curious minds project is based on the observation that learning takes place in social interaction and that learning is a dynamic, socially situated process (Hirsch-Pasek, Golinkoff, Berk, & Singer, 2009; Van Geert & Steenbeek, 2005). Therefore, the Curious Minds project uses good practice examples of interactions, by means of Curious Minds video clips which show a triangular interaction between a child and an adult while working on a science- and technology task (www.talentenKracht.nl). It is important that adults gain insight in this interaction process and that they learn how children learn from them (Alfieri, Brooks, Aldrich and Tenenbaum, 2010). For this to happen, it is crucial that adults are capable of seeing the dynamic relationship between child, adult and talent eliciting tasks as an expression of the child's science and technology talents on the one hand, and the adults' potential to stimulate these talents on the other hand.

The aim of the current study, 'Having eyes, giving eyes, receiving eyes', is to gain insight in the adult's perception of the child's science and technology talents. The adults that we will focus on in this study are the children's parents and the children's teachers, i.e., the professional educators. Among this group of professional educators a distinction is made between day care teachers and primary school teachers, and among the primary school teachers a further distinction is made between experienced teachers and student teachers. Since these groups have different positions and obligations in the educational process as a whole, for instance they differ in terms of experience, training, responsibilities and in terms of the age of the children put under their care, we expect to find different views or sensitivities with regard to children's talent among the four groups (Steenbeek & Van Geert, 2009).

The study's main objective was to define the respondent groups' current level of knowledge about the properties of young children's talented science and technology reasoning. This current level of knowledge can be used as a reference level for the development of future interventions which aim at further educating the respondent groups in how to recognize and stimulate talented behaviour of children. In this study 'see' is defined as what respondents notice in the observable behaviour that accompanies the science and technology talent of children in the dynamic interaction process (Steenbeek & Van Geert, 2009).

The research questions are: Which properties of children's talented science and technology reasoning do professional educators and parents see in Curious Minds video? Which properties of the adult's stimulating this talent do professional educators and parents see in those video clips? Are there differences or/and similarities between the groups of professional educators and parents?

This study is exploratory, and no specific expectations are formulated about differences between the respondents and the given description of talent. However, differences are expected in the answers of the respondent groups, because of their different positions and obligations in the educational process as a whole.

METHOD

Participants

Participants were teachers in primary education, student teachers in primary education, day care teachers and parents (n = 142). Table 1 shows the details of the respondents.

Table 1. Number of respondents per target group by gender and work experience

	N	Gender	Average years work experience
Teachers	43		14
		Male	7
		Female	36
Student teachers	25		Not relevant
		Male	5
		Female	20
Day care teachers	33		12,5
		Male	0
		Female	33
Parents	41		Not relevant
		Male	9
		Female	32

Qualitative interview method

In this study, qualitative interviewing was used to establish the expertise level of professional educators and parents. The expertise level is defined in terms of their ability to notice fundamental aspects of science and technology talents of children expressed in the children's actual behaviour. A semi - structured interview design was used to obtain descriptions of spontaneous observations from the respondents in an optimally valid and reliable way (Saunders, Lewis & Thornhill, 2008). The semi - structured interview had mandatory questions, but the interviewers were free to deviate from the interview protocol dependent on the answers of the respondents.

Procedure

The interviews usually took place at the office of the respondents or in their homes. The interview started with a short introduction about the Curious Minds project, the aim of the project, and an explanation of the procedure. To avoid that the respondents would give socially desirable responses, it was explicitly indicated that responses would not be qualified, i.e., there were no right or wrong answers. In order to guarantee the reliability of this study, a manual was used as a guide for conducting the interviews.

The average duration of the interviews was 15 minutes. The three Curious Minds video clips (Uittenbogaard & Feijs, 2006, 2007). were shown on a laptop brought by the interviewer. Selection of the video clips was based on differences in science and technology content and in the interaction between the child and the adult (Appendix I: Extensive description of the clips). Together, the three selected video clips showed a representative range of science and technology subjects, and child adult interaction forms. In the first clip, two children were asked to solve a difficult counting task (count marbles in a jar). The second video clip focused on the principle of light reflection in a system of mirrors (a periscope). The third clip asked the child to recognize and continue a spatial pattern in the form of a wall consisting of two types of blocks, the beginning of which was constructed by the adult

The aim of the interview was to find out what the respondents notice when watching the video clips. The respondents were asked what they noticed in the video clips by means of the following questions:

- What did you notice in the child's behaviour?
- What did you notice in the adult's behaviour?
- What did you notice about the material?

These questions were repeated after each of the three video clips.

Analysis

The interviews were audio recorded with permission of the respondents to ensure that all information was captured. All the audio recordings of the interviews were transcribed using InqScribe (2009) and analysed with ATLAS.ti (Mortermans, 2001; Murh, 1997). The methodology used is based on the grounded theory approach (Glaser & Strauss, 1967). In this approach, the transcripts of interviews are analysed in an inductive manner, with the aim of obtaining the best possible view of the respondents' own ideas. To achieve this aim, the codes are derived from what was said in the interviews (bottom - up) instead of having been formulated in advance, on the basis of (theory-driven) expectations (top-down). In this way the interviews are coded as openly as possible, so that the view of the respondents will come forward.

Two interviews of each target group were used for designing a codebook. The codes in this codebook were named in concordance with the properties mentioned by the participants. After assigning the codes, further analysis showed that some codes could be grouped, or "clustered" based on their content, under different domains of talented behaviour of children and talent- stimulating behaviour of adults. In table 2 a description of the clusters and examples of accompanying codes and properties are presented. Note that the clusters based on the respondents' spontaneous answers turned out to be more or less similar to the domains that are mentioned in the Curious Minds definition of talented behaviour of children (as mentioned in the Introduction). These domains refer to performance, experiential, motivational, cognitive and help-seeking aspects of behaviour. Note that experience refers to the way in which the child or the adult perceived or evaluated her or his own performing or guiding of the task.

In order to statistically test whether the differences in answer frequencies between the clusters were due to chance or not, a Monte Carlo procedure was used, more precisely a random sampling of answers based on the null hypothesis that the cluster probabilities were similar for all clusters and all respondent groups. Statistical randomisation procedures are recommended if the data structure is idiosyncratic (meaning that it is determined by the respondents' initiatives) and the samples of respondents are small (Todman & Dugard, 2001). The Monte Carlo Simulation was used in two ways, namely to search for significant results within the four respondent groups and between the respondent groups. A significant result within a respondent group (i.e., a within-group analysis) means that the respondents within this group mention a particular property significantly more than can be expected on the basis of chance alone. If applied to differences between clusters, the within group comparison tests whether a particular cluster is mentioned significantly more often than another cluster by a particular group of respondents, for instance the parents. A significant result between the respondent groups (i.e., a between-group analysis) means that the differences in the frequencies with which different groups have chosen a particular cluster are significantly greater than can be expected on the basis of chance.

Table 2. Clusters and description of the clusters

Clusters	Description	Examples of codes	Examples of properties
Experience	Refers to the experience of performing the task (child) or guiding the task (adult)	Enthusiasm Having fun	"I think it is important that they have fun in whatever they are doing" (experience of a child) "She is very enthusiastic" (experience of an adult)
Motivation	Refers to the willingness of the child to know 'how it works'	Eager to learn Curious	"That the child is willing to learn"
Cognition	Refers to the child's (cognitive) thought process	Concentration Insight	"And they were really focused"
Act	Refers to actively doing the task (child) or guiding the task (adult)	Investigating Directive	"Yes, investigating. All the time busy with trying to do better" (action of a child) "I find her a little too directive (acting of an adult)
Attitude	Refers to the adults mental attitude towards the child during the task execution	Openness Negative approach	"The open attitude towards the child"
Asking	Refers to the adult's verbal behaviour	Open questions Right questions	"He poses the right questions, so the children start thinking"

To ensure that different researchers assigned the same codes to the content of quotations, the interrater reliability was calculated. Two interviews were separately coded by three raters. The percentage agreement between interviewers was found to be 0.83, using the proportion reliability method, and this agreement was found to be higher than chance agreement ($p<0.05$). Thus, the agreement satisfies the arbitrary set criterion of 0.80 for interrater proportion reliability.

RESULTS

The results will be summarized in accordance with the 2x2x4 design of the study: two content conditions (properties of talented science and technology behaviour in children on the one hand, and properties of talent stimulating behaviour of adults on the other hand), two coding conditions (one based on specific property codes, the other on clustered codes), and finally four respondent conditions (teachers, student teachers, day care teachers, and parents).

Properties of the child's behaviour

The distribution of the three most frequently assigned codes for each respondent group is given in table 3.

The code Smart was assigned significantly more often than chance in the day-care teachers group ($p = 0.04$), in the students group ($p = 0.01$) and in the parents group ($p = 0.04$). However, for the teachers Smart does not occur in their top three.

Teachers and students are similar in that only one of the three categories is significantly different from chance. It is striking to see that the code that teachers mentioned more often than chance refers to Having fun in the science and technology activity. Day care teachers and parents are similar in that all three of the most frequent codes are statistically significantly different from chance. Day care teachers and parents also agree on their choice of the code Investigating as one of the three most frequently mentioned codes. They differ in their choice of a Good cooperation (day care teachers) versus Convinced of themselves (parents).

Table 3. Most frequently assigned codes of child's behaviour for each respondent group

Respondent group	Code	Significance (p)
Day care Teachers	Smart	0.04*
	Good co-operation	0.04*
	Investigating	0.09**
Teachers	Having fun	0.07**
	Impulsive	0.22
	Enthusiastic	0.33
Students	Smart	0.01*
	Good co-operation	0.23
	Impulsive	0.28
Parents	Smart	0.04*
	Investigating	0.05*
	Convinced of themselves	0.09**

* Significant at 0.05 level
** Significant at 0.10 level

In figure 1 four clusters are presented which are formed on the basis of the content of the assigned codes (properties) mentioned by the respondents. Summed over all respondent groups, the cluster Cognition has the highest score (e.g. ""I was surprised at the child's insight ... fantastic'), the cluster Act scores second highest (e.g. "Investigative ... constantly looking for improvement'), the cluster Motivation third (e.g. "He really wants to find out"), and the cluster Experience (e.g. "The initial unbelief of the child when it says: That cannot be! ") is scored the least.

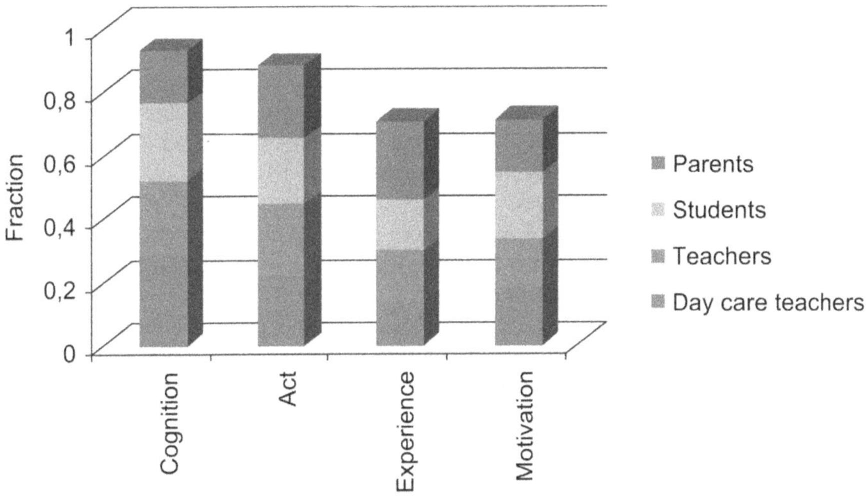

Figure 1. Cluster scores of codes of child's behaviour, by all four respondent groups.

The question now is whether – per individual cluster - respondent groups differ in the amount of codes assigned. In other words, does one of the respondent groups contribute significantly more (or less) than the other respondent groups to this particular cluster?

In general, the four respondent groups are relatively similar in their choice patterns. A chi square analysis showed that only the teachers and the day care teachers differ significantly in the composition of the choice pattern as a whole. The clusters differed in the respondent groups which choose them most frequently: Experience for the parents, Motivation and Action as equally frequently occurring clusters with the students, Action for the teachers and Cognition for the day care teachers.

The preference of teachers for the Experience cluster differed significantly from the preference for Experience from the three other respondent groups. Although the Cognition cluster found its highest frequency in the group of day care teachers, its occurrence was equal in the students and the teachers, and only the difference with parents was statistically significant. Hence, students, day care teachers and teachers agree in their preference for Cognition as an important property of talented science and technology behaviour in children, and differed as a group of professional educators with the parents. Note that these results do not say which was the most preferred cluster in each of the respondent groups, they only state in which respondent group a particular cluster found its highest frequency of choice.

The second and probably also more interesting question is whether – per respondent group – differences can be found over the clusters. That is, does a respondent group significantly mention one of the clusters more often than the other clusters? (In statistical terms: are differences within a respondent group, significantly bigger or smaller than can be expected by chance?)

For the student group, none of the differences between the cluster frequencies was statistically significant. This means that the students' most frequently chosen category, which was Cognition, was not statistically significantly more frequently chosen than any of the other three clusters. In the group of parents, the most frequent cluster was Experience, the second frequent cluster was Action. These two clusters differ significantly from the two least frequently chosen clusters, namely Cognition and Motivation.

In the teachers, Cognition and Action were chosen with almost equal frequency. This choice differed significantly from the choice for the almost equally chosen categories Experience and Motivation.

In the day care teachers group, the choice for the most preferred category, Cognition, differs significantly with the choices for the other three clusters. The second frequent choice, for Action, differs significantly with the least frequent choice, for Experience.

3.2. Properties of adult behaviour

The distribution for the adult's behaviour is given in table 4.

Table 4. Most frequently assigned codes of adult's behaviour for each respondent group

Respondent group	Code	Significance (p)
Day care Teachers	Not directive	>0.01*
	Directive	0.04*
	Calm	0.06**
Teachers	Directive	0.01*
	Not directive	0.02*
	Calm	0.08**
Students	Directive	0.02*
	Not directive	0.09**
	Bring things	0.11
Parents	Directive	0.05*
	Supporting	0.10**
	Coaching role	0.12

*Significant at 0.05 level
** Significant at 0.10 level

Directive behaviour is highest in three of the four respondent groups, and the frequencies are significantly greater than chance (p=, 0.01, 0.02, and 0.05, respectively). An example of a property of Directive behaviour is 'Asking the right questions...helping the child'. The day care teachers score Directive (p = 0.04) as second element with Not Directive (p > 0.01) as their first. The teachers and students score Not Directive (p = 0.02, resp. p-0.09) as second. An example of a property of Non-Directive behaviour is 'He (the adult) asks open questions all the time'.

Both day care teachers and teachers mention Calm (e.g. "the adult remained very calm and composed") as one of the properties, whereas students mention Initiation'(e.g. 'I notice the adult presenting new points of view'), and parents mention the Coaching role of the adult (e.g. 'I really notice the adult complimenting the child").

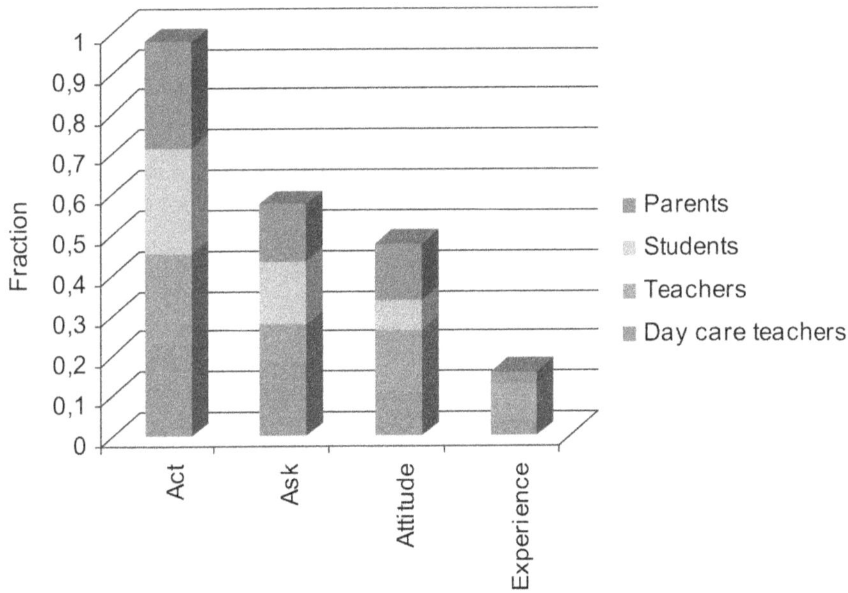

Figure 2. Cluster scores of codes of adult's behaviour, by all four respondent groups

In figure 2 four clusters are presented which could be made on the basis of the content of the codes assigned to the properties mentioned by the respondents. Notice that a new cluster occurs, namely the cluster Asking, referring to the adults' questions (For the description of the clusters, see table 2.) In terms of the overall choice, Act (58%) is the largest cluster, and Experience is the smallest. The teachers mentioned properties belonging to the Act cluster (which refers to actually doing or guiding the task) significantly less ($p < 0.01$) than should be expected on the basis of chance. This lack of interest for the behavioral aspect differs statistically significantly from the choices of students, parents and kindergarten teachers (p = 0.04, 0.06, and 0.02 respectively). However, the lack of preference for Experience in contrast to Act is primarily due to the choices of the professionals, and not the parents.

We compared the overall structures of the choice patterns of the four respondent groups by means of a chi square analysis. It showed that only the difference between parents and teachers is statistically significant, implying that the overall qualitative similarity in choice patterns among the respondent groups is quite high. Then the question was whether – per individual cluster - respondent groups differed in the amount of codes assigned. In other words, does one of the respondent groups contribute significantly more (or less) than the other respondent groups to this particular cluster? The only interesting differences concerned, first, the choice of the Attitude cluster in the teacher group, which was significantly lower than its frequency in the other three respondent groups, and second, the choice of the Ask cluster (referring to the adults' questions) in the parent group, which was significantly lower than the choice for Ask in the professional groups (students, teachers, day care teachers).

A more interesting question is whether – per respondent group – differences can be found over the clusters. That is, does a respondent group significantly mention one of the clusters more often than the other clusters?

The students' first choice concerned Act, the second was Ask. These two choices differed significantly from the choice for the least preferred clusters, namely Experience and Attitude.

The parents showed the clearest preference for Act, and this choice differed significantly from their choice for the three other clusters. The second most frequent cluster, Attitude, differed significantly from the two least frequently chosen clusters, Ask and Experience.

The teachers' most frequent choice was Act and this differed significantly from the other three clusters. The teachers were the only group for whom the difference in preference for the four clusters was significant for each possible combination.

The most frequent choice of the day care teachers was Act, and this differed significantly from the other three clusters. The clusters Ask and Act were chosen about as frequently, this choice was statistically different from the choice for Experience.

Task object

In total 11 codes were assigned to properties that the respondents mentioned about the task material in the video clips. Compared to the other elements of the interaction dynamics- the child (623 properties) and the adult (511 properties) - task object properties were mentioned the least in all respondent groups (299). The distribution of the three most frequently assigned codes for each respondent group is given in table 5.

All respondents have Positive judgement in their top three of most frequently assigned codes, and this preference differs statistically significantly from a choice based on chance. The day care teacher group and parent group both have significant results for Positive judgement ($p = 0.01$ and $p = 0.02$, respectively). The students' score on Good to work with ($p = 0.01$) is significantly greater than a preference based on chance alone. None of the other results is significant. No further similarity between respondents is observed in the results.

The combined scores on Good to work with, Positive judgement and Defiant are high compared to the other eight codes. No argumentation or details on these high scores are available from the interview.

Table 5. Most frequently assigned codes of task material for each respondent group

Respondent group	Code	Significance
Day care Teachers	Positive judgement	0.01*
	Defiant	0.07**
	Simple	0.21
Teachers	Positive judgement	0.08**
	Good to work with	0.20
	Defiant	0.31
Students	Good to work with	0.01*
	Positive judgement	0.12
	Common	0.27
Parents	Positive judgement	0.02*
	Defiant	0.06**
	Simple	0.11

*Significant at 0.05 level
** Significant at 0.10 level

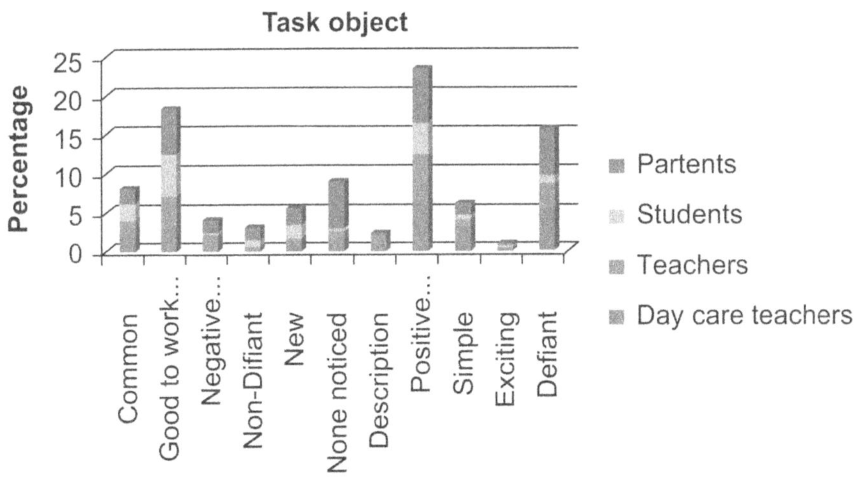

Figure 3. Frequency scores by codes of the task object, by all four respondent groups.

Because the 11 most frequently assigned codes of the task object have little similarity in terms of content properties, further clustering of these codes has little additional significance.

CONCLUSIONS

The aim of this study was to find an answer to two questions, first, what are the properties that professionals and parents assign to talented science and technology behaviour in young children, and second, what do they think are the properties of adult behavior that can stimulate science and technology talents in young children. The respondents were asked to describe the properties they observed while watching three video clips of science and technology behaviour that emerged in the interaction between a child, an adult and a particular task object.

These questions were presented to respondents from four groups, parents and professional educators, which were subdivided according to their experience (student teachers versus active practitioners) and to the age of their pupils (teachers versus day care teachers). Thus, a second aim of this study was to find whether there are any differences in opinions and observations between these four respondent groups.

Child

The most frequent observation of day care teachers, students and parents is that the child is smart. Referring to the child, the teachers most frequently mentioned single property is having fun. This is a somewhat counter-intuitive finding in that having fun is more easily associated with preschool age than with the more serious activities that one expects from primary school children. The fact that it is a somewhat unexpected property at this age, namely that children have fun with primarily cognitive science and technology activities, might explain why teachers and not kindergarten teachers see this as a typical property of talented science and technology behavior.

When we examine the codes as summarized in clusters, all the respondents seem to focus on the cognition part of talent behaviour the most, whereas they focus the least on experience and motivation.

When we examine this in detail, it is striking that only the group of day care teachers indeed chose cognition significantly more often than they chose other clusters. In the student group, cognition was not statistically significantly more frequently chosen than any of the other three clusters. In the group of parents, the most frequent cluster was experience, whereas in the teachers group, cognition and action were chosen with almost equal frequency.

Adult

The respondents' most frequently chosen single property is that the adults in the video clips are directive or not directive. The day care teachers mention the non-directive aspect the most, whereas the students and teachers seem to focus on the directive aspect. The parents do not mention the non-directive aspect in the top three of most frequently assigned codes. Instead they focus on the coaching and supporting role of the adult.

All groups mention the cluster Act the most and the cluster Experience the least. So, the respondents focus more on how adults act in guiding a child in a task than on how adults experience the guiding of the child. However, the lack of preference for experience in contrast to acting is primarily due to the choices of the professionals, and not the parents. This conclusion is entirely in line with a distinction we have found with regard to the child's behaviour, where the parents typically selected the experience-related aspects, in contrast to the professionals' preference for cognitive aspects.

In terms of clustered scores, the four respondent groups are similar in their preference for the Acting component over the components relating to asking questions, experience and Attitude.

Task object

The respondents mentioned properties of the task object noticeably fewer than properties of the child and the adult. However, the parents make many comments on the task object relative to the other respondent groups. All the respondent groups mention positive judgement in their top three of most frequently assigned codes. In addition, the students mention good to work with also significantly often. No further similarity between respondents is observed in the results. Thus, the task object seems to be the element of the dynamic interaction process most undervalued by the respondents.

Differences between professional educators and parents

With regard to the child's talented science and technology behaviour as well as with regard to the adult's role of talent stimulation, parents mentioned properties relating to experience significantly more than the professionals do. The professionals on the other hand focus more on cognition. The training that the professionals have followed can eventually explain the difference between parents and professionals, in that they are trained to work with children in a result-oriented way. Their result-oriented way of working is further strengthened by the fact that professionals are used to work with a larger number of children than parents.

DISCUSSION

Comparison between Curious Minds definition of talent and respondents' answers

The basic assumption underlying the Curious Minds concept is that talent of children in science and technology manifests itself and develops in the dynamic interaction between child, adult, and task object. This interaction triangle results in a talent-evoking situation (constituted by adult and task-object) and expression of talent by the child. A child is seen as talented if the child is intrinsically motivated, cognitively active and exhibits a level of performance leading to further learning (Van Geert & Steenbeek, 2007; Steenbeek, Van Geert and Van Dijk, 2011). An adult is seen as adequately stimulating this talent when providing an environment such that talent can optimally develop. That is, it is important that the adult 'sees' cognitive, behavioural, motivational, and experiential aspects of the child's behaviour. In addition, it is important that the adults gain insight in this process and that they learn how children learn from them. For this to happen, it is crucial that adults are capable of seeing the dynamic relationship between child, adult and talent eliciting tasks as an expression of the child's science and technology talents on the one hand, and the adults' potential to stimulate these talents on the other hand.

Recommendations for the development of training programs

This study was carried out in order to gain insight in the talent recognition skills of parents and professionals. This insight can be used as a baseline for developing a training program for professionals and parents in order to further develop their talent recognition skills. Based on the results of this study, several recommendations can be formulated.

A first recommendation is to work with video clips as a means for starting a discussion about children's talented behaviour and the role adults can play in stimulating this behaviour. In this study we found that all respondent groups enjoyed seeing the video clips and that these clips yielded more than enough inspiration for conversations about talented science and technology behaviour, of both child and adult.

Second, different respondent groups validate, or at least see, different aspects of talented behaviour more than other groups. The professional educators primarily see the cognitive aspects, whereas parents focus more on the experiential aspects. Therefore, we recommend that for parents the focus of a training program needs to be on recognizing the other aspects of talented behaviour as well, such as the role cognition plays in talented science and technology behaviour. We recommend that for professional educators the focus of a training program needs to be on other aspects than the cognitive one, which they already recognize spontaneously.

A last recommendation is that the role of the task objects and their affordances should be stressed in a training program for all respondent groups. That is, it was noticed that all respondent groups mention aspects of the task object the least, whereas the affordances that these materials offer (or not) greatly contribute to the richness of the talent stimulating environment in which children can explore, play, and show talented science and technology behaviour (Foo and Hedberg, 2005).

Talent; a retrospective or a prospective standpoint?

Within Curious Minds, the development of talent is seen from a prospective standpoint. Talent is described as a process that develops over time, with the child actively interacting with the adult and its environment during the task execution.

The majority of the interviewee responses describe talent not as a process but as an operational developmental level reached by the child. Hence, the interviewees in general have a retrospective view of talent, i.e., given an observable result (the excellent performance). It appears as if the interviewees look back on the history of the child as having a talent as a property that produces or generates the observed excellent behaviour.

Any program to be developed for the field of science and exploration learning behaviours should account for this retroactive view on talent currently present in the respondent groups. One can use this as a starting point for making them aware of the possibility of taking a prospective view in line with our Curious Minds definition of talent. That is, adults must learn to see themselves as having an important function in getting the talent process off the ground, in such a way that a positive developmental talent-spiral can develop over time (Steenbeek & Van Geert, 2009).

Henderien Steenbeek
Groningen University, The Netherlands
h.w.steenbeek@rug.nl

Paul van Geert
Nena Hageman
Rutger Meissner
Sabine van Vondel
Groningen University, The Netherlands

Kees Broekhof
Hans Cohen de Lara
Sardes Educational Services, The Netherlands

APPENDIX I: DESCRIPTION OF THE VIDEO CLIPS

The video clips showed in the interviews are to be found on the website www.talentenkracht.nl. These three specific clips are selected on base of the different task materials and the interaction between the child and the adult. The following descriptions are also to be found on this website (for this report the descriptions are translated form Dutch in English).

Marijn, Maurits and the marbles (shown from 0.00 till 2.53)

The first video shown was about Marijn, Maurits and the marbles: On a table is a jar of marbles. Marijn (age 5.9) and Maurits (age 4.11) have been asked how many marbles the jar contains. "More than one hundred" estimates Marijn. "Maybe two hundred" says Maurits. "No, not that many" Marijn argues, "It doesn't seem more than one hundred." The tutor asks the children how they could come to know the answer. They agree that the marbles have to be counted. They apply many different strategies but none seems to work since the marbles roll away. Finally, the tutor intervenes, handing them a towel and asking if this might come in handy. "Yes" says Marijn "then it is not that slippery." Marijn and Maurits start counting the marbles.

Jop and the periscope (shown from 0.00 till 3.10)

The second video shows that Jop (age 4.8) is presented with a periscope. At first he does not know what to do with it; he wants to throw something inside. The tutor tells him to look through the periscope. He sees a painting with a frog in front of him, but if he looks without the periscope the frog is gone. Subsequently he discovers the painting is behind him. He seems to understand that one can see behind oneself with the periscope. Searching for an explanation Jop discovers the mirrors inside the periscope. How the mirrors contribute to the fact that one can look behind still seems to puzzle Jop.

Quinten and the walls (shown from *1.55 – 3.15)*

The last video shown is about Quinten (age 5.3). The adult asked to continue building a wall made of blocks differing in size. He already finished one wall successfully when this video clip starts. The tutor pronounces the third and last wall as being a little bit more difficult than the other two. Quinten says: "this is going to be fun!" he follows the actions of the tutor closely and easily completes the task.

APPENDIX II: INTERVIEW PROTOCOL

Part 1: Open questions about three video clips (questions are asked every time after showing the clips)

2.1 What is your opinion about the video clip?
2.2 What did you notice in the video clip?
2.3 What did you notice in:
   a. the child's behaviour?
   b. the adult's behaviour?
   c. the task object?
   d. the interaction between the adult and the child?
   e. the course of the task?
2.4 Would you behave the same way as the adult did in guiding the child?
   a. Why would you behave the same way?
   b. Why wouldn't you behave the same way?

Part 2: Final questions

3.1 Have you seen the video clips before?
3.2 What do you know about the Curious Minds project?
3.3 Did you check out the website of the Curious Minds project?
   (Following questions are not asked to parents)
3.4 How long do you have experience in working with children?
3.5 Did you take any courses which can be compared to the Curious Minds project?
3.6 Do you have any further questions?

PART II

PROFESSIONAL DEVELOPMENT AND ATTITUDES TOWARDS SCIENCE AND TECHNOLOGY

JULIETTE WALMA VAN DER MOLEN

7. INTRODUCTION TO PART II

Attitudes towards science and technology

Role of this Part in the book

Achieving a positive attitude towards science and technology is seen as a highly important factor within the VTB-Pro program. Besides a coherent offer of knowledge and skills in science and technology and insights into inquiry-based learning, the VTB-Pro program calls for explicit attention to a more positive attitude towards science and technology as an underlying *and* overarching component in the professionalization of primary school teachers. This part of the book describes the scientific research that was conducted within the VTB-Pro program on the concept of primary school teachers' attitude towards science and technology and on underlying factors that may contribute to the enhancement of these attitudes.

The importance of positive teacher attitudes

The study on attitudes towards science and technology has received considerable societal and research attention over the last decades. The reason for this interest is probably twofold. A major short-term concern for many countries is the fallback in the number of students pursuing a career in technology or science, while at the same time society is increasingly dependent on these disciplines (Osborne, Simon, and Collins, 2003; Osborne and Dillon, 2008). A second, more general concern is the increasing gap between the generally low scientific and technological literacy of the population and a negative attitude towards these topics on the one hand, and the advancement of the technical disciplines on the other hand (Osborne and Dillon, 2008). Several interventions in different countries to change this did not succeed in changing students' attitudes towards science and technology, mainly because they failed to take the *teachers'* attitudes into account (Haney, Czerniak and Lumpe, 1996).

Research shows that children develop their interest in and attitudes towards science before the age of 14 (Osborne and Dillon, 2008). While the lack of interest in science often manifests itself during secondary school when young people must choose the subjects they will study, most students have already excluded the choice of scientific or technological subjects long before, i.e., during their years in primary school (Osborne & Dillon, 2008, Tai, Qi Liu, Maltese, & Fan, 2006;

Marc J. de Vries, Hanno van Keulen, Sylvia Peters, and Juliette Walma van der Molen (Eds.),
Professional development for primary teachers in science and technology.
The Dutch VTB-Pro project in an international perspective.

Young & Kellogg, 1993). Science education should therefore pay explicit attention to improving students' interest in and attitudes towards science, and this should take place beginning at the primary school level (Haney, Czerniak, & Lumpe, 1996; Tobin, Tippins, & Gallard, 1994; Van Driel, Beijaard, & Verloop, 2001).

Primary school teachers are therefore crucial in determining positive attitudes of students (Harlen and Holroyd, 1997; Skamp, 1991; Palmer, 2001; Osborne, Simon, & Collins, 2003). However, it has been shown that (pre-service) primary school teachers' scientific and technological literacy is low and that their attitude towards science and technology is mostly negative (Palmer, 2004; Trumper, 1998). These negative attitudes often originate from negative experiences that teachers had during their own primary and secondary education and these attitudes persist during their pre-service teacher training (Jarret, 1999; Mulholland & Wallace, 1996; Palmer, 2001; Tosun, 2000; Young & Kellogg, 1993). Studies investigating the attitudes of primary teachers towards science have indicated that teachers with less positive attitudes share a number of characteristics. They have lower confidence and self-efficacy beliefs about teaching science (Skamp, 1991; Tosun, 2000; Yates & Goodrum, 1990); they spend less time discussing and teaching these topics in their classrooms (Goodrum, Hackling, & Rennie, 2001; Harlen & Holroyd, 1997); they rely more on standardized methods and top-down instruction (Appleton & Kindt, 1999; Harlen & Holroyd, 1997; Jarvis & Pell, 2004; Plonczak, 2008); and they are less able to stimulate the attitudes of their students (Harlen & Holroyd, 1997; Jarvis & Pell, 2004; Osborne et al., 2003; Van Driel et al., 2001; Weinburgh, 2007).

The situation described above is unfortunate, and it is important that the self-esteem, interest, and enthusiasm of teachers in science and technology be stimulated and supported. Research shows that such an endeavour can have a positive effect, not only on the teachers themselves, but also on the knowledge level and attitudes of elementary school students towards science and technology (e.g., Jarvis, 2004; Palmer, 2004). In addition to an integrated offering of knowledge and skills within the domain of science and technology and in addition to more insights in inquiry-based learning, a very important (and often overlooked) element of the professionalization of elementary school teachers should thus be the explicit focus on teachers' own attitudes towards science and technology. The goal of that part of teachers' professionalization should be that teachers and aspiring teachers become aware of their own thoughts, values, feelings, and behaviour in the field of science and technology.

Issues in research into primary teachers' attitudes towards science and technology

The research described in the chapters in this part of the book is based on the contention that as a basis for any educational change project or pre- or in-service training, one needs to investigate the attitude of teachers. Such attitudes have been investigated in numerous scientific studies, but it is difficult to compare these studies or to generalize their results due to several reasons (Gardner, 1995; Bennet,

Rollnick, Green, and White, 2001). Most importantly, many studies only marginally theoretically define the construct of attitude. In the relevant psychological research literature, the concept of "attitude" is construed as an internal, personal, psychological tendency to evaluate a particular object or construct in a positive or negative manner (Eagly & Chaiken, 1993). This personal tendency can persist for a longer or shorter period of time and, according to the tripartite model, involves cognitive, affective, and behavioural components. In such a manner, a positive attitude with respect to, for example, learning or studying can consist of thoughts or opinions about the importance of learning for the attainment of a good job or a good future, personal feelings of pleasure derived from learning, or actual behaviour in the form of studying hard or the intention to undertake a particular study.

Within the field of social psychology, attitude is traditionally seen to be one of the most important motives behind numerous processes and clearly related to motivation and interest. In keeping with this, attention to the notion of attitude in the literature on science education has strongly increased over the past few years. Critical in this regard, however, is the distinction between a scientific attitude and attitudes towards science (see Osborne, 2003). A scientific attitude manifests itself in the form of scientific thinking: curiosity, creativity, perseverance, critical reflection, and so forth. Such a scientific attitude consists of necessary scientific process skills. However, a different set of thoughts, values, feelings, and behaviours can be understood under *attitudes towards science and technology* — thoughts, values, feelings, and behaviours that address — for instance — one's thoughts about the level of difficulty characteristic of the sciences and technology, the value attached to science and technology for society, feelings of pleasure or interest with regard to science and technology, and the desire or intention to learn more about science and technology. Many studies in the field of science education fail to make this distinction and focus on a mixture of attitude concepts without specification. In addition, many educational change projects or professionalization programs do not focus specifically on improving attitudes towards science. In most cases, only scientific attitudes are taken into account and it is assumed that people's attitudes towards science will improve as a by-product of other interventions. As was outlined in the theoretical framework of the VTB-Pro program, however, more explicit attention should be given to interventions designed to improve teachers' attitudes towards science, for example through the use of debating and reflection techniques.

A second major issue in the international research on teachers' attitudes towards science concerns the measurement of attitude as an underlying construct in general and, more specifically, the manners in which measurement instruments on teachers' attitudes towards science have thus far been constructed. Attitude is a difficult multidimensional construct to measure. In most cases, self-report survey instruments were developed to measure teachers' attitudes towards science and technology. Many researchers, however, used poorly designed measurement instruments or inadequate methods of analysis and failed to conduct pilot testing, validation, and evaluation of their measurement instruments. Several researchers (e.g., Blalock et al., 2008; Coulson, 1992; Gardner, 1995; Osborne, 2003; Reid, 2006) have pointed at the methodological and psychometric shortcomings of most attitude survey-instruments that have been used to date. Shortcomings that were highlighted in these reviews include: poor construction of items that measure a multitude of non-specified attitude objects, using total attitude scores even though different underlying attitude dimensions are measured by the instrument, using questionable response categories such as a 'neutral' midpoint, using item wording that is not clear enough or that leads to skewed or socially desirable answer patterns, etc. Because of the theoretical and methodological problems that have thus far adhered to research into primary teachers' attitudes towards science, research in this area has thus far progressed only on a very limited scale.

Chapters in this part

The scientific research that was conducted in the realm of the VTB-Pro program and that is described in the following chapters, takes the above-described issues into account. It focuses on the possible underlying dimensions of primary teachers' attitudes towards science and technology and tries to disentangle this concept in a theoretically and methodologically sound manner. In addition, it describes attempts to develop psychometrically reliable and valid attitude instruments and it investigates which aspects of teacher training might be most successful in enhancing attitudes towards science and technology. In chapter 8, Asma, Walma van der Molen, and van Aalderen-Smeets describe a focus group study that was conducted to validate the development of a new comprehensive survey instrument to measure primary teachers' attitudes towards science and technology. The results of that study led the researchers to formulate a new theoretical framework that describes the main underlying components of primary teachers' attitudes in this domain, which is presented in the last part of their chapter.

In chapter 9, Denessen, Vos, Damen, Koch, Louws, and Wigboldus discuss the difficulties that relate to questionnaire ratings of attitude and report on the development of two *implicit association* (IAT's) measures to investigate teachers' general attitudes towards science and technology and their gender-stereotyped attitudes towards the construct. Although more research needs to be done to determine the construct validity of these new implicit association tests in teacher research, the study generates a promising route for the future, in which a combination of both explicit and implicit testing of teachers' attitudes might prove to be very fruitful.

In chapter 10, van Cleynenbreugel, de Winter, Buyse, and Laevers describe an intervention study that was set up specifically to improve primary teachers' attitudes and competencies in relation to science and technology. Apart from measuring attitudes towards science and technology, the researchers also measured teachers' attitudes towards learning by design and towards inquiry learning. A rich dataset was obtained in which questionnaires, interviews and observations were used at teacher and pupil level. Results showed that, especially in the technology domain, teachers' self-efficacy and enjoyment were improved after the intervention and that this had a positive influence on pupils' perceptions of the importance of science and technology, their enjoyment of it, and their intention to invest time in these domains.

Finally, in chapter 11, de Wilde, Sjoer, and de Vries report on a study among pre-service primary teachers in which they investigated (a) what their definition and (stereotypical) image of science and technology is and how their attitudes are related to this, and (b) how their image and attitudes changed after a special project on waterworks. The study combined qualitative and quantitative research methods and showed that the concept or image that pre-service teachers have of science and technology is rather limited and did not improve after the intervention. Similarly, their attitudes did not progress after the intervention. Results are discussed in light of explicit criteria that should be incorporated into interventions in order to obtain the desired progress in concept, image, and attitudes.

In sum, the different chapters in this part of the book provide an overview of the opportunities that lie ahead to investigate, measure, and stimulate primary teachers' attitudes towards science and technology. Such opportunities are certainly not only applicable to the situation in The Netherlands. On the contrary, both explicit and implicit measurement techniques and intervention strategies can and should be applied internationally and we would very much like to take up that challenge with our international readers!

Juliette Walma van der Molen
University of Twente, The Netherlands
j.h.walmavandermolen@utwente.nl

LIEKE ASMA, JULIETTE WALMA VAN DER MOLEN, AND SANDRA VAN AALDEREN-SMEETS

8. PRIMARY TEACHERS' ATTITUDES TOWARDS SCIENCE AND TECHNOLOGY

Results of a focus group study

INTRODUCTION

The study on primary teachers' attitudes towards science and technology has received considerable research attention over the last decades. However, if one looks at the extant literature in this domain, a major problem that becomes apparent is the lack of consistency in the conceptualization of what is meant by teacher attitudes. Attitude is a complex and multidimensional construct and a clear definition and thorough theoretical understanding are essential for research in this area. The present chapter is intended to shed more light on this construct and to present the results of a focus group study amongst pre- and in-service primary teachers that investigated both their personal attitudes towards science and technology and their attitudes towards teaching science and technology at primary school level.

Primary teachers' attitudes towards science have been investigated in a range of scientific studies world wide, but scientific progress in this field is slow due to several major theoretical and methodological issues (Bennett, Rollnick, Green, & White, 2001; Gardner, 1995; Kind, Jones, & Barmby, 2007; Osborne, Simon, & Collins, 2003). Most importantly, both in research and in educational change projects, the concept of an attitude towards science is often poorly articulated (Barmby, Kind, & Jones, 2008; Bennett et al., 2001; Coulson, 1992; Osborne et al., 2003; Pajares, 1992). Many studies do not, or incompletely, define the construct of attitude, do not explicate the different components of attitude that they measured, or do not make a distinction between attitudes towards science and related concepts, such as opinions or motivation. It is, therefore, difficult to determine what precisely was measured or investigated and, as a consequence, to determine if it actually is attitude that was investigated in these studies (Blalock et al., 2008). Furthermore, most researchers do not offer explanations for the choices they have made as to what components of attitude or attitude objects they selected to measure. These choices often seem to be based on pragmatic or convenience arguments.

The above-described lack of consistent definition and theoretical underpinning has led to a great variety of measurement instruments aimed at investigating teachers' attitudes towards science and technology. For example, some studies

Marc J. de Vries, Hanno van Keulen, Sylvia Peters, and Juliette Walma van der Molen (Eds.), Professional development for primary teachers in science and technology. The Dutch VTB-Pro project in an international perspective.

focused mainly on the affective components of attitude (e.g., Hartshorne, 2008; Ramey-Gassert, Shoyer, & Staver, 1996), while others focused exclusively on self-efficacy (e.g., Palmer, 2006; Wenner, 1993, 1995). In addition to (and probably also because of) the poor theoretical definition of what constitutes primary teachers' attitudes towards science and technology, many studies used measurement instruments that were poorly designed and that were not tested according to current methodological standards (Blalock et al., 2008; Coulson, 1992; Gardner, 1995; Reid, 2006). As a consequence, the results of the different studies cannot be compared nor replicated (Pardo & Calvo, 2002).

In an attempt to remedy these issues, the overall research project, of which the current chapter is part, was focused on disentangling the construct of primary teachers' attitudes towards science and technology. In a related review article (Van Aalderen-Smeets, Walma van der Molen, & Asma, 2011) we aimed to explicate and structure the range of underlying components or dimensions of primary teachers' attitudes towards science and technology and we related these components to general psychological attitude theories. Based on this theoretical exercise, we developed a new measurement instrument: the Dimensions of Attitudes towards Science questionnaire (DAS). After construction of the first version of the DAS, we investigated its validity and reliability by means of a quantitative survey study and a qualitative in-depth focus group study, which is reported in this paper. Although the qualitative study was primarily intended to examine some important theoretical assumptions underlying the DAS, the results of the study proved to be essential for further improvement of the survey instrument. As will be shown in the results and discussion sections of this chapter, interesting implications of teachers' beliefs and emotions were discerned that would not have been uncovered by means of a quantitative survey only. Before describing the method and results of our focus group study, we will first present the reader with an overview of the theoretical considerations that form the basis of the DAS and that were evaluated in the focus group study.

Theoretical Considerations

Attitude is not a single unitary concept, but a construct consisting of multiple subcomponents and attributes. The separate evaluations of each of these attributes contribute in varying degrees towards the overall attitude towards the object (Ajzen, 2001). Or, as Pajares (1992, p. 314) describes: "When clusters of beliefs are organized around an object or situation and predisposed to action, this holistic organization becomes an attitude". The overall psychological construct of attitude is often divided into three components: cognition, affect and behaviour (Eagly & Chaiken, 1993).

The cognitive component of attitude encompasses evaluative thoughts and beliefs a person has about the attitude object. In our case, this could be the belief that science is hard to understand, the belief that men are more interested in and better suited for a career in technology, or the belief that science is of essential economic value. These beliefs may range from a positive to a negative evaluation of attributes and everything in between.

The second component of attitude is affect. This component consists of feelings and moods a person experiences in relation to the attitude object. For instance, feelings of anxiety when confronted with science teaching or feeling a sense of insecurity when being asked questions about a technological problem. A positive attitude is characterized by the experience of positive physical reactions and emotions when confronted with the object, while a negative attitude is accompanied by negative affective reactions.

The third component of attitude is behaviour, which constitutes the behavioural response or action a person engages in when confronted with the attitude object. This response can be overt, in which case the person is actually acting out the behavioural response or action (e.g., teaching science and technology in class every Tuesday at 10 am). But the response may also be covert, in which case the person is intending to act out the behaviour or action but the action is not yet taking place. This covert response, labelled behavioural intention, is more frequently measured in surveys of primary teachers' attitudes towards science than the overt behaviour. Whether or not the intention to engage in certain behaviour is carried out is dependent on different circumstantial conditions at that time and place (Ajzen, 2001). The three components of attitude refer to the attitudinal responses a persons has in relation to the attitude object (cognitive response, affective response, and behavioural response), but also to the different processes underlying the formation of an attitude.

Apart from the necessity of a distinction between different components or dimensions of attitude, it is important to distinguish between different objects of attitude. The attitude object is the entity or thing with respect to which an attitudinal evaluation is made; it is the object of evaluation. At first hand, this seems a simple description, but in many studies there is a lack of clarity on this subject. Many studies measure multiple objects of attitude (e.g., general attitude towards science, attitudes towards science in high school, or attitudes towards effective teaching of science) as if it were one attitude object, without even mentioning possible theoretical distinctions. In addition, different attitude objects are often blended to yield one overall score of attitude towards science. One common example of the blending of attitude objects is when questions on teachers' personal attitude towards science and questions on their professional attitude towards teaching science are intermixed.

In our view, personal attitude towards science and technology refers to the attitude a person has as a citizen, independent of that person's profession (in this case teaching at primary school level). Examples of this general attitude are: someone's beliefs about the historical or economic relevance of science for society or daily life and the general interest in or affect towards informing oneself about science and technology via different media. In contrast, a teacher's professional attitude towards the teaching of science and technology in primary school refers to the beliefs and feelings that he/she may have with respect to teaching these topics within the school context, such as beliefs about the appropriateness and importance of science and technology for children at primary school level, or feelings of joy or anxiety towards teaching these topics.

To our knowledge, thus far only three studies attended to a possible distinction between primary teachers' personal and professional attitude towards science and technology (Atwater, Gardner, & Kight, 1991; Harty, Samuel, & Anderson, 1991; Koballa, 1986). Although these studies indicate that the two concepts should indeed be treated as distinct, albeit related, the measurement instruments that were used are problematic or were not reported in the articles. Further research, using both qualitative and quantitative methods, should therefore investigate how primary teachers consider these two objects of attitude and how the two might be related. To that end, the DAS is divided into two parts that each measures the different dimensions of attitude and related constructs. The first goal of the present focus group study was to validate this distinction and to explore how primary teachers perceive the two objects of attitude and possible relations between them.

A second goal of the focus group study was to find out which components of the tripartite model of attitude (affective, cognitive, and/or behavioural) primary teachers refer to when discussing their personal attitudes towards science and technology and their attitudes towards teaching science and technology. Furthermore, we explored how Bandura's concept of self-efficacy is related to the cognitive, affective, and behavioural components of attitude. Self-efficacy refers to the perceived capability, or confidence, a person has to perform a particular behaviour that may contain difficult and stressful elements (Bandura, 1997). A large number of studies that investigated attitudes of primary teachers towards teaching science and technology focused on the concept of self-efficacy (e.g., Coulson, 1992; Palmer, 2001, 2006; Ramey-Gassert et al., 1996; Skamp, 1991; Tosun, 2000; Wenner, 1993, 1995; Yates & Goodrum, 1990). In most of these studies, however, self-efficacy was not explicitly related to other dimensions of attitude. Although many primary teachers' indeed seem to feel insecure about their teaching of science and technology, it should be investigated how the concept of self-efficacy is related to primary teachers' beliefs, feelings, and behaviour towards (the teaching of) science and technology and whether other, perhaps previously neglected factors, might play a role.

METHODS

Participants

Participants were in-service and pre-service primary teachers from the Netherlands. In total, 84 teachers participated in the study (12 male, 72 female; 38 in-service, 46 pre-service). The group of in-service teachers was mixed: some just started their careers as teachers, others were very experienced. On average, the teachers had 17 years of experience. The pre-service teachers were in their third year of study and had teaching experience through internships. In the relevant literature we found no indication that there is a difference between pre- and in-service teachers on their attitudes towards science and technology and teaching science and technology. Therefore, results of in-service and pre-service teachers were summed, with the exception of some instances were clarification of results made a distinction necessary.

The total group of in-service and pre-service teachers was divided into three groups, based on their level of training in teaching science and technology. The first group consisted of elaborately trained primary teachers who had participated in a training program on science and technology education. Within this group, the in-service teachers had previously participated in the national VTB-Pro program aimed at the professionalization of primary teachers into teaching science and technology. The pre-service teachers in this group had taken part in a minor on science and technology education as part of their study. The second group of teachers had previously received some training (between one and five lessons) on science and technology education. The third group consisted of teachers who received no prior training on teaching science and technology. Five focus group discussions were conducted for every level of training; which led to a total of fifteen focus group discussions. By dividing the teachers in groups based on their level of prior training, we were able to compare their insights about the objects and components of their attitudes towards science and technology and examine possible differences between the three groups.

Procedure and protocol

We decided to conduct focus group discussions to examine the attitudes of teachers towards (the teaching of) science and technology because this research method is well-suited for exploratory research. Focus groups may provide a natural setting in which people normally state and form their opinions and attitudes. Therefore, discussions took place at a familiar setting: either at the participating primary schools or at the participating colleges for teacher training.

We used a semi-structured protocol for every focus group discussion to make sure that differences between focus groups were minimized and that the same procedure was followed in every discussion. The protocol consisted of four parts: an introduction on focus groups and the topic of discussion, an explanation of what is understood by science and technology in the present study, questions on the attitudes teachers have towards science and technology, and concluding remarks on the most important aspects of the discussion. The questions on attitudes were focused on the difference between professional and personal attitude and the different components and concepts that may play a role in either the negative or positive attitudes that pre- or in-service teachers may experience towards (the teaching of) science and technology.

The discussions were guided by one or two of the authors. The first focus group discussion was conducted by two researchers to streamline the procedure and minimize the differences in interview styles. The group interviews lasted between half an hour and one hour.

Data-analysis

The focus group discussions were audio taped and notes were taken. Subsequently, a detailed report was made of each focus group discussion. For each focus group, all statements relating to science and technology were marked in the report. These statements were listed in a data file in order to obtain a comprehensive list of the relevant statements and issues from the total number of group discussions. This list was used to score the occurrences of statements for all discussions. When a statement occurred it was given the score '1', if it did not occur it was scored as '0'. Scores for the five focus groups within one level of training were summed. Remarks were subdivided between personal and professional attitude and categorized as relating to the different components of attitude that were mentioned in the introduction.

RESULTS

Below, we start our results section with a broad overview of teachers' own thoughts about a possible distinction and/or relation between personal and professional attitude that we observed in our study. Subsequently, we will examine teachers' remarks on professional and personal attitudes more closely, in order to find out which components and underlying attributes of attitude and related concepts were exactly referred to in the discussions. Overall, we found that teachers predominantly talked about their professional attitude and that relatively little information were provided on their personal attitude towards science and technology. Because of the relatively large amount of remarks on professional attitude, in our presentation of results, we provide separate overviews for positive and negative remarks, self-efficacy and other factors. Our presentation of teachers' accounts of their personal attitude will be kept simpler.

Professional versus personal attitude

One major goal of our study was to find out whether teachers perceive a distinction between their personal and professional attitude towards science and technology. In *every* focus group discussion, teachers claimed to experience a difference. Some teachers stated to have a positive attitude towards teaching science and technology, but a negative attitude towards science and technology in daily life. Furthermore, teachers said that they thought it was possible to have no personal interest in science and technology whatsoever, but to be aware of the importance of teaching it to primary school children. From these results, we infer that primary teachers' personal and professional attitudes towards science and technology should indeed be treated as separate constructs. In the remainder of the results section, we will show that responses to both attitude objects differ considerably for professional and personal attitude.

When discussing a possible relation between personal and professional attitude, teachers were less unanimous. Some teachers who received no or only little prior training in science and technology stated that the two do not influence each other.

Examples of such responses are: 'If you have to teach it, you will' and 'in your own time you can do whatever you like, but not at school'. However, most of the less-trained teachers made contradictory statements that actually did reveal a perceived (causal) relation. Many teachers especially ventured the belief that personal attitude influences the way science and technology is taught. 'Teachers with a positive personal attitude teach in a more enthusiastic way' and 'teachers with a positive attitude show how things work in real life, not through pictures or videos.' Comparable comments, on the influence of personal attitude on professional attitude, were also made by other more elaborately trained teachers. Some statements were: 'If you are interested in something, you will teach that subject more frequently, and in a better and more extensive manner' and 'in free teaching hours, teachers will spend time on what they are interested in.' Some elaborately-trained teachers mentioned that to them, a positive personal attitude resulted in less fear of failure and a more constructive way of dealing with frustrations in teaching science and technology.

In addition to a perceived influence of personal attitude on professional attitude, teachers also reported that they experienced influences in the opposite direction. Teachers that had received elaborate prior training stated that their training, although mainly focused on professional attitude, indirectly influenced their personal attitude as well. They observed that the training had stimulated them to be more conscious about science and technology in their daily life. It should be noted, however, that such heightened interest and consciousness was mainly directed at the possibilities of every-day science and technology phenomena that they might use in their classes at primary school.

Professional attitude

The focus of the pre- and in-service teachers during our discussions was largely on their professional attitude. Table 1 lists the remarks made on the different attitude components that pertained to professional attitude, for teachers who received no, little, or elaborate prior training in science and technology.

Positive remarks Overall, teachers in the three training groups were about equally positive about teaching science and technology. Often, teachers stated that they enjoyed teaching science and technology and that they experienced enjoyment among their pupils as well (affective component). Furthermore, some teachers explicitly referred to the importance of gender equality and the stimulation of both boys and girls in science and technology (cognitive component). In addition, most teachers stated that science and technology are relevant to children and that they thought it was an important topic to teach at primary school level (cognitive component). However, it should be noted that teachers who had only little or no prior training in science and technology had difficulty explaining *why* science and technology at primary level are important. In contrast, the elaborately trained teachers were able to provide explanations. Their remarks were predominantly focused on the contribution of science and technology to children's 'personal knowledge', 'increased consciousness of their environment', 'safeguarding against consumer society' and 'increased vocabulary'.

Table 1. Number of remarks on professional attitude towards science and technology

Attitude components	*No training*	*Little training*	*Elaborate training*
Positive cognitive factors	4	6	5
Positive affective factors	8	9	8
Total positive	**12**	**15**	**13**
Negative cognitive factors	9	11	0
Context factors	21	21	5
Self-efficacy	1	2	0
Total negative	**31**	**34**	**5**

Negative remarks As listed in Table 1, the three training groups did differ considerably in their accounts of negative aspects. Elaborately-trained teachers made few negative remarks about teaching science and technology, while teachers who received no or little training made numerous negative remarks. As listed, responses predominantly pertained to negative cognitive factors and remarks about perceived contextual circumstances that might hinder the teaching of science and technology (such as perceived lack of materials or support from the school board).

Teachers who received no or only little prior training stated that they experienced a lack of knowledge about science and technology in general and about teaching science and technology in particular. As a consequence, they believed that science and technology were difficult to teach (cognitive component). In addition, the untrained pre-service teachers said that there was little attention for the topic in their teacher education. Although a fair number of teachers who received no or little prior training did state that science and technology are relevant and important to primary school children (see section on positive remarks), they also believed that mathematics, reading, and writing should have higher priority in primary education. Overall, although teachers responded positively to the importance of teaching science and technology, teachers that received no or only little prior training perceived a range of obstacles that, in their view, hinder implementation at primary school level. Because these beliefs are qualitatively different from other cognitive attitude components, we have labelled these beliefs under the heading 'context factors'.

Context factors From both the literature and our own practical experience, we knew that many primary teachers are inclined to mention contextual circumstances when they are questioned about their intention to teach science and technology. At the onset of our study, however, we did not anticipate that teachers' emphasis on context factors would be so prominent. In many of our focus group discussions, teachers mentioned that they did not succeed in implementing science and technology into their teaching on a structural basis due to lack of school support and insufficient school organization. In addition, teachers mentioned that they experienced other external obstacles, such as lack of materials, lack of time, and lack of money and a good method to teach science and technology. Also, many of them thought that teaching science and technology takes too much time to prepare.

Interestingly, however, our results showed a considerable difference between, on the one hand, pre- and in-service teachers that had received no or only little prior training in science and technology and on the other hand, teachers who were trained in a more elaborate way (see Table 1). Overall, teachers in the two former groups felt that teaching science and technology on a regular basis is too complex, due to the external obstacles mentioned above. The more elaborately trained teachers, however, seemed to have overcome such obstacles and expressed a very different perception. Some illustrative statements made by elaborately-trained teachers were: 'If they [teachers who perceive external obstacles in teaching science and technology] would understand the broadness of science and technology, they would know that teaching science can be done in a much simpler way, with less time investment and less expensive materials.' And: 'Many teachers are not prepared to teach science and technology and cannot see the possibilities; they only see the mess and the extra work'. Also, many of the well-trained teachers explained how they integrated science and technology lessons into other subjects in class, which makes it less time-consuming and more related to compulsory math and reading lessons. They also stated that by using the right inquiry-learning didactics, teachers might invest less preparation time and leave a larger proportion of the lessons to pupils.

Self-efficacy As listed in Table 1, teachers did not disclose a lot of personal notes on their own feelings of insecurity or perceived lack of capability. Some teachers confessed that they felt insecure or were anxious about potential difficult questions by their pupils. But, overall, teachers did not refer to such 'internal obstacles', not even when they were questioned about it explicitly. In the focus groups with elaborately trained teachers, self-efficacy was a topic of discussion, but these teachers referred to the low self-efficacy of their untrained colleagues (Table 1 only lists remarks that teachers made about their own believes and emotions, not about the perceived attitudes of others). According to well-trained teachers, low self-efficacy plays a large role in teachers' reluctance to teach science and technology. In their view, however, low self-efficacy can be conquered if one learns to develop a different attitude. Examples of observations made by elaborately-trained teachers were: 'Many teachers are afraid of failure, but one needs to learn that in science and technology education not everything has to succeed', 'Many teachers need structure and something to hold on to, while science and technology education needs space', and 'Teachers need to learn to deal with the chaos and mess that science and technology education can cause'. The elaborately-trained teachers disclosed that, prior to their training, they also experienced cold feet but that the training diminished their fear and reluctance. Interestingly, one teacher stated that after the training the bar was *less* high. 'We understand now that we do not have to know everything about science and technology and that we can just start with what we already are familiar with.'

Personal attitude

As mentioned before, teachers expressed relatively few thoughts, beliefs, and feelings that related to science and technology in their personal lives. Table 2 lists the components of personal attitude that were discussed.

Table 2. Number of remarks on personal attitude towards science and technology

Attitude components	*No training*	*Little training*	*Elaborate training*
Positive cognitive factors	1	0	0
Positive affective factors	9	4	9
Total positive	**10**	**4**	**9**
Negative cognitive factors	7	5	0
Negative affective factors	0	3	0
Total negative	**7**	**8**	**0**
No opinion	**2**	**0**	**2**

As listed, almost all positive comments on personal attitude towards science and technology were on affective factors. Irrespective of type of prior training, in all groups some teachers mentioned that they enjoyed watching or reading about science and technology and found it fun and interesting. It should be noted, however, that in many of these positive remarks, teachers referred to technical applications that directly affected their personal life, such as enjoyment in working out the functions of their mobile phone. Elaborately-trained teachers stated that their increased focus on science and technology had made them wonder about the world around them.

Negative remarks were predominantly focused on cognitive factors, such as difficulty and, to a smaller extent, gender differences. Teachers who had received little or no prior training mentioned that in their personal lives they were not interested in science and technology and some even said they thought it was boring. Furthermore, these teachers mentioned to experience a lack of knowledge and some of them believed that men are more interested in science and technology than women are. In addition to these specific beliefs, several teachers also stated that they simply had a negative attitude towards science and technology in their personal life and that they felt so little affinity with the topic that they felt unable to form an opinion about our questions on personal attitude. For the same reason, they disclosed that they had difficulty answering the survey questions on personal attitude that we pilot-tested among them.

As was found for professional attitude, differences between groups of teachers with different levels of prior training were observed only with respect to negative remarks. On average, teachers from all groups were about equally positive about science and technology in daily life. However, teachers with little or no prior training in science and technology also held negative personal beliefs, whereas teachers that were more elaborately trained did not mention any negative aspects.

DISCUSSION

The results of the present study confirmed our assumption that primary teachers perceive a difference between their professional and personal attitude. When asked explicitly, the majority of the (pre-service) primary teachers in our sample stated to experience a difference between the two attitude objects. In addition, they reported that in their view the two constructs could develop independently. Most notably, when probed about the different underlying components and sub-attributes of their attitudes, teachers predominantly talked about their professional attitudes towards teaching science and technology and reported having very little feelings for or beliefs about these topics in their daily life. Furthermore, some teachers clearly had a more positive attitude towards teaching science and technology than towards science and technology in their personal life. Apparently, the two attitude constructs can develop to a different extent and in different directions. In our view, teachers' distinctive approach towards their personal and professional attitude towards science and technology underscores the importance of treating personal and professional attitude as two separate objects of attitude in future qualitative and quantitative studies.

Apart from investigating a possible distinction between teachers' personal and professional attitude towards science and technology, a second goal of our study was to find out which components and sub-attributes of the tripartite model of attitude (affective, cognitive, and/or behavioural), and the related concept of self-efficacy primary teachers refer to when discussing their attitudes towards science and technology. What are the aspects of attitude they spontaneously come up with in an informal conversational setting? The results of our study showed that respondents focused mainly on the cognitive and affective components of attitude. No remarks that could be categorized as relating to the behavioural component of attitude were made. Furthermore, teachers hardly commented on their self-efficacy. Instead, the less-trained teachers referred largely to external obstacles in their teaching environment. In the remainder of this discussion, the results that relate to the tripartite model of attitude are discussed in more detail. Subsequently, we will examine the theoretical implications of the results for self-efficacy and context factors.

Reconsidering the tripartite model

Initially, we set out to investigate teachers' attitudes as consisting of cognitive, affective and behavioural components. During our focus group discussions, however, teachers only mentioned their beliefs and feelings and did not come up with behavioural attitudinal aspects. This may suggest that the pre- and in-service teachers in our sample do not view their behaviour and behavioural intention as part of their underlying attitude. This finding corresponds with the way attitude is considered in the influential Theory of Planned Behaviour (TPB) (Ajzen, 1991). According to the TPB, attitude (together with subjective norm and perceived behavioural control) determines behavioural intention, which is the immediate motivational factor for behaviour itself. According to this theory, attitude consists of a cognitive, or evaluative, component and an affective component. Behavioural intention is viewed as a direct outcome of these two attitude dimensions and not as a component of attitude itself.

The sub-attributes of cognition and affect that we found in our present focus group study are highly similar to the attitude attributes that we put forward based on our theoretical examination of previous literature in this domain (see Van Aalderen-Smeets et al., 2011). Most importantly, the teachers in our sample did not refer to any unexpected aspects of attitude that deviated from the attributes that we found in our large scale literature review. Furthermore, many of the attributes found in the literature were also spontaneously mentioned by the teachers during our focus group discussions. This implies that the attributes of the cognitive and affective components that we hypothesized based on our examination of the literature are indeed corroborated by the discussions that we had with the pre- and in-service teachers on their attitudes towards science and technology. The cognitive aspects of professional attitude that were discussed are: relevance of science and technology for pupils, gender differences in enjoyment of or achievement in science and technology, and difficulty of teaching science and technology. The cognitive remarks that related to personal attitude were scarce and focused mainly on difficulty. The affective responses for both professional and personal attitude were focused on enjoyment or dislike of (teaching) science and technology.

Reconsidering self-efficacy and perceived influence of context factors

For self-efficacy the results were somewhat different than we expected. To our surprise, the teachers in our sample hardly mentioned their own feelings of insecurity or low confidence in teaching science and technology. This contrasts with several previous studies that did find that many teachers display low self-efficacy in this domain (e.g., Tosun, 2000; Yates & Goodrum, 1990). One possible explanation may be related to the research method that we used: it might be that teachers felt reluctant to disclose their uncertainties in front of colleagues or classmates during our focus group discussions.

However, an alternative explanation may be sought in the fact that many previous studies did not measure self-efficacy in combination with other components of attitude or related concepts. As a consequence, in previous studies teachers have been unable to display their beliefs and emotions on related constructs and aspects of attitude and were forced to focus exclusively on their (lack of) self-efficacy. Because self-efficacy has been measured as the only indicator of teachers' attitude towards science in quite a number of studies (and in some ways has even been equated with attitude), it could be argued that its role has perhaps been somewhat overestimated. Thus, even if teachers felt that their attitude towards science and technology depended on other aspects as well, they were unable to express this in the closed survey items that were focused primarily on self-efficacy.

In contrast, during our focus group discussions, teachers were able to talk freely about their attitude and related concepts and were not restrained by a limited survey. Interestingly, even when asked explicitly about their personal self-efficacy, many teachers reported that they did not experience low self-efficacy per se and were quite confident about teaching science and technology. In our view, this implies that in order to genuinely find out which aspects of attitude may hinder teachers to teach science and technology, it is important to examine self-efficacy in relation with other components and attributes of attitude.

Our results show that the teachers in our sample did not necessarily perceive self-efficacy as the main reason for their reluctance to teach science and technology. Rather than disclosing internal obstacles, such as insecurities about teaching science and technology or fear of difficult questions by pupils, the teachers in our study referred to external obstacles, such as lack of suitable materials, lack of a structured science and technology teaching-method, and lack of preparation time. This implies that context factors may play a large role in teacher attitudes towards science and technology. However, this aspect has thus far been hardly examined in conjunction with other attitudinal aspects in empirical studies on teacher attitudes towards science and technology (Van Aalderen-Smeets, et al. 2011) and its influence on actual teaching behaviour thus remains largely unclear. In the present study, we observed that the less-trained teachers focused on external obstacles to quite a large extent, while the elaborately trained teachers held very different perceptions on the importance of structured materials and methods. This interesting difference is reflected on in more detail below, where we will show that teachers' differential perceptions of internal and external factors that may hinder (or foster) their teaching of science and technology could be integrated in a new theoretical framework of teachers' attitudes towards science and technology.

Implications for further research

The results of the present study have led to two implications for our overall research project and further research in this area. First, the results of the present focus group study contributed to our understanding of the underlying dimensional structure of primary teachers' attitudes towards science and technology, which is reflected in our formulation of a new theoretical framework. Second, it provided valuable insights for the development of an integrated attitude survey instrument that we would not have gained if we would have tested the instrument by means of paper-and-pencil methods only. These implications are clarified below.

Theoretical framework for attitude

Our large-scale review of the literature and the present focus group study instigated us to formulate a new theoretical framework for primary teachers' attitudes towards science and technology, which consists of three components and seven underlying attributes (see Figure 1). In our view, the proposed framework is suitable for describing both the professional and personal attitude of primary teachers towards science and technology. As can be seen in Figure 1, the framework integrates previous theoretical considerations as formulated in the Theory of Planned Behaviour, findings from previous studies on teachers' attitudes towards science and technology, and our present findings that relate to teachers' perceptions of self-efficacy and their perceived dependency on contextual factors. In our new framework, the cognitive and affective component of the original tripartite model remain and consist of sub-attributes that were found both in the present study and in previous studies (although in previous studies these attributes were not investigated in conjunction). However, the original tripartite model was revised in two ways: based on the results of our focus group study and in line with the Theory of Planned Behaviour, we excluded the behavioural component as part of the underlying construct of attitude and added a new third component that we labelled 'Perceived Control'.

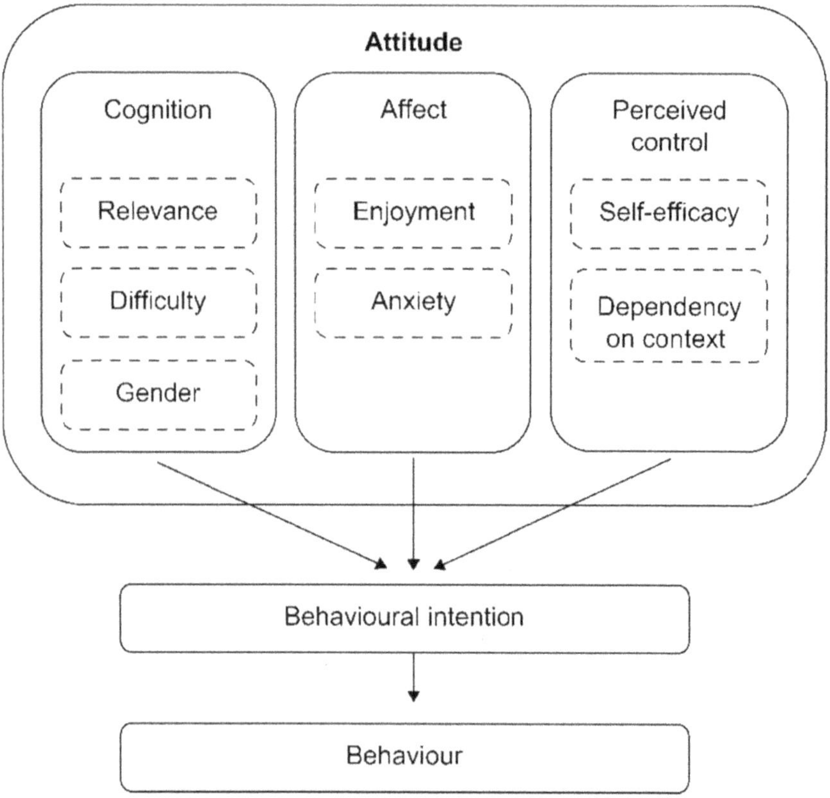

Note: The same framework is suggested for both personal and professional attitude towards science and technology

Figure 1. New framework for primary teachers' attitudes towards (teaching) science and technology

Results from our focus group study showed that teachers made no comments on behaviour or behavioural intention as part of their attitudes towards (teaching) science and technology. Based on these results and on our re-evaluation of the extant literature on the psychological concept of attitude, we agree with Ajzen (1991) that behaviour and behavioural intention are conceptually different from attitudes and that this component should not be part of the construct of attitude itself. Instead, attitudes should be viewed as antecedents of behavioural intention, which among other things determines actual behaviour. Therefore, in our new framework, behaviour and behavioural intention are included as outcomes of attitude and not as part of the theoretical construct of attitude itself.

Instead of behaviour and behavioural intention as a third component of attitude, we propose a new third component, of which self-efficacy is part. Our review of the literature showed that many researchers have studied self-efficacy as a main attitudinal predictor of the teaching of science and technology at primary school level (Van Aalderen-Smeets et al., 2011). Although the pre- and in-service teachers in the present study did not disclose a lack of self-efficacy, we agree that self-efficacy should be included in a complete framework of primary teachers' attitudes towards science and technology. However, if we want to form a complete picture of the role that different aspects of attitude might play, self-efficacy should not be investigated in isolation but in conjunction with other components and attributes. Theoretically, we suggest that self-efficacy is neither part of the affective nor the cognitive component of attitude, because self-efficacy is qualitatively different from both. In our view, self-efficacy consists of both cognitive and affective aspects that are focussed on people's internal beliefs about and feelings of being in control to execute particular behaviour or not.

The results of the present study have shown that apart from internal beliefs and feelings that adhere to self-efficacy, an additional aspect that seems to be closely related to teachers' sense of 'being in control' concerns the beliefs and feelings that teachers have about external, or contextual factors. In our view, teachers' perceived dependency on context factors (such as their belief that they can only teach science and technology when their school takes care of the right materials and enough preparation time) cannot be left out of a complete theoretical framework of primary teachers' attitudes towards science and technology. On the contrary, we believe that together with self-efficacy (which reflects teachers' internal perception of control) it should be included in a dimension that we call 'Perceived Control'. It should be noted that we deliberately refer in this label to teachers' *perception* of being in control. Similar to the cognitive and affective dimensions of attitude that are necessarily subjective, this new component reflects someone's subjective beliefs and feelings about internal and external obstacles, not the factual presence of such obstacles.

Ajzen (2002) proposed that perceived external, contextual obstacles should be viewed as a part of self-efficacy. However, the results of the present study suggest that teachers' dependency on context factors may differ from their self-efficacy. Overall, the teachers in the present study felt to have sufficient self-efficacy, but a large group still perceived low control in teaching science and technology because they felt that they could not teach science and technology without the right materials or methods provided to them. Measuring both aspects as one attitudinal attribute would make it impossible to deduct these clearly different aspects and would hinder a distinction between different combinations of the attributes. In our view, theoretically teachers could be (a) low on self-efficacy and highly dependent on context factors, (b) high on self-efficacy, but still highly dependent on context factors, or (c) high on self-efficacy and not dependent on context factors. These three categories reflect increasing levels of confidence or a sense of being in control. A fourth possibility, in which teachers would be low on self-efficacy and not dependent on context factors at the same time, seems less realistic. Because both constructs are clearly related but in theory do not always need to point in the same direction, unlike Ajzen (2002), we included them as separate sub-attributes of the dimension that we labelled perceived control.

Integrated attitude instrument

Our revised framework for primary teachers' attitudes towards science and technology provided us with valuable insights for the development of a new and integrated attitude instrument, the Dimensions of Attitude towards Science questionnaire (DAS). The DAS is an instrument that consists of two separate questionnaires: one that measures the professional attitude of teachers and another that measures their personal attitude. Unlike most previous studies, in which attitude items on different attitude objects were intermixed (see Van Aalderen-Smeets, et al., 2011), we believe that measuring professional and personal attitude separately leads to a more reliable and valid investigation of teacher attitudes. For both questionnaires, sub-scales were included that were directly derived from the seven underlying attributes of the new theoretical framework.

Although some previous survey instruments measured certain aspects of teachers' attitudes, self-efficacy, or dependency on context factors, thus far no attitude instrument was developed that measures the total construct of primary teachers' attitudes towards science and technology, including our newly proposed component of perceived control. The STEBI, developed by Enoch and Riggs (1990), is a much-used instrument measuring self-efficacy. It combines the measurement of self-efficacy with items on teacher outcome expectancy, but does not include the measurement of teachers' perceived dependency on context factors or other components of attitude. Lumpe, Haney, and Czerniak (2000) developed the CBATS (Context Beliefs About Teaching Science) to measure the influence and occurrence of context factors. But, this instrument also does not combine the measurement of perceived external obstacles with other aspects of attitude or self-efficacy. As mentioned previously in this discussion, in the above-described studies teachers were thus unable to display their beliefs and emotions on related constructs and aspects of attitude and were forced to focus exclusively on one or some aspects of attitude. Because of this, primary teachers' interrelated beliefs about and feelings towards different components and attributes of their attitudes towards science and technology still remain unclear and, as a consequence, interventions that are aimed at increasing teachers' attitudes still fall short on including the complete range of attitudinal aspects.

Concluding remarks

To summarize, we believe that the present qualitative focus group study uncovered important insights that would not have been uncovered with the use of survey research only. The present study has important implications for both the creation of a reliable and valid attitude instrument and the theoretical development of the concept of primary teachers' attitudes towards science and technology. Based on the theoretical framework and the present study, the DAS was revised and the underlying components and attributes can now be studied with large groups of in-service and pre-service teachers. In future research, we may now further investigate the distinction between teachers' personal and professional attitudes towards science and technology and we may develop professionalization interventions that are targeted at specific attitude attributes and measure the effects of these interventions with the appropriate sub-scale of the DAS and evaluate the outcomes relative to scores on other sub-scales.

Lieke Asma
Juliette Walma van der Molen
j.h.walmavandermolen@utwente.nl
Sandra van Aalderen-Smeets
University of Twente, The Netherlands

EDDIE DENESSEN, NIENKE VOS, TOM DAMEN, SEVERINE KOCH, MONIKA LOUWS, AND DANIEL WIGBOLDUS

9. EXPLICIT AND IMPLICIT MEASURES OF TEACHER ATTITUDES TOWARDS SCIENCE AND TECHNOLOGY

INTRODUCTION

In this chapter, teacher attitudes towards science and technology and the way these attitudes can be accessed in empirical research are discussed. The study of teacher attitudes towards science and technology is highly relevant, since teacher attitudes seem strong predictors of the quality of their teaching and of student attitudes and performance (Osborne, Simon, & Collins, 2003).

Motivation of students can very well be positively stimulated by teachers' enthusiasm about the subjects they teach. Teachers' enthusiasm, however, does vary across subjects, especially concerning subjects related to science and technology (Jarvis & Pell, 2004). In contrast to some other countries, and unlike Dutch *secondary* school teachers, Dutch primary teachers do not need to specialize in specific subject areas. Primary teachers provide teaching in all subject areas, ranging from the basics (reading, writing, mathematics), to social and creative subjects (citizenship education, art). It is therefore likely that teachers vary in the degree to which they feel comfortable teaching different subjects. Previous studies have shown that many primary teachers experience difficulties teaching science and technology and that, in general, they hold less positive attitudes towards teaching science and technology than towards other school subjects, and that this is particularly true for female teachers (Tiedemann, 2000).

Previous studies on teachers' general attitudes towards teaching science and technology revealed that these attitudes cannot be treated as one single construct, but that several attitude dimensions need to be discerned (see for example, Jarvis & Pell, 2004; Johnston & Ahtee, 2006; Osborne, Simon, & Collin, 2003). The following four attitude dimensions have been identified in previous studies: (1) teachers' enjoyment when teaching science and technology, (2) their perceived relevance of teaching science and technology (3) their perceived levels of competence and efficacy related to teaching science and technology, and (4) their motivation to invest in their science and technology teaching competences.

Apart from attitudes towards their teaching of science and technology, teachers may also vary with respect to their attitudes about their students' interest, motivation and ability in learning science and technology. Specifically, teachers

Marc J. de Vries, Hanno van Keulen, Sylvia Peters, and Juliette Walma van der Molen (Eds.), Professional development for primary teachers in science and technology. The Dutch VTB-Pro project in an international perspective.

appear to have differential attitudes concerning these factors towards boys and girls in their classroom. (Tiedemann, 2000).

In general, teachers seem to hold more favorable attitudes concerning boys' aptitude and interest in science and technology (Tiedemann, 2000). Such gender-biased attitudes may cause self-fulfilling prophecy effects (Brophy& Good, 1970; Rosenthal, 1994), since positive expectations can have positive effects on student attitudes and academic achievement (Jussim & Harber, 2005).

As has also been suggested in the chapter by Asma, Walma van der Molen, and van Aalderen-Smeets of this book, when studying teacher attitudes, it thus seems fruitful to distinguish teacher attitudes towards their teaching of science and technology and their (in particular gender stereotypical) attitudes related to students' interest and capacities to learn science and technology. In both respects, teacher gender seems to be important to consider as well when investigating those attitudes.

Measurement of teacher attitudes towards science and technology

Usually, teacher attitudes towards science and technology are measured with the use of questionnaires. Those questionnaires include statements about science and technology and teachers have to rate their level of agreement (Van Aalderen-Smeets, Walma van der Molen, & Asma, 2011). Since the construction of the Science Attitude Scale (Shrigley, 1973), many questionnaires concerning teacher attitudes towards science and technology have been developed, mostly to measure specific sub-constructs of teacher attitudes, such as teachers' self-efficacy in teaching science or teachers' perceived level of competence (Atwater, Gardner, & Kight, 1991; Bleicher, 2007; Coulson, 1992; Johnston & Ahtee, 2006; Tosun, 2000; Walma van der Molen, 2009). Some examples of questionnaire items are 'I don't know enough about science to do it with children', 'Science is a worthwhile subject', and 'I have never done well in science, so I will probably not teach it as well as other subjects' (Coulson, 1992; Tosun, 2000).

Teachers' ratings of their agreement with these types of questionnaire items provide some insight in teacher attitudes towards science and technology, although measurement validity problems and a lack of clear definitions make direct comparisons of these studies and generalizing their results difficult (Gardner, 1995; Van Aalderen-Smeets, et al., 2011). Especially the multidimensional nature of teacher attitudes results in measurement problems. Numerous adjectives can be attributed to the teaching of science and technology (joyful, relevant, important, difficult, interesting, pleasant, dull, easy, et cetera) and it can be questioned whether attitude measures using different adjectives result in a similar picture of teachers' attitudes (see for example Gardner, 1995). In various studies, different numbers of attitude dimensions have been reported and as a result, we still remain puzzled about the number of attitude dimensions that can be discerned in this respect.

In addition, it can be doubted whether teacher attitudes can be fully captured with self-reports using rating scales (see for example Nosek, 2007). Self-reported attitude ratings may lead to biased responses due to social desirability issues and strategic response behavior. Especially with respect to sensitive subjects, like group stereotypes or ratings of attitudes in normative contexts, this type of response behavior might be elicited more strongly. Respondents, then, are more likely to report attitudes that are congruent with a social norm (Gawronski, Lebel, & Peters, 2007). With respect to teachers' attitudes towards science and technology, it may be the case that teachers are not very willing to report negative attitudes towards science and technology, especially when the local school context is being perceived as strongly science and technology minded. Teachers may feel obliged to report positive attitudes with which they show their support for the school's policy regarding its science and technology curriculum. Also, in a context in which one strives for gender equity, as is the case in The Netherlands, teachers may be unwilling to report gender stereotyped attitudes in a questionnaire.

To overcome these validity problems, implicit attitude measures can be used, such as implicit association tests (Nosek, Greenwald, &Banaji, 2007). With the use of an implicit association test (IAT), attitudes are derived from response latencies when sorting pictograms or words that are presented on a computer screen. Respondents are asked to sort pictograms or words in categories as fast as possible. Two categories share the same response key, whereas one or two opposing categories use a different response key. The underlying assumption of an IAT is that a shared response key leads to quicker responses when a person strongly associates both categories (e.g., insects and negative), and to slower responses when a person does not associate both categories (e.g., flowers and negative) (Greenwald, McGhee, & Schwartz, 1998).

IAT-scores seem to have a stronger predictive value than questionnaire ratings concerning discriminative behavior towards specific social groups (Poehlman, Uhlmann, Greenwald, &Banaji, 2004). In a cross-national comparative study on the gender stereotyped attitudes towards science, Nosek et al. (2009) reported that, in general, science is more strongly associated with males than with females. By relating TIMMS 2003 national differences in 8th grade science performance between boys and girls with national estimates of gender stereotyping, they found that national implicit stereotyping of science as male was strongly related to the gender differences in science performance across countries. From another IAT-study on gender stereotypic attitudes towards math it appeared that teachers, irrespective of their own gender, held gender stereotypic attitudes, and that women demonstrated more negativity toward math relative to arts and language on implicit measures (Nosek, Banjani, & Greenwald, 2002).

In this chapter, we present two newly developed single target science and technology IATs, which we have tested in a sample of Dutch primary school teachers. We also present some analyses of the psychometric value of this IAT by relating IAT-scores to questionnaire ratings and by comparing IAT scores of male with those of female teachers.

METHOD

Participants

The data for this study were obtained between October and December 2009. In total, 139 Dutch primary school teachers participated in this study. Teachers from Grade 3, 4, and 5 (43 male, 95 female, on average 16 years of teaching experience) were selected from a sample of 36 primary schools. All schools were participating in a national network for professionalization of science and technology in primary education (VTB-program, see Walma van der Molen, de Lange, & Kok, 2009).

Measures

Explicit measures: Attitude questionnaires Teachers' attitudes towards science and technology were measured with a rating questionnaire consisting of 20 items. The questionnaire items were selected from previous measures of attitudes, such as the Revised Science Attitude Scale (Bitner, 1994), science attitude items from the PISA-study (OECD, 2006), and items from a previously validated Dutch questionnaire (Walma van der Molen, 2009). With the questionnaire we aimed at measuring four dimensions of attitudes towards science and technology that have been identified in previous studies (see for example, Jarvis & Pell, 2004; Johnston & Ahtee, 2006; Osborne, Simon, & Collin, 2003): (1) Teachers' enjoyment for teaching science and technology (5 items, for example 'I feel comfortable when teaching science and technology', Cronbach's alpha = .84), (2) their perceived relevance of teaching science and technology (5 items, for example 'I think it is important to teach science and technology', Cronbach's alpha = .80), (3) their perceived levels of competence and efficacy related to teaching science and technology (5 items, for example 'For me teaching science and technology is very difficult', Cronbach's alpha = .70), and (4) their motivation to invest in their science and technology teaching competences (5 items, for example 'I am willing to invest in the development of my abilities to teach science and technology', Cronbach's alpha = .85). Teachers were asked to rate their agreement with the statements on a four-point Likert scale.

Four items were formulated to measure teachers' gender stereotyped attitudes towards science and technology. Teachers were asked to rate whether they thought a statement was more applicable for boys than for girls. Teachers rated four statements: 1) enjoying to repair things is…, 2) to be competent as a scientific or technical professional is…, 3) to study science or technology is…, 4) to experience difficulties with science and technology is…. The statement were to be rated on a five point Likert scale (1 = mostly applicable to boys, 2 = more applicable to boys than to girls, 3 = equally applicable to boys and girls, 4 = more applicable to girls than to boys, 5 = mostly applicable to girls). After recoding of the ratings of the fourth statement, the reliability of the scale was moderate (Cronbach's alpha = .56).

Implicit measures Two IATs were developed to measure teachers' attitudes towards science and technology. The first IAT was aimed at teachers' positive and negative associations with science and technology. The scores on this IAT were taken as indicators of teachers' general attitude towards science and technology. A second IAT was developed for the measurement of teachers' gender stereotyped attitudes (associations between science and technology and gender).

As mentioned above, the IAT is a computerized task with which response times are registered when participants sort words or pictograms related to specific categories. The underlying assumption is that responses are faster when categories that share the same response key are strongly associated.

In contrast to previously developed IATs, that include arts or humanities as opposing category to science or math (see Nosek et al.2002; Nosek et al., 2009), we did not define a category opposite to science and technology. The IATs that were developed for this study were so-called single-target IATs (Wigboldus, 2003). According to Wigboldus et al., the standard IAT has the disadvantage that it is a relative measure, which means that associations with one category are defined in comparison to association with an opposing category. It then is difficult to identify whether a strong association is the result of a positive association with one category or a negative association with the opposing category, or both. To gain less ambiguous association scores, Wigboldus et al., suggest to use a single target IAT with which associations can be measured with a category without the need of an opposing category. For the IATs in this study, we thus did not define a category opposite to science and technology.

In the first IAT, positive and negative pictograms as well as science and technology related pictograms were presented on a computer screen (see Figure 1 for an overview of the pictograms). The science and technology pictograms were selected from a library of approximately 2000 pictograms that were collected on the Internet. The science and technology pictograms were chosen in such a way that they were related to the five domains (systems) of science and technology that were distinguished in PISA-studies: (1) physical systems, (2) living systems, (3) earth and space systems, (4) technical systems, and (5) mathematical systems (OECD, 2006). After attributing the pictograms to the five domains, five of the authors of this chapter then independently rated the pictograms for inclusion in the IAT. The highest rated pictogram within each domain was selected for inclusion in the IAT. At the start of the IAT, the participants were prompted with the science and technology pictograms and the positive and negative pictograms of the IAT (see Figure 1 for the overviews presented). In the IAT, two response keys were defined, the 'A' and the 'L' keys on the keyboard. On either side of the screen the category labels were presented. Two test blocks were designed. In the first test block the left side of the screen contained the category labels 'negative' and 'science and technology', whereas the right side on the screen contained the category 'positive'. Participants were instructed to press the key that corresponded with the category ('A' for negative or science and technology pictograms; 'L' for positive pictograms) as quickly as possible after the pictogram was presented. In the second test block the science and technology category was projected on the right side of the screen to be sorted with the same key ('L') as the positive pictograms. When the response times of the first block (science and technology and negative pictograms on the one side, positive pictograms on the other) were faster than the response times of the second block (negative pictograms on the one side, science and technology and positive pictograms on the other), the participant is suggested to show a negative attitude towards science and technology. Each test block contained 40 trials, presented in a random order.

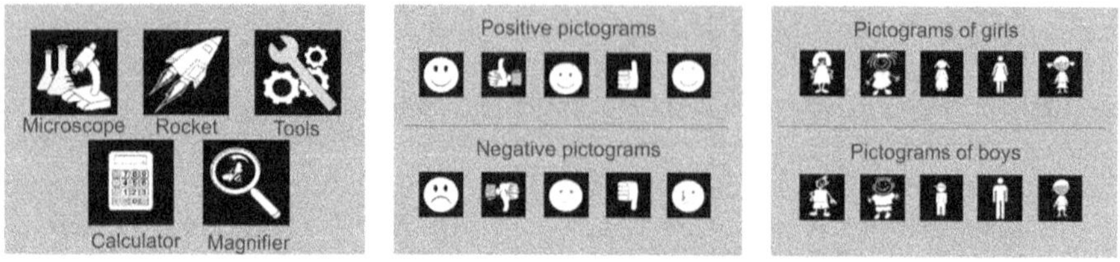

Figure 1: Overview of science and technology pictograms, positive and negative pictograms, and boys and girls pictograms as elements of IATs for the assessment of teacher attitudes towards science and technology.

In the second IAT, concerning gender stereotyped attitudes, pictograms of boys and girls were presented in combination with the science and technology pictograms (see Figure 2 for a screenshot of the test). In the first test block the left side of the screen contained the category labels ‘girls’ and ‘science and technology’, whereas the right side of the screen contained the category ‘boys’. Participants were instructed to press the key that corresponded with the category (‘A’ for girls or science and technology pictograms; ‘L’ for boys pictograms) as quickly as possible after the pictogram was presented. In the second test block the science and technology category was projected on the right side of the screen to be sorted with the same key (‘L’) as the boys pictograms. When the response times of the first block (science and technology and girls pictograms on the one side, boys pictograms on the other) were slower than the response times of the second block (girls pictograms on the one side, science and technology and boys pictograms on the other), the participant is suggested to show a biased attitude towards science and technology in favor of boys. Again, each test block contained 40 trials, presented in a random order.

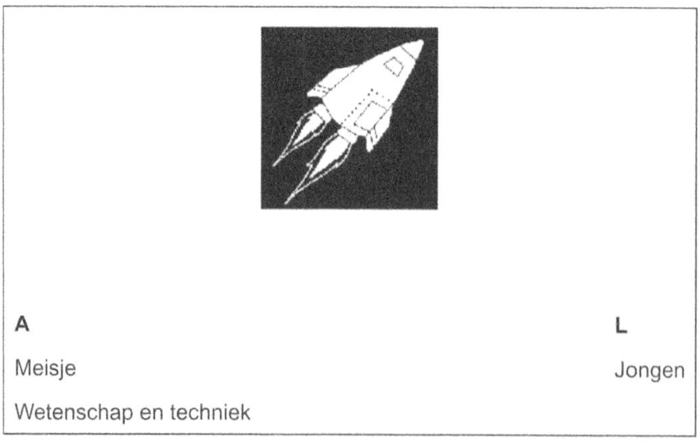

Figure 2: Screenshot of the first test block of the gender-science and technology IAT

Because IATs do not include items, reliability estimates in terms of Cronbach's alpha coefficients cannot be calculated. To estimate the reliability of the IAT test scores odd-even split halves reliability estimates were computed. Based on the correlation between both halves of the IATs, the reliability estimate for the general science and technology IAT was .78; for the gender stereotype IAT the reliability estimate was .64.

Analyses

The present study was aimed at the construct validity of the scores on two science and technology-related IATs. Measurement validity was studied by relating the IAT scores with questionnaire ratings and by comparing the scores of male teachers with those of female teachers.

RESULTS

Explicit measures

In Table 1 the descriptive statistics of teachers' questionnaire ratings are presented. Given the possible score range of 1-4, the teachers in this sample seem on average to report quite positive attitudes towards science and technology, although there were some differences between men and women. Male teachers reported significantly higher scores on enjoyment for teaching science and technology than female teachers ($t(113) = 3.66$, $p < .001$) as well as higher levels of perceived competence ($t(113) = 2.99$, $p < .001$). Male and female teachers rated their perceived relevance of teaching science and technology and their motivation to invest in their science and technology teaching competences equally high.

From the analyses of gender stereotyped attitudes, it appeared that only one teacher rated one of the items on the right side of the rating scale (4 and 5) indicating that the science and technology statement was more applicable to girls than to boys. Therefore, we decided to recode the scale to a scale ranging from 1 to 3. A score of 1 on this scale (indicating a constant rating of the science and technology items as being equally applicable to boys and girls) was provided by the majority of the teachers in the sample (n = 69, 60.5%). The mean score of 1.22 on this scale indicated that teachers only slightly showed gender stereotyped attitudes.

Table 1: Descriptive statistics of explicit measures of teacher attitudes towards science and technology (range 1-4) and teachers' gender stereotyped attitudes (range 1-3).

		Enjoyment		*Relevance*		*Perceived competence*		*Motivation to invest*		*Gender stereotyped attitude*	
	n	*m*	*sd*	*m*	*sd*	*m*	*sd*	*m*	*sd*	*m*	*sd*
Male	37	3.23	.43	3.24	.36	2.96	.39	2.79	.53	1.25	.40
Female	77	2.86	.53	3.23	.38	2.72	.43	2.74	.50	1.21	.30
Total	114	2.98	.53	3.24	.37	2.80	.43	2.76	.51	1.22	.33

Table 2 shows the inter-correlations among the scales of the attitude questionnaire. These correlations indicate that the dimensions of teachers' attitudes towards science and technology were related. However, attitude scores on the four dimensions of teacher attitudes towards science and technology did not correlate with the gender stereotyped attitude scores.

Table 2: Correlations among explicit measures of teacher attitudes towards science and technology and their gender stereotyped attitudes (n = 114).

	Enjoyment	*Relevance*	*Perceived competence*	*Motivation to invest*	*Gender stereotyped attitude*
Enjoyment	1.00				
Relevance	.45*	1.00			
Perc. competence	.67*	.25*	1.00		
Motivation to invest	.55*	.45*	.36*	1.00	
Gender stereotyped attitude	–.07	–.06	–.11	–.17	1.00

* $p < .05$

Implicit measures

Table 3 contains the descriptive statistics of the implicit measures of teacher attitudes. The general attitude towards science and technology was computed by taking the difference in response times between the two test blocks of the IAT. A positive mean general attitude IAT score indicates a positive association with science and technology. The overall mean IAT score of 23.49 did differ statistically from a neutral attitude score of zero ($t(114) = 2.49$, $p = .014$), which indicates that teachers have slightly positive implicit attitudes towards science and technology. The differences in general attitude scores of men and women were not significant ($t(112) = 0.06$, $p = .953$).

A positive gender stereotyped IAT score indicates an association of science and technology with boys. The overall mean score of 33.88 was statistically higher than the midpoint of zero ($t(114) = 4.13$, $p < .001$), which points at biased attitudes in favor of boys. Also with respect to the gender stereotyped IAT scores, the mean difference between male and female teachers was not significant ($t(112) = 1.48$, $p = .142$). The correlation between the general IAT scores and the gender stereotyped IAT scores was .07 ($p = .464$), which indicates that general attitude scores did not correlate with gender stereotyped attitudes.

Table 3: Descriptive statistics of the IAT scores related to teachers' general attitudes towards science and technology and their gender stereotyped attitudes.

		General attitude		*Gender stereotyped attitude*	
	n	*m*	*sd*	*m*	*sd*
Male	34	24.34	93.04	52.51	120.01
Female	88	23.13	103.67	25.96	70.10
Total	114	23.49	100.21	33.88	88.27

Relations between questionnaire ratings and IAT scores

To gain insight in the relations between explicit and implicit measures of teacher attitudes, correlation between the scores on both measures were computed (see Table 4). As the figures in Table 4 show, the correlations between the general attitude IAT-scores and the questionnaire ratings of teachers weakly point at a positive relation between IAT-scores and each of the four questionnaire scales, although none of the correlations reached statistical significance.

Table 4: Correlations between explicit and implicit measures of teacher attitudes.

	Explicit measures									
	Enjoyment		*Relevance*		*Perceive competence*		*Motivation to invest*		*Gender stereotyped attitude*	
Implicit measures	*r*	*p*	*r*	*p*	*r*	*p*	*r*	*p*	*r*	*p*
Attitude IAT ($n = 103$)	.20	.059	.17	.115	.15	.164	.20	.063	.04	.688
Gender IAT ($n = 104$)	.10	.377	–.03	.818	.04	.683	–.03	.819	.24	.021

The relation between questionnaire ratings and the IAT scores regarding teachers' gender stereotyped attitudes were statistically significant (see Table 4), which means that higher scores on the gender stereotyped IAT were provided by teachers who reported higher levels of gender biased attitudes in the questionnaire.

To further explore the convergent validity of both measures of attitudes, regression analyses were performed to calculate the shared variance of the two attitude measures (see Table 5). These regression analyses show that there was approximately 6 percent shared variance among the explicit and implicit measures of teacher attitudes. These results indicate that the information about teacher attitudes that has been gained with questionnaire ratings shows marginal overlap with the information about teacher attitudes that has been gained with implicit association tests.

Table 5: Shared variance among the explicit and implicit measures of teacher attitudes towards science and technology and their gender stereotyped attitudes (R^2, results of regression analyses, n=103).

	General attitudes towards science and technology IAT	*Gender stereotyped attitude IAT-scores*
General attitudes towards science and technology questionnaire	.062	.008
Gender stereotyped attitude questionnaire	.002	.052*
Total	.063	.067

* $p < .05$

DISCUSSION

The present study focused on the value of implicit measures of teacher attitudes towards science and technology. Two implicit association tests were presented in order to measure automated associations that teachers may have with science and technology. Below we will discuss the results of our study for both IATs separately.

General attitude towards science and technology IAT

The first IAT was developed to measure teachers' general positive and negative associations with science and technology. In contrast to the literature and results of research on questionnaires of science and technology attitudes, we have assessed teacher attitudes with this IAT as a one-dimensional construct. The question is whether the IAT scores refer to a more global (or overall) association of teachers' attitudes towards science, or to specific dimensions (or aspects) of teacher attitudes. From correlation analyses it appeared that IAT scores did not strongly relate to the questionnaire ratings, although weak relations between teachers' IAT scores and their ratings of the four attitude aspects that have been distinguished in the questionnaire have been reported. In addition, regression analyses pointed at an overlap of both measures of approximately 6 percent. From these reports it may be tentatively concluded that IAT scores may capture an overall attitude towards science and technology.

Low correlations between questionnaire ratings and IAT scores mean that the information obtained by both measures is somehow different. It could mean that different aspects of attitudes have been captured, but it could also mean that one measure or the other is a more valid measure of teacher attitudes. With the results of this study, we cannot be conclusive about the correct interpretation of these correlations. We should consider, however, that low correlations between questionnaire ratings and IAT scores may have several causes. First, the overall character of the IAT may inhibit the assessment of specific teacher attitudes and thus result in a too broad picture of teacher attitudes. This may cause some loss of information. We saw, for example that male and female teachers rated two specific questionnaire scales differently (enjoyment and motivation to invest), with men reporting more positive attitudes than women. Gender differences were not found in the IAT data. The absence of those gender difference may be interpreted as a validity problem of the IAT. For future research we need more external criteria to provide more conclusive evidence for the validity of the IAT scores.

A second cause of low correlations among measures may be sought in the nature of the attitudes that have been assessed. Contrary to IATs, which measure automated more implicit (or intuitive) positive and negative associations with science and technology, questionnaire items are rated more in a more deliberate way. Different cognitive processes may result in differences between questionnaire ratings and response latencies in the IAT (Greenwald et al., 1998). The implicit aspects of teacher attitudes yield assessment problems, particularly when using rating scales (Donaghue, 2003). Especially in teacher research, IATs may well be suited for capturing concepts of tacit knowledge and beliefs (Shulman, 1987; Berliner, 2004; Schwarz, 2008).

A third cause of low correlations among measures can be found in some technicalities with respect to the IATs that have been developed for this study. In contrast to many other IATs, we have used pictograms instead of verbal stimuli (i.e. words). The use of pictograms as well as the kind of pictograms that we have selected for the IAT may result in less measurement validity. Since not much research has been done on the validity of IATs using pictograms, it would be interesting to include IATs with verbal and visual stimuli in a combined study to explore measurement validity of pictogram use in IATs in more detail. In addition, we have used a single target IAT. IAT studies that we have referred to in the introduction of this chapter used an opposing category to science and technology, such as arts and humanities. With a double target IAT, relative preferences are measured: do respondents have more positive associations with science and technology than with arts and humanities? Since an opposing category was not included in our IAT, results of measures may diverge. It would also be interesting to combine a single target and a double target IAT in one study to get more information about the validity of both types of IATs.

Another possible factor that may cause low relations between explicit and implicit measurement scores is the selectivity of the sample in this study. The teachers in our sample worked at schools that were members of a national network of schools for professionalization of science and technology in primary education (VTB-program, see Walma van der Molen, de Lange, & Kok, 2009). This selective sample may have caused less variation in questionnaire ratings. It can, for example, be claimed that teachers' scores on their perceived relevance of teaching science and technology and their motivation to invest in their science and technology teaching competences is uniformly high at the schools in this network. We therefore suggest performing a future IAT study on a broader sample of teachers.

Gender stereotyped attitude towards science and technology IAT

The second IAT was developed to measure teachers' associations of boys and girls with science and technology. We also developed a short questionnaire to measure teachers' gender stereotyped attitudes. Teachers' questionnaire ratings showed that more than half of the teachers rated themselves as being unbiased towards boys or girls. In contrast, the scores on the gender stereotyped IAT showed that teachers associated science and technology more with boys than with girls.

The correlation between questionnaire ratings and IAT scores showed that there was some overlap (of approximately 5 percent) between both measures, although the correlation was small. This small, though statistically significant, correlation between measures of gender bias may have different causes than the low levels of correlation between teachers' general attitude scores that we discussed before. With respect to gender stereotypes, the issue of social desirability is more likely to lead to biased responses to questionnaires. Although teachers may provide biased ratings of their general science and technology attitudes due to self-representation, it can be assumed that self-representation and social desirability is stronger when it comes to gender bias, because gender bias can be perceived as a social sensitive issue. This is in line with the results of a meta-analysis of correlations between implicit and explicit measures of attitudes (Hofman, Gawronski, Gschwendner, Le, & Schmitt, 2005), from which the authors concluded that correlations are higher for relatively mundane topics (e.g., consumer preferences) than for socially sensitive topics (e.g., prejudice against minority groups).

We should also note that some methodological problems may have caused low levels of correlation between questionnaire ratings and IAT scores. Besides IAT technicalities discussed above, the small amount of variance of gender stereotype questionnaire ratings negatively affects correlation among measures. Low levels of variance of questionnaire ratings have been reported frequently in research on gender stereotypes with respect to mathematics. Studies in which the Fennema-Sherman 'mathematics as a male domain'-scale have been used, usually show that the majority of respondents claims to be unbiased, which leads to low mean scores and small variance of the self-ratings (Brandell, Nyström, Sundqvist, 2004; Forgasz, Leder, & Gardner, 1999). A clear benefit of the use of a gender stereotyped IAT is that it evokes more teacher variance than the questionnaire. It seems that this IAT generates more nuances with respect to teachers' gender stereotyped attitudes.

Future research

A lot of research still needs to be done on the measurement validity of implicit association tests in teacher research. The implicit association test is a relatively new instrument. Research designs that are aimed at the construct validity of IAT scores are scarce. One reason for this lack of validity studies is that it is difficult to define a golden standard to which IAT scores are to be related for validation purposes. Validation of IAT scores is usually defined in terms of predictive validity. The predictive value of IAT scores has been shown in many previous studies (Nosek, Greenwald, & Banaji, 2005; Nosek et al., 2009). With respect to implicit measures of teacher attitudes it has been shown that teachers' implicit prejudiced attitude scores were predictive towards ethnic performance gaps in their classrooms, whereas questionnaire ratings were not (Van den Bergh et al., 2010). A previous study of Nosek et al. (2009) has shown that IAT scores related to attitudes towards science correlated with science performance differences between boys and girls. In countries where strong associations between science and boys were reported, boys showed higher levels of science performance than girls did. These studies point at strong predictive values of IATs.

In summary, we are not yet able to be conclusive concerning the value of the IATs presented in this study. To gain more information about the quality of the IATs we will relate this sample's IAT scores to some external criteria, such as teachers' differential expectations of future careers of the boys and girls in their classrooms, teachers' differential behavior towards boys and girls when teaching science and technology related subjects, and the attitude development of the students in their classroom during one school year. These data have yet to be gathered and analyzed. From those analyses we expect to present some positive results concerning the predictive value of the IATs that were developed for this study.

Eddie Denessen
E.Denessen@pwo.ru.nl

Nienke VosTom Damen
Severine Koch
Monika Louws
Daniël Wigboldus
Radboud University Nijmegen, The Netherlands

CHARLOTTE VAN CLEYNENBREUGEL, VEERLE DE WINTER, EVELIEN BUYSE, AND FERRE LAEVERS

10. UNDERSTANDING THE PHYSICAL WORLD: TEACHER AND PUPIL ATTITUDES TOWARDS SCIENCE AND TECHNOLOGY

An intervention study in primary education

INTRODUCTION

In this study, an intervention was set up aimed at improving both teachers' attitudes and competences in relation to science and technology, and teaching these subjects. The present report focuses specifically on teacher attitudes and how these evolve throughout the school year. A pre- and posttest design was used to evaluate teacher attitudes in relation to science and technology (and teaching these subjects) before and after the intervention trajectory. Additionally, children's attitudes were investigated to explore the relations between teachers' attitudes (and how these evolve) on the one hand, and (changes in) their pupils' attitudes on the other. At last, we explored the intervention inputs for critical tools/aids to change attitudes.

Trying to foster positive attitudes towards science and technology is a worthy goal in itself. Moreover, a positive attitude towards a certain topic is considered a necessary condition for the development of a broad and in-depth understanding of that topic in several areas (e.g., Cheung, 2009; Ho & Kuo, 2010; Ogbuehi& Fraser, 2007). More positive attitudes towards science and technology may concord with higher involvement in and deeper understanding of the physical world (Laevers, 1993; Laevers, 1998; Walma van der Molen, 2007).

In international literature, attitudes have consistently been described as multidimensional constructs, consisting of a cognitive, affective, and behavioural dimension. The cognitive dimension refers to perceptions and views, the affective component covers feelings with regard to a certain topic and the behavioural dimension captures the intentions to undertake actions in a particular field (e.g., Ajzen, 2005; Vazquez-Alonso, Manassero-Mas, & Acevedo-Diaz, 2006; Walma van der Molen, de Lange, & Kok, 2009). In relation to education, teachers' feelings of self-efficacy have often been described as an important additional dimension of teachers' attitudes. Teachers' feelings of self-efficacy have been shown to be powerful predictors of positive teaching behaviour in the classroom, and are therefore also linked with pupil outcomes, such as achievement and motivation (e.g., Pajares, 1996; Tschannen-Moran, Woolfolk Hoy, & Hoy, 1998).

Marc J. de Vries, Hanno van Keulen, Sylvia Peters, and Juliette Walma van der Molen (Eds.), Professional development for primary teachers in science and technology. The Dutch VTB-Pro project in an international perspective.

It may not surprise that positive attitudes in teachers towards specific *teaching practices* to address certain topics can also be considered important antecedents of their actual teaching behaviour and related pupil outcomes. Although Shrigley and Johnson already acknowledged this in 1974, attitudes towards teaching have not often been included in research. Chen (2006), for example, stated that "although teachers' positive attitudes towards teaching are essential for instruction to succeed, a related attitude survey is not available in the literature" (p. 804). With regard to technological and scientific topics in primary classes in particular, recent efforts have been made to stress the importance of positive attitudes towards active teaching strategies to address these topics. Active strategies focus heavily on the development of pupils' understanding of science and technology in 'co-construction'. Children are thought to learn the most about technological and scientific issues by designing and actively inquiring things together in class. The active strategies encompass four pedagogical strategies with a positive effect on learning: collaborative, contextual, reflective and project-based learning (Sidawi, 2009). The teachers' role can then be defined as one of guiding and supporting pupils' learning processes (Kemmers, Klein Tank, & Van Graft, 2007). According to Jalil, Sbeih, Boujettif and Barakat (2009), teachers who act as acoach with minimal interference and who offer children autonomy in their knowledge building, cause a significant positive attitude shift on pupil level. These teaching strategies also have a positive impact on pupils' motivation (Barak & Zadok, 2009). The strategies create opportunities for creative thinking and offer pupils the possibility to obtain a better self-image and view glimpses of their potential (Lewis, 2009). The active teaching strategies considered in this study, are intertwined, but address different underlying questions. In 'learning by design' a solution for a technical problem is searched; in 'inquiry-based learning' a broad and in-depth understanding of science topics is at the centre of attention (Van Graft & Kemmers, 2007).

Positive attitudes towards a certain topic and related teacher practices have been found to foster more positive attitudes in children as well (e.g., Abulude, 2009; Jarvis, 2006; Pell and Jarvis, 2003). Based on these findings, we hypothesise more positive (changes in) teacher attitudes to foster (positive changes in) pupil attitudes in the field of science and technology in particular. Therefore, in the present study, we also examined the relation between teacher and pupil attitudes in these fields.

In the study, an intervention trajectory was set up to promote teachers' attitudes and competences in relation to the domain of science and technology. The 'teacher profile', that served as a basis for the interventions, contains five dimensions: (1) openness to and interest in science and technology (attitude); (2) intuitive understanding of physical phenomena (competence); (3) the capacity to extract knowledge from experience (conceptualisation); (4) the ability to create learning environments where children engage in intense mental activity (didactics enhancing involvement); and (5) the competence to identify the cognitive load of activities, i.e. the developmental domains triggered in pupils through the activity and the mental operations that can be challenged by the activity (critical view on content and material). While teachers implement a module on science and technology in their class during the school year, they are exposed to four specific intervention inputs. These were designed to support development in the five above-mentioned dimensions and respond to the experiential view on learning and development (Laevers, 1993; 1998).

A first input consisted of a half day workshop titled 'The Eye and the Fire' (introduction session). In the first part, the teachers responded to pictures (taken in outside areas) which made them aware of the paramount presence of physical phenomena and technological applications. Developing 'an eye' for these phenomena can be regarded as one of the main objectives in the domain of science and technology. A broadened view on science and technology gives teachers the opportunity to facilitate transfer between daily life experiences and science and technology activities. This awareness increases teachers' observational skills: it will be easier for them to recognise and appreciate children's talents in the science and technology domain. The second part of the workshop offered a framework in which children's level of involvement is presented as an indicator for the power of learning environments. Here, the active ingredients of an approach that elicits interest and fascination in children are considered. Particular attention is paid to the 'open framework model' (Laevers, 2006a) in which both children and teacher take initiative in co-construction of learning. As highlighted above, co-construction is also a key concept in inquiry-based learning and learning by design.

Secondly, a visit to a Science Centre was organised, where an interactive trajectory consisting of ten selected exhibits was laid out for the teachers (interactive visit science centre). With the support of key questions, in groups of four, the teachers concentrated on the perceptions, thoughts and questions that arose while they were experimenting with the materials. They were also invited to formulate a 'thoughtful' explanation of the phenomena. This exercise does not only help teachers be receptive to the technological and scientific dimensions in their environment (develop 'The Eye') but also helps them identify mental processes within themselves and articulate these in a dialogue with others. This way, they encounter their own limitations in knowledge and insights and try to overcome these with the help of their colleagues. They are put in the position of the learner and therefore experience for themselves how an environment can affect a person's interest and cognition.

A third input consisted of a training session containing an in-depth analysis of video recordings of children dealing with science and technology activities. These were developed by the Dutch project Talenten Kracht [Curious Minds][ii] (in-depth session Curious Minds). The guide to view the clips is based on the PaLe (Laevers, 2006b), a tool designed to make a process-oriented analysis of learning environments. Here, well-being, involvement and mental activity are at the centre of attention. Furthermore, the activity is held against several developmental domains to identify the cognitive load, i.e. the domains that are mobilised during the viewed episodes. These domains are not limited to an understanding of the physical world, but may also, for example, refer to social competence, or meta-competences such as entrepreneurship or creativity. Finally, the context factors to enhance involvement in the observed situation are explored, including adult style dimensions (sensitivity, support of autonomy, stimulation), the richness of the materials at hand, and which specific layers of understanding of the physical world are inherent to the material and can hence be triggered by interaction with it.

A fourth intervention consisted of a coaching session, grafted on the science and technology projects in which teachers were already engaged. Reflections in groups started with a specific practice example teachers were proud of, thereby inspiring colleagues and creating a more positive self-concept. Starting from these strong points, weaker points and personal obstructions were explored, followed by an attempt to overcome these in co-construction with teachers who encountered the same challenges.

A closer analysis of the interventions allows to identify which dimensions of teachers' attitudes are likely to be addressed. In general, there is a particular focus on the behavioural dimension (supporting change at the level of planning actions), the affective dimension (enjoying the experimentation and exploration of the physical reality) and the cognitive dimension (offering teachers a view on science and technology that is more linked to daily life experiences, which makes science and technology less difficult and more compelling. Furthermore, feelings of self-efficacy may be mobilised (the feeling to be able to successfully engage in teaching science and technology). Finally, rooted in Experiential Education, the intervention inputs emphasise the importance of a rich learning environment, where activities evolve in interaction or co-construction with the children (Laevers, 2006a). This may lead to more positive teacher attitudes towards learning, also by design and inquiry-based learning, which both adhere to the same essential learning principles.

In sum, four research goals are addressed. Firstly, we evaluated the impact of the intervention on change in teacher and pupil attitudes. Secondly, we investigated the change in attitude scores over the school year for teachers who followed the intervention trajectory, compared to control teachers. We hypothesised a larger growth in every dimension of the attitudes towards science and technology for the teachers who followed the intervention trajectory, as compared to the control group. Thirdly, we evaluated the impact of changes in teacher attitudes (for the teachers who followed the intervention trajectory) on pupil attitudes towards science and technology. We expected a positive change in teacher attitudes resulting in a positive change in their pupils' attitudes as well. Fourthly, we made an exploratory analysis in order to find out which 'tools/aids' offered by the intervention inputs can cause a growth in teacher attitudes. Therefore, at the end of each input, teachers were asked to indicate what they had learned and to rate their level of involvement during each session.

METHOD

Participants

The study involved schools from Flanders (Belgium) and the Netherlands. In a first stage, teachers were recruited to participate in the intervention, making them partof the 'trajectory group'. In Flanders, these teachers were involved in an on-going technology project, called 'Dorp op school' ['Village at School']. Similarly, the Dutch teachers worked at schools that were engaged in integrating science and technology more in their programme (so-called 'VTB schools'). The 39 colleagues of these (26) teachers, who did not follow the intervention, were assigned to the 'control group'.

Table 1. Division of participating teachers over Flemish and Dutch schools

	# Teachers	
	Trajectory group	*Control group*
Flemish schools (n = 9)	15	29
Dutch schools (n = 6)	11	10
	26	39

As presented in Table 1, 44 Flemish teachers participated as opposed to 21 Dutch teachers. The trajectory group consisted of 58% Flemish teachers, whereas the proportion of Flemish teachers equalled 74% in the control group. However, these proportions did not differ significantly ($\chi^2(1) = 1.98$, *ns*). There were no significant differences detected between the trajectory and control group with regard to other available background features. Firstly, the proportion of male teachers, equalling 15% and 10% in the trajectory and control group respectively, did not differ between both groups ($\chi^2(1)= 0.38$, *ns*). Secondly, seniority of the teachers (defined as a categorical variable with 1 = < 10 years, 2 = 10-15 years, and 3 = > 15 years of teaching experience respectively) did not differ between both groups ($\chi^2(1) = 1.48$, *ns*).

For the trajectory teacher group, scores on attitude scales were also available for the pupils in their class. In total, 489 children completed the questionnaires, of which 327 were Flemish and 233 were boys. The proportion of boys in Flanders (46%) and the Netherlands (50%) did not differ significantly ($\chi^2(1) = 0.74$, *ns*).

Procedure

The intervention offered to the teachers of the 'trajectory group' consisted of four inputs, described in the introduction. Each input was concluded with an evaluation by the teachers (see 'Instruments').

During an intake with the participating teachers and their principals, the schools received the teacher and pupil attitude questionnaires for Pretesting. These were completed by (a) the 'trajectory teachers'; (b) the 'control teachers'; and (c) the pupils of the teachers following the intervention trajectory. The 'Pretest questionnaires' were handed back before the first intervention input. All intakes took place between October 2009 and January 2010. In June 2010, after the final intervention input, schools received the attitude questionnaires for Posttesting, which were completed by the same three groups as described for the Pretest.

Instruments

Teacher attitudes: To measure teachers' attitudes towards science and technology, we used a Dutch questionnaire, developed by VTB. Previous research established that this questionnaire is a comprehensive instrument, which operation alises attitude as a set of thoughts, feelings, and behaviours (Walma van der Molen, 2009). We also included the VTB questionnaire on attitudes for pupils, because of its similar design. This would enable us to explore the connection between teacher and pupil attitudes towards science and technology.

The pupil questionnaire consists of scales on two domains, i.e. 'attitudes towards technology' and 'attitudes towards science'. Two other domains were added to the teacher questionnaire, referring to their attitudes towards designing and inquiring as classroom practices to foster learning in the field of science and technology. More specifically, the scales on 'attitudes towards learning by design' and 'attitudes towards inquiry-based learning' were selected from the Oberon study (2009).

Within each domain, a cognitive, affective, and behavioural dimension of attitudes are distinguished, in accordance with the attitude concept. The cognitive component is further divided into two subcomponents: the evaluation of the *difficulty* and the *importance* of science and technology (and inquiry-based learning and learning by design in the teacher form). The affective subscales refer to the *enjoyment* teachers and pupils experience in relation to science and technology (and related practices). The behavioural subscales cover the *intentions* to invest time and energy in these topics. In addition, Oberon (2009) defined a fifth subscale on the teacher questionnaire, *self-efficacy*, which we also included. Here, teachers rate how they perceive their ability to act appropriately in a given area. The respective scales within each domain will be referred to as DIFFICULTY, IMPORTANCE, ENJOYMENT, INTENTIONS, and SELF-EFFICACY.

We checked the quality of the teacher attitude questionnaire, using factor[iii] and reliability analysis. Based on the findings, we made a few (small) changes in the composition of the scales. Like Oberon (2009), we omitted the IMPORTANCE and SELF-EFFICACY scales for attitudes towards science and the DIFFICULTY scale for attitudes towards inquiry-based learning and learning by design. We also omitted the DIFFICULTY scale for attitudes towards technology. More details on these analyses can be found in Table A.1 and in the internal report (De Winter & Van Cleynen breugel, 2010).

Table A.1. Overview of scales, example items and study results concerning the teacher questionnaire on attitudes towards science, technology, and related practices

		Oberon study[(a)]		*Present study*	
Subscales	*Example item*	*# items*	*α*[(b)]	*# items*	*α*[(b)]
Attitude Technology		$R^{2\,(c)} = 59\%$		$R^{2\,(c)} = 57\%$	
Difficulty	The technology course can only be given by specially trained teachers.	2	.67	/[(d)]	/

Table A.1 (Continuation). Overview of scales, example items and study results concerning the teacher questionnaire on attitudes towards science, technology, and related practices

		Oberon study[a]		*Present study*	
Subscales	*Example item*	*# items*	*α*[b]	*# items*	*α*[b]
Importance	Technology is important for the community.	7	.75	6[e]	.77
Enjoyment	I like to repair things myself.	6	.87	6	.89
Intentions	I would like to learn more about technology in primary school.	4	.81	4	.78
Self-efficacy	I wonder whether I have the necessary skills to teach technology.	5	.78	7[d]	.84
<u>Attitude Science</u>		R^2 [c] = 59%		R^2 [c] = 57%	
Difficulty	Science is complicated.	2	.60	2	.67
Importance	Researchers do important work.	/[f]	.46	/[f]	.52
Enjoyment	I like to invent things.	4	.79	4	.72
Intentions	I like to read about new inventions, for example in the newspaper or on the Internet.	4	.72	4	.68
Self-efficacy		/[g]		/[g]	
<u>Attitude Learning by design</u>		R^2 [c] = 55%		R^2 [c] = 58%	
Difficulty	It seems difficult to me to apply learning by design in primary school.	/[f]	(.57)	/[f]	(.22)
Importance	Learning by design in primary school is necessary to prepare children for secondary school.	3	.62	3	.71
Enjoyment	It appeals to me to let children solve technical problems.	5	.84	2[h]	.70
Intentions	I would like to learn more about how to conduct children with learning by design.	/[h]	/	3[h]	.85
Self-efficacy	When children can't figure out a technical problem, I can help them.	3	.62	3	.68
<u>Attitude Inquiry-based Learning</u>		R^2 [c] = 60%		R^2 [c] = 62%	
Importance	It is important that even young children learn how to do research (at their level).	4	.75	4	.77
Enjoyment	It appeals to me to let children unravel things.	3	.74	3	.75
Intentions	I would like to learn more about inquiry-based learning at primary school.	3	.78	3	.84

Difficulty	It seems difficult to me to apply inquiry-based learning in primary school.	/[f]	(.55)	/[f]	(.49)
Self-efficacy	I know how to motivate pupils for inquiry-based learning.	3	.60	3	.64

[a] Extracted from the interim report on an effect study VTB-Pro (school year 2008-2009) by Oberon (2009).
[b] Cronbach's alpha coefficients.
[c] Explained variance in attitudes towards technology, science, learning by design, and inquiry-based learning respectively, based on factor analysis.
[d] Based on factor and/or reliability analysis, the items of the DIFFICULTY scale, i.e., *'The course technology can only be taught by specially trained teachers'* and *'To be able to be a good teacher in technology, you need a specialised training'* were added to the 'self-efficacy' scale. This adaptation is theoretically defendable.
[e] Compared to the Oberon study, the item *'The government should spend more money on technology'*, was excluded.
[f] Based on factor and/or reliability analysis, this subscale was excluded.
[g] In the interim report of Oberon (2009, p. 6), it is stated that the 'self-efficacy' scale is not applicable for attitude towards science.
[h] Other than in the Oberon study, the items on the INTENTIONS scale were not added to the ENJOYMENT scale.

Each scale of the final teacher questionnaire contains a series of items to be rated on a five point scale from 'I completely disagree' to 'I completely agree'. Items formulated in a negative sense were inverted and subsequently, scale scores were computed by averaging item scores. Higher scores on each scale indicate that teachers perceive the respective domains (i.e., technology, science, and related practices) as more difficult (DIFFICULTY scales), more important (IMPORTANCE scales), and more enjoyable (ENJOYMENT scales); that they intend to invest more time and energy in the domain (INTENTIONS scales); and that they have more confidence in their own capacities (SELF-EFFICACY scales).

Correlations between the different attitude scales for the four domains were calculated and reported in the internal report (De Winter & Van Cleynen breugel, 2010) and in Tables A.2, A.3 and A.4. Furthermore, the Pretest did not show any differences in attitude scores between countries. At the end of the school year, 2 out of the 15 attitude scales showed a significant difference, i.e. with regard to IMPORTANCE, learning by design ($t(22) = 2.17$, $p < .05$) and ENJOYMENT, inquiry-based learning ($t(13.07) = 3.19$, $p < .01$). More specifically, Flemish teachers describe learning by design as significantly more important than their Dutch counterparts[iv]. Furthermore, Flemish teachers report significantly more enjoyment in inquiry-based learning than Dutch teachers[v]. Although no significant differences were detected between seniority groups for any of the Posttest attitude scales, these were found for INTENTIONS, learning by design ($F(2,23) = 4.28$, $p < .05$) and inquiry-based learning ($F(2,23) = 4.75$, $p < .05$) on the Pretest, with the youngest group scoring significantly lower than the oldest[vi, vii].

Table A.2. Correlations between teacher attitude scales concerning science and technology (Posttest, n = 25-26)

	2.	3.	4.	5.	6.	7.
Attitude Technology						
1. Importance	.06	.18	–.09	–.29	.11	.05
2. Enjoyment		.34	.49*	–.15	.70**	.62**
3. Intentions			.36	–.20	.42*	.37
4. Self-efficacy				–.47*	.61***	.43*
Attitude Science						
5. Difficulty					–.50**	–.20
6. Enjoyment						.59**
7. Intentions						

*$p < .05$. **$p < .01$.

Table A.3. Correlations between technology related teacher subscales (Posttest, n = 23-24)

	Attitude Learning by design			
	Importance	*Enjoyment*	*Intentions*	*Self-efficacy*
Attitude Technology				
Importance	.47 *	.48 *	.08	–.07
Enjoyment	.38	.24	.23	.05
Intentions	.18	.27	.26	.14
Self-efficacy	.03	–.04	.22	.33

*$p < .05$.

Table A.4. Correlations between science related teacher subscales (Posttest, n = 24)

	Attitude Inquiry-based learning			
	Importance	*Enjoyment*	*Intentions*	*Self-efficacy*
Attitude Science				
Difficulty	–.16	–.04	–.16	–.34
Enjoyment	.07	.42 *	.35	.28
Intentions	.23	.16	.19	.09

*$p < .05$.

Pupil attitudes: In accordance with the teacher questionnaire, the pupil questionnaire consists of the DIFFICULTY, IMPORTANCE, ENJOYMENT and INTENTIONS subscales. The GENDER differences subscales were added for attitudes towards science and technology because perception of gender differences, i.e. the idea that boys are better in technology and/or science, is considered an additional element of the cognitive aspect of attitudes (Walma van der Molen, 2009). Similar to the analyses of the teacher questionnaire, a factor[viii] and reliability analysis was perform edon the pupil questionnaire. All the original scales were included, except for the science DIFFICULTY scale. More details on these analyses can be found in Table A.5 and the internal report (De Winter & Van Cleynen breugel, 2010).

Table A.5. Overview of scales, example items and study results concerning the pupil questionnaire on attitudes towards science and technology

		Design study[a]		Present study	
Subscales	*Example item*	*# items*	*α*[b]	*# items*	*α*[b]
Attitude Technology		$R^{2\,(c)}$ = 51%		$R^{2\,(c)}$ = 54%	
Difficulty	I find it hard to learn about technology.	4	.50	3[d]	.66
Importance	Technology has a big influence on people.	7	.73	7	.70
Gender	Boys are better car-mechanics than girls.	3	.76	3	.77
Enjoyment	I enjoy fixing things myself.	6	.78	6	.80
Intentions	Later, I want to follow a technical profession.	3	.92	3	.84
Attitude Science		$R^{2\,(c)}$ = 56%		$R^{2\,(c)}$ = 56%	
Difficulty	I find science difficult.	3	.60	/[e]	(.36)
Importance	People who figure out new ideas are important to society.	7	.70	7	.76
Gender	Boys are better scientists than girls.	3	.85	3	.88
Enjoyment	I like to figure out new ideas.	7	.88	7	.82
Intentions	Later, I would like to have a job in science.	3	.84	3	.83

[a] Extracted from the design report of the pupil attitude monitor (Walma van der Molen, 2007).
[b] Cronbach's alpha coefficients.
[c] Explained variance in attitudes towards science and technology respectively, based on factor analysis.
[d] Compared to the design study, the item *'Technology is only for smart people'*, was excluded.
[e] Based on factor and/or reliability analysis, this subscale was excluded.

Each scale of the final pupil questionnaire contains a series of items to be rated on a four point scale from 'I completely disagree' to 'I completely agree'. Items formulated in a negative sense were inverted and subsequently, scale scores were computed by averaging item scores. Higher scores for the DIFFICULTY, IMPORTANCE, ENJOYMENT and INTENTIONS scales must be interpreted similarly as explained for the teacher questionnaire. Additionally, higher scores on GENDER reflect pupils' beliefs that boys are better in technology and/or science than girls.

Correlations between the different attitude scales in science and technology were calculated and reported in the internal report (De Winter & Van Cleynen breugel, 2010) and the Appendix (Table A.6). Moreover, the Pretest score of Flemish pupils was significantly higher than the ENJOYMENT and INTENTIONS score for science and technology of the Dutch pupils. However, Flemish pupils perceive technology as significantly more difficult than their Dutch counterparts. On the Posttest, all significant differences between the two countries disappear, except for ENJOYMENT and INTENTIONS for science. As on the Pretest, both mean attitude scores are significantly higher in Flanders. Detailed information on these group differences can be found in Table A.7. Both on pre- and Posttest, pupil gender has a meaningful influence on their attitudes. All differences are significant (see Table A.8), except for the scores on IMPORTANCE of technology in the Pretest. Boys enjoy both science and technology more, they find it easier and generally more important, and they are more interested in learning more about it than girls. Furthermore, boys perceive themselves as better at both domains, while girls do not report these gender differences.

Table A.6. Correlations between pupil attitude scales (Posttest, n = 480-487)

	2.	3.	4.	5.	6.	7.	8.	9.
Attitude Technology								
1. Difficulty	-.07	.00	-.36***	-.18***	-.09*	-.06	-.26***	-.18***
2. Importance		-.02	.29***	.32***	.65***	.07	.27***	.17***
3. Gender			.02	.13**	.07	.76***	.03	.11*
4. Enjoyment				.45***	.27***	.10*	.68***	.28***
5. Intentions					.34***	.21***	.41***	.45***
Attitude Science								
6. Importance						.15***	.43***	.33***
7. Gender							.12*	.17***
8. Enjoyment								.52***
9. Intentions								

*$p < .05$. **$p < .01$. ***$p < .001$.

Table A.7. Comparison of mean attitude scores between Flemish and Dutch pupils

	Flanders (N = 320-327)		The Netherlands (N = 158-163)		t-test[(a)]		
	M	SD	M	SD	df	t	p
PRETEST							
Technology							
Difficulty	2.19	.65	1.86	.64	487.00	5.32***	.000
Importance	2.90	.44	2.90	.49	488.00	–0.13	.895
Gender	2.17	.87	2.33	.98	291.96	–1.69	.092
Enjoyment	3.32	.52	3.17	.58	488.00	2.85**	.005
Intentions	2.25	.82	2.03	.86	485.00	2.75**	.006
Science							
Importance	2.94	.47	2.90	.59	265.67	0.63	.531
Gender	1.82	.87	1.94	1.00	282.24	–1.32	.188
Enjoyment	3.22	.55	2.97	.70	260.19	4.00***	.000
Intentions	2.10	.80	1.87	.80	482.00	2.93**	.004
POSTTEST							
Technology							
Difficulty	2.02	.66	1.93	.65	483.00	1.47	.141
Importance	2.81	.46	2.88	.56	278.07	–1.26	.208
Gender	2.18	.91	2.01	.98	483.00	1.82	.069
Enjoyment	3.18	.63	3.09	.65	483.00	1.50	.134
Intentions	2.15	.83	2.03	.83	481.00	1.56	.119
Science							

Importance	2.85	.49	2.84	.63	260.27	0.25	.804
Gender	1.77	.83	1.86	.98	277.00	−0.96	.338
Enjoyment	3.05	.62	2.89	.71	485.00	2.65**	.008
Intentions	2.04	.76	1.88	.79	482.00	2.04*	.042

* $p < .05$. **$p < .01$. ***$p < .001$.

[(a)] two-tailed independent samples *t*-test for mean differences.

Input evaluation: At the end of each intervention input, all participants were asked to rate their involvement during the session on a five point scale. Score 1 means "I was bored, I only stayed because I had to" and score 5 means "I was interested almost constantly. I felt seriously involved, challenged to think about it and/or engage myself to work with it". The participants were also asked to report in an open form what they had gained from the sessions for their personal development and/or what they could transfer to their classroom practices.

Table A.8. Comparison of mean attitude scores between boys and girls on the pupil questionnaire

	Boys (N = 230-233)		*Girls (N = 250-256)*			*t-test*[(a)]	
	M	*SD*	*M*	*SD*	*df*	*t*	*p*
PRETEST							
Technology							
Difficulty	1.97	.69	2.18	.63	486.00	-3.51**	.001
Importance	2.94	.43	2.86	.48	487.00	1.87	.063
Gender	2.74	.89	1.75	.63	414.65	14.04***	.000
Enjoyment	3.40	.53	3.15	.54	487.00	5.11**	.001
Intentions	2.48	.88	1.90	.70	486.00	7.98***	.000
Science							
Importance	3.01	.53	2.85	.48	484.00	3.361**	.001
Gender	2.40	.98	1.36	.49	333.06	14.61***	.000
Enjoyment	3.23	.62	3.05	.59	484.00	3.31**	.001
Intentions	2.19	.89	1.88	.69	432.70	4.26***	.000
POSTTEST							
Technology							
Difficulty	1.84	.69	2.12	.60	457.83	-4.65***	.000
Importance	2.90	.52	2.78	.47	482.00	2.57***	.000
Gender	2.58	.98	1.70	.65	393.83	11.54***	.000
Enjoyment	3.29	.64	3.02	.62	482.00	4.77***	.000
Intentions	2.37	.89	1.87	.70	435.24	6.79*	.011
Science							
Importance	2.97	.53	2.74	.53	484.00	4.81***	.000
Gender	2.27	.96	1.36	.51	348.39	12.79***	.000
Enjoyment	3.12	.64	2.88	.64	484.00	4.15***	.000
Intentions	2.14	.84	1.83	.67	442.33	4.49***	.000

* $p < .05$. **$p < .01$. ***$p < .001$.

[(a)] two-tailed, independent samples *t*-test for mean differences.

Data-analysis

To evaluate change in teacher and pupil attitudes after the intervention (first research goal), Pre- and Posttest scores of trajectory teachers and their pupils were compared by means of paired samples t-tests. For comparison, similar tests were performed on the scores of control group teachers. While we expected these t-tests to indicate significantly positive[ix] changes in attitude scores for trajectory teachers and pupils, we did not expect this for the control teachers. In order to obtain a straightforward comparison between trajectory and control teachers (second research goal) Posttest attitude scores were predicted by the variable 'belonging to the trajectory vs. control group' (with 0 = control teacher and 1 = trajectory teacher), controlling for Pretest scores. As we expected the intervention to have a beneficial effect on teacher attitudes, we hypothesised the regression coefficient of the dummy coded predictor to be positively significant for each outcome (except for DIFFICULTY).

Thirdly, multilevel analyses were performed to predict (changes in) pupil attitudes by means of changes in teacher attitudes. Preliminary, the variance in pupil attitude scores was partitioned into a component at both class and pupil level. Class level variables (i.e. teacher attitude scores) were only added as potential predictors (Hox, 2002; Snijders & Bosker, 1999) for outcomes where a significant amount of variance was situated on class level.

Finally, the hypothesised influence of the intervention inputs on attitude change was explored through the evaluating information gathered after each input (fourth research goal). On the one hand, regression analyses were performed to predict the Posttest attitude scores based on the involvement scores related to each input, controlling for Pretest attitude scores. On the other hand, the qualitative information obtained from the teachers with regard to their learning gains on a personal and professional level, was clustered and summarised.

RESULTS

Changes in Teacher and Pupil Attitudes over the School Year

Table 2. Paired samples t-tests for mean differences in teacher attitudes for the trajectory group (n = 24-26)

	Posttest scores				*Pretest scores*				*t-test*[(a)]		
	M	*SD*	*Min.*	*Max.*	*M*	*SD*	*Min.*	*Max.*	*df*	*t*	*p*
Attitude Technology											
Importance	4.37	.35	3.83	5.00	4.28	.35	3.50	4.83	24	1.09	.286
Enjoyment	3.85	.70	2.33	4.83	3.60	.84	1.67	4.83	24	1.89	.071
Intentions	3.86	.45	3.00	4.75	3.69	.52	2.75	5.00	25	2.41*	.024
Self-efficacy	3.33	.65	1.71	4.29	3.17	.74	1.57	4.86	23	1.73	.097
Attitude Science											
Difficulty	3.25	.74	1.50	4.00	3.56	.68	1.50	5.00	25	-2.96**	.007
Enjoyment	3.76	.54	2.50	4.50	3.68	.55	2.75	5.00	24	0.56	.578

Intentions	3.47	.66	2.00	4.50	3.34	.73	2.00	5.00	25	1.33	.195
Attitude Learning by design											
Importance	4.18	.47	3.00	5.00	4.03	.41	3.00	5.00	23	0.76	.454
Enjoyment	4.19	.41	3.50	5.00	3.96	.34	3.00	5.00	23	2.04	.053
Intentions	3.99	.44	3.00	4.67	3.96	.34	3.00	5.00	23	0.29	.775
Self-efficacy	3.35	.59	2.33	4.33	3.14	.66	2.00	4.67	23	1.46	.158
Attitude Inquiry-based learning											
Importance	4.22	.45	3.25	5.00	4.10	.37	3.25	4.75	22	1.05	.307
Enjoyment	4.22	.45	3.67	5.00	3.96	.36	3.33	5.00	22	2.49*	.021
Intentions	4.02	.50	3.00	5.00	4.06	.48	3.00	5.00	23	-0.33	.747
Self-efficacy	3.43	.48	2.00	4.33	3.23	.56	2.33	4.67	**22**	**1.27**	.217

*$p < .05$. **$p < .01$.
[(a)] two-tailed paired samples t-tests for mean differences.

For the 'trajectory teachers', the means, standard deviations, and ranges of attitude scores concerning technology, science, and related practices are presented in Table 2, for the Pre- as well as Posttest scores[x]. Table 2 further reveals that the Posttest scores for the 'trajectory teachers' are beneficially higher[xi] than the Pretest scores on all comparable scales, with one exception (i.e. slightly reduced intentions to include more inquiry-based learning in their school practices). Paired samples t-tests for mean differences revealed that three of these changes in mean attitude scores over the year reached significance. In more detail, we found that after the intervention trajectory teachers significantly (a) intend to learn more about technology; (b) perceive science as less difficult; and (c) find inquiry-based learning more pleasant, as compared to the Pretest measure. Additionally, three mean differences are borderline significant, indicating that after the intervention, teachers also tend to (a) enjoy technology more; (b) enjoy learning by design more; and (c) feel more self-efficient in the technology domain.

For the pupils, the means, standard deviations, and ranges of attitude scores concerning science and technology are presented in Table 3, for the Pre- as well as the Posttest[xii].

Table 3. Paired samples t-tests for mean differences concerning pupil attitudes (n = 478 – 485)

	Posttest scores				*Pretest scores*				*t-test[(a)]*		
	M	*SD*	*Min.*	*Max.*	*M*	*SD*	*Min.*	*Max.*	*df*	*t*	*p*
Attitude Technology											
Difficulty	1.99	.66	1.00	4.00	2.07	.66	1.00	4.00	483	-2.75*	.006
Importance	2.84	.50	1.14	4.00	2.90	.46	1.43	4.00	484	-2.80*	.005
Gender	2.12	.94	1.00	4.00	2.23	.91	1.00	4.00	484	-2.83*	.005
Enjoyment	3.15	.64	1.00	4.00	3.27	.55	1.00	4.00	484	-4.50**	.000
Intentions	2.11	.83	1.00	4.00	2.17	.84	1.00	4.00	479	-1.70	.089
Attitude Science											
Importance	2.85	.54	1.00	4.00	2.93	.51	1.00	4.00	483	-3.12*	.002
Gender	1.80	.89	1.00	4.00	1.86	.92	1.00	4.00	484	-1.85	.065

Enjoyment	3.00	.65	1.00	4.00	3.14	.61	1.00	4.00	483	-4.92**	.000
Intentions	1.99	.77	1.00	4.00	2.02	.81	1.00	4.00	477	-0.98	.327

* p< .01. ** p< .001.
(a) two-tailed paired samples t-tests for mean differences.

As for the teachers, pre- and Posttest attitude scores were also compared with paired samples t-tests for the pupils in the 'trajectory classes' (see Table 3). In accordance with the expectations, we found a decrease in stereotypical thinking about gender in technology. Although not significant, a similar result was found for gender stereotypes in science. Pupils also find technology less difficult at the end of the school year, compared to the beginning of the year. However, unexpectedly, we also detected significant decreases in ENJOYMENT and IMPORTANCE of both science and technology over the course of the school year. Pupils' intentions towards science and technology also decreased over the year, but not significantly.

Comparison of Trajectory and Control Group Teachers

Preliminary analysis on Pretest scores, comparing the initial attitudes of trajectory and control group teachers, only revealed significant initial differences in mean scores for two scales (see Table 4). On average trajectory teachers score higher on the ENJOYMENT scales for technology as well as learning by design[xiii]. No significant differences were detected between the control and trajectory teachers for the other 13 attitude scales.

Paired sampled t-tests to evaluate changes in attitude scores over the year for the control group of teachers only revealed a significant change in positive attitudes for the DIFFICULTY scale of science, indicating that also control group teachers perceive science as less difficult over the course of one school year (with $t(38) = 2.71, p < .05$).

Table 4. Pretest comparison of mean attitude scores between trajectory and control group teachers

	Control group (N =38 -39)				*Trajectory group (N = 25-26)*				*t-test*[(a)]		
	M	*SD*	*Min.*	*Max.*	*M*	*SD*	*Min Min.*	*Max.*	*df*	*t*	*p*
Attitude Technology											
Importance	4.12	.39	3.00	5.00	4.28	.35	3.50	4.83	62	1.71	.092
Enjoyment	3.35	.78	1.83	4.67	3.60	.84	1.67	4.83	63	1.23	.222
Intentions	3.45	.54	2.25	4.75	3.69	.52	2.75	5.00	63	1.82	.074
Self-efficacy	2.91	.64	1.71	4.57	3.17	.74	1.57	4.86	62	1.49	.141
Attitude Science											
Difficulty	3.60	.58	2.00	5.00	3.56	.68	1.50	5.00	63	-0.29	.776
Enjoyment	3.40	.51	2.25	4.25	3.68	.55	2.75	5.00	62	2.10*	.039
Intention	3.01	.74	1.00	4.50	3.34	.73	2.00	5.00	63	1.74	.087

Attitude Learning by design											
Importance	3.91	.32	3.00	4.67	4.03	.41	3.00	5.00	63	1.22	.227
Enjoyment	3.68	.52	2.50	5.00	3.96	.34	3.00	5.00	63	2.63*	.011
Intentions	3.89	.47	2.00	5.00	3.96	.34	3.00	5.00	62	0.70	.484
Self-efficacy	3.22	.52	2.33	4.00	3.14	.66	2.00	4.67	62	-0.53	.597
Attitude Inquiry-based learning											
Importance	4.03	.30	3.25	4.50	4.10	.37	3.00	5.00	62	0.88	.382
Enjoyment	3.86	.42	2.67	4.67	3.96	.36	3.00	5.00	62	0.94	.351
Intentions	3.96	.45	3.00	5.00	4.06	.48	3.00	5.00	63	0.83	.413
Self-efficacy	3.30	.53	2.00	4.00	3.23	.56	2.00	4.67	62	-0.52	.605

* $p < .05$.
[a] two-tailed, independent samples t-test for mean differences.

However, to obtain a straightforward comparison between trajectory and control teachers, regression analyses were performed, including type of intervention (i.e. intervention followed vs. not followed) as well as Pretest scores as predictor variables for Posttest scores. Results of the regression analyses, comparing estimated Posttest scores for control and trajectory teachers, are reported in Table 5. Estimated Posttest scores are significantly higher for the trajectory teachers than for the control teachers, for the same five attitude scales where the trajectory teachers have grown (marginally) significantly over the course of the trajectory school year. Furthermore, it must be noted that the significant decrease in perceived difficulty of science over the year, which we detected for both trajectory and control teachers, does not differ significantly between both groups.

Table 5. Predicting Posttest attitude scores by Pretest scores and 'control vs. trajectory group' (n = 61-65)

		Regression coefficient[a]		*Estimated Posttest score*	
	$M_{Pretest}$	β_{group}	*P*	*Control group*	*Trajectory group*
Attitude Technology					
Importance	4.18	.19	.078	4.15	4.31
Enjoyment	3.45	.18*	.027	3.43	3.70
Intentions	3.55	.23**	.008	3.51	3.76
Self-efficacy	3.01	.22*	.014	2.99	3.26
Attitude Science					
Difficulty	3.58	-.05	.585	3.34	3.27
Enjoyment	3.51	.10	.379	3.54	3.65
Intentions	3.14	.11	.262	3.19	3.35
Attitude Learning by design					
Importance	3.96	.21	.095	3.96	4.16
Enjoyment	3.79	.28*	.021	3.80	4.11
Intentions	3.92	.17	.153	3.81	3.97
Self-efficacy	3.19	.08	.477	3.26	3.36

Attitude Inquiry-based learning					
Importance	4.05	.25	.055	4.01	4.21
Enjoyment	3.90	.26*	.035	3.95	4.19
Intentions	4.00	.22	.090	3.78	3.90
Self-efficacy	3.27	.11	.354	3.31	3.43

[a]Regression coefficient of the dummy coded variable, with 0 = control group and 1 = trajectory group.

Associations Between (Changes in) Teacher and Pupil Attitudes

Table 6 shows the estimation and significance of the class level variance in pupil attitude scores for Pretest, Posttest and changes over the school year. Overall, the proportion of variance on class level for the outcomes as presented in Table 6 ranges between 0% and 13%. Class level features matter the most for gender stereotypical attitudes (significant class level variances ranging between 9% and 13%).

Table 6. Estimation of class level variance in attitude scores, concerning the pre- and Posttest measurement, and changes over the school year (n = 478-490)

	Pretest scores			*Posttest scores*			*Change scores*		
	B	*SE*	*p*	*B*	*SE*	*p*	*B*	*SE*	*p*
Attitude Technology									
Difficulty	.03*	.02	.033	.00	.00	1.00	.02	.01	.093
Importance	.03*	.01	.025	.02*	.01	.023	.01	.01	.208
Gender	.04	.02	.087	.09*	.04	.014	.06*	.03	.022
Enjoyment	.01	.01	.181	.05*	.02	.017	.02	.01	.057
Intentions	.04*	.02	.046	.02	.02	.189	.02	.02	.286
Attitude Science									
Importance	.01	.01	.113	.02*	.01	.040	.01	.01	.588
Gender	.08*	.03	.016	.10**	.04	.008	.05*	.02	.018
Enjoyment	.02	.01	.060	.04*	.02	.026	.02	.01	.079
Intentions	.02	.02	.154	.01	.01	.470	.02	.01	.261

*$p < .05$. **$p < .01$.

To explain significant class level variances in pupil attitude scores, teacher attitude scores were added as predictors of pupil scores. More specifically, teacher scales of technology and learning by design were used as predictors of pupil scores on attitudes towards technology, and teacher scales of science and inquiry-based learning as predictors for pupil attitudes towards science. Pretest, Posttest, and change scores of teachers were included as predictors of Pretest, Posttest, and change scores of pupils, and were not mixed.

No significant predictors were found concerning Pretest scores. With regard to Posttest scores, pupils' enjoyment of technology can be predicted by the self-efficacy teachers experience for technology ($\beta = -0.15$, $p < .01$), and for design-based learning ($\beta = -0.20$, $p < .001$). However, the associations are negative, indicating that higher feelings of self-efficacy in teachers at the end of the school year are associated with lower feelings of enjoyment in children with regard to technology.

The Impact of the Intervention Inputs: An Exploratory Analysis

As shown in Table 7, self-reported mean involvement scores during the inputs range from 4.13 to 4.61 on a five-point scale. Significant correlations are found between involvement scores for the introduction session on the one hand, and the science centre visit and the in-depth session about talents on the other hand. Scores for the latter also significantly correlate with involvement scores for the coaching session.

Table 7. Descriptive statistics and correlations of the involvement scores (n=23-46)

	Involvement scores				*Correlations*			
	M	SD	*Min.*	*Max.*	*1.*	*2.*	*3.*	*4.*
1. Introduction session	4.13	.61	3.00	5.00		.57*	.69**	.35
2. Interactive visit science centre	4.61	.50	4.00	5.00			.38	.42
3. In-depth analysis Curious Minds	4.52	.51	4.00	5.00				.59*
4. Coaching session	4.21	.57	3.00	5.00				

$^{*}p < .05$. $^{**}p < .01$.

Using regression analysis, involvement scores for each input were linked to changes in teacher attitude scores. Involvement during the introduction session was found to operate as a predictor of growth in SELF-EFFICACY with regard to technology ($\beta = 0.38$, $p < .05$). Secondly, involvement during the interactive visit to a science centre predicted the growth in INTENTIONS with regard to technology ($\beta = 0.31$, $p < .05$). No other attitude changes could be predicted based on involvement for the different intervention inputs.

Finally, Table 8 contains an overview of the learning experiences participants report to have encountered for each of the four intervention inputs. Although the content of the different inputs highlights different aspects of science and technology and teaching it (see introduction), most of the learning experiences teachers mentioned hold to some extent for several inputs.

Over the different inputs, teachers' specific learning experiences can be combined into three categories. The first category concerns a broader view on science and technology, which (a) helps teachers recognise science and technology in daily life; and (b) helps see connections with their own knowledge and image of the physical world more easily. Related to this, teachers indicate to have learned about how specific science and technology issues can be translated into classroom practices. In this regard, they refer to the importance of a rich learning environment, with room for exploring and experimenting, based on reality. In a second (related) category, teachers indicate to have learned about specific teacher competences that are highly relevant in guiding children's learning processes with regard to science and technology. Here, providing autonomy and stimulating thoughts, communication and creativity are specifically stressed. Furthermore, according to what teachers have learned, attentiveness to pupils' involvement, competences and talents is an important teaching skill in this domain.

Table 8. Overview of self-reported learning experiences for each intervention input

Introduction Session (N[1]=88)	*%*	*(N)*
Broader view on science and technology/recognising science and technology in daily life	23.86%	(21)
Self-knowledge: strong and weak points, personal obstructions	20.45%	(18)
Providing autonomy and being sensitive to the child's exploratory drive	12.50%	(11)
Stimulation: the teacher as source of enrichment in communication and engagement in activities	11.36%	(10)
Domains of development: the development of fundamental schemes and the importance of intuition	12.50%	(11)
Importance of a rich learning environment: based on reality, challenging to different levels of competence, with focus on process instead of result	10.23%	(9)
Attentiveness to pupils' involvement, their competences and talents	9.09%	(8)
Interactive visit science centre (N=101)		
Importance of a rich learning environment: with room for experimenting (in co-construction), starting from wonderment, with focus on process instead of result	26.73%	(27)
Self-knowledge: strong and weak points, personal obstructions	22.77%	(23)
Knowledge about science and technology and inspiration for the classroom practices	17.82%	(18)
Broader view on science and technology/recognizing science and technology in daily life	12.87%	(13)
Enjoyment of science and technology	5.94%	(6)
Attentiveness to the nature of cognitive processes mobilised through the activity	5.94%	(6)
Stimulation: the teacher as source of enrichment in communication and engagement in activities	2.97%	(3)
Importance of science and technology for pupils	2.97%	(3)
Attentiveness to pupils' involvement, their competences and talents	1.98%	(2)
In-depth session about talents (N = 60)		
Providing autonomy and being sensitive to the child's exploratory drive	30.00%	(18)
Stimulation: the teacher as source of enrichment in communication and engagement in activities	28.33%	(17)
Attentiveness to pupils' involvement, their competences and talents	23.33%	(14)
Importance of a rich learning environment: challenging to different levels of competence, with focus on the process instead of the results	8.33%	(5)
Attentiveness to the nature of cognitive processes mobilised through the activity	6.67%	(4)
Broader view on science and technology/recognizing science and technology in daily life	3.33%	(2)
Coaching session (N=54)	*%*	*(N)*
Self-knowledge: strong and weak points, personal obstructions	38.89%	(21)
Inspiration for classroom practices	22.22%	(12)
Confirmation of self-efficacy	12.96%	(7)
Providing autonomy and being sensitive to the child's exploratory drive	11.11%	(6)
Broader view on science and technology recognizing science and technology in daily life	5.56%	(3)
Attentiveness to pupils' involvement, their competences and talents	5.56%	(3)
Attentiveness to the nature of cognitive processes mobilised through the activity	3.70%	(2)

[1] Total number of remarks of the participants.

Finally, knowledge of different developmental domains and the capability to recognise (possible) mental loads in an activity is considered an important additional competence for teaching science and technology. In a third category, personal feelings and notions can be grouped together. Teachers pointed out that they gained better insight into themselves through the input sessions. They mentioned their strong and weak points and the personal obstructions they encountered. This was, for example, translated into a more accurate and overall more positive perception of their self-efficacy. Apart from the three main content categories teachers mentioned, particularly with regard to the interactive science centre visit, teachers indicated that they experienced enjoyment during activities in science and technology domains.

DISCUSSION AND CONCLUSION

The main purpose of this article was to examine the impact of an intervention trajectory on changes in teacher and pupil attitudes towards science and technology. Furthermore, we explored whether, and if so, which elements of the intervention inputs could provide an explanation for these changes.

With regard to the effectiveness of the intervention, our goal to enhance positive teacher attitudes has especially been reached in relation to the technology domain. By the end of the year, compared to the beginning, teachers who followed the intervention (a) intended to learn more about technology; (b) felt more self-efficient about their capacities in the field, and (c) enjoyed it more. Their attitudes in these three aspects evolved to a significantly stronger extent than was the case for the teachers who did not follow the intervention. For two out of these three effects, self-attributed involvement scores of teachers evaluating the inputs can offer some insight into the specific active ingredients of the intervention. Firstly, higher involvement scores for the visit to the science centre predicted a larger growth in intentions with regard to technology. The interactive visit to a science centre primarily aimed at allowing the teachers to experience what a wondering child experiences in relation to science and technology in a rich and open context. The qualitative analysis on the evaluation sheets showed that teachers especially reported to have learned from this visit that a rich environment is an important basis for children to learn about science and technology in co-construction. Furthermore, confronting teachers with rich environments, i.e., the exhibits at the centre, offers a lot of inspirational material that teachers can translate into their classroom practices. For teachers who were more involved during the visit, this may consequently evoke more intentions to implement more science and technology activities in their classrooms. Secondly, higher involvement scores during the introduction session predicted a larger growth in feelings of self-efficacy with regard to technology. In the introduction session, participants were confronted with the principles of experiential education and the link to a science and technology implementation that starts from a rich context and daily life experiences. Furthermore, teachers receive the message that an intuitive understanding is very important and that as a teacher, you can make a difference by stimulating and coaching your pupils, and hence finding a solution in co-construction with your pupils. This may give teachers more grip on how to conduct a science or technology activity and may make it more achievable. In accordance with the aims of the input, many teachers mentioned these items in the evaluation forms as learning experiences: they gained a broader view on science and technology and better self-knowledge. Teachers also mentioned the importance of'providing autonomy' and 'stimulation' to children.As teachers may experience these aspects of good teaching behaviour as feasible, the combination of the above elements may lead to higher estimated feelings of self-efficacy. With regard to the growth in enjoyment of technology over the year, the involvement during the inputs does not operate as a significant predictor. However, the enjoyableness of science and technology is mentioned by six teachers during the visit to a science centre. Part of the goal of this visit was getting teachers to wonder (again). Moreover, the possibility to operate in active co-construction with their pupils during the science and technology module in their classroom, might have given teachers the chance to become acquainted with a more enjoyable way of working with science and technology. Additionally, the intervention had similar effects on teachers' enjoyment of teaching practices related to science and technology (learning by design and inquiry-based learning), which are stressed as good practices in these domains throughout the intervention inputs. Also with regard to these two attitude scales, teachers who followed the intervention trajectory have grown significantly more than their control group counterparts.

With respect to pupil scores, attitudes with regard to gender stereotypes and the perception of difficulty of science and technology improved over the year, whereas pupil attitudes decreased concerning (a) their perceptions of importance of both domains; (b) their enjoyment in both domains; and (c) their intentions to invest more time and effort in these domains. The latter results were unexpected and deserve future research attention. As there were no data available on pupil attitude scores in the control group, findings are however hard to interpret. However, the pupil results suggest that although intervention efforts may aim at improving teacher attitudes and related teacher practices, this does not necessarily improve pupil attitudes directly. This idea was also confirmed by the (striking) finding that more feelings of self-efficacy in teachers (with regard to technology and design-based learning) are related to less enjoyment of technology by pupils at the end of the school year. As one possible explanation, it may not be unlikely that through the intervention, teachers become 'too' self-confident and consequently tend to impose new content on pupils, rather than building knowledge in co-construction with them. In line with experiential education, this lack of autonomy and self-initiative for children may result in less enjoyment of tasks related to technology (Laevers, 2005). Alternatively, as innovation in teacher practices does not go without adaptation, this may cause some friction and drops in satisfaction and enjoyment for the pupils confronted with it. Perhaps if the intervention was spread over a longer period, and/or we conducted a follow-up measure on pupil attitudes, we might have discovered an increase in enjoyment again. Therefore, finding significant improvements in positive attitudes over the period of only one year in pupils, and especially in teachers, holds a strong promise for the future. However, further (in depth) study is certainly necessary based on the unexpected findings on pupil attitudes, as additional and/or alternative explanations may be equally valuable. Perhaps intentions to learn more about science and technology as well as enjoyment in both domains drop by the end of the school year, because it is traditionally a period with a lot of testing, close to the holidays, for example.

With respect to gender, it is also interesting to note that gender-stereotypic beliefs may be determined by class and/or teacher features, as the amount of class level variance was meaningful. Fairly straightforward, it is possible that equivalent teacher attitudes about gender stereotypes play a significant role here. This could however not be evaluated, as this scale was omitted in the version of the teacher questionnaire we used. Other (possibly related) determinants of the detected decrease in gender stereotypic beliefs of pupils may lay in the 'active ingredients' of the intervention we imposed, such as a focus on hidden talents and intensified working with science and technology. This offers pupils and teachers a chance to appreciate the skills of their feminine counterparts, even more when using co-construction and working in a rich environment that provides many opportunities to explore and experiment. Further study would be useful to determine tools and aids to positively affect pupils' views on gender in relation to science and technology, as this is a major issue, especially in the Netherlands (Joukes, 2010).

In general, three remarks are at stake. First, comparison of effects between trajectory group and control group teachers may hold an underestimation of effects, as the control group teachers belonged to the same schools as the teachers who followed the intervention. As these teachers shared the same school environment, this may have led to communication and interaction about the interventions, restoring an influence on the attitudes of the control teachers as well. However, on five attitude scales, we still detected a significant difference in growth on attitude scales in favour of the trajectory group teachers, underscoring the effectiveness of the intervention. Second, it may not surprise that the intervention seems to have a more meaningful influence on attitudes towards technology vs. attitudes towards science. After all, interventions were imbedded in a *technology* module which was implemented through the school year. Third, concerning attitudes towards teacher practices in the field of science and technology, only teachers' enjoyment was fostered through the intervention. Yet, as the correlation analysis showed, both for teachers and pupils, enjoyment of science and technology are associated with all other attitude scales, except for the gender scales. This may indicate that enjoyment in science and technology activities hold a key to improve attitudes on both domains, not only concerning the affective component, but also concerning the cognitive and behavioural components in an indirect way as well.

Finally, we note that next to the fact that effects on (teacher and especially pupil) attitudes may be underestimated due to the limited period of active intervention mentioned above, some other limitations have to be articulated with regard to the present study. First, the aim of the larger study was broader, and hence, future analyses including different outcomes, may show a larger impact of the intervention on the actual competence, behaviour and involvement in science and technology practices of pupils and/or teachers, compared to the impact on their attitudes. As gathering all data through questionnaires, interviews, and observations during the intervention year was very time-consuming, the limited number of participating teachers in the study has to be acknowledged as a downside as well. Consequently, if in future studies the statistical power could be enhanced by enlarging the teacher sample, this may lead to more significant effects on a broader variety of subscales concerning teacher attitudes. We may expect this, because the changes in teacher attitude scores after the intervention year, although not statistically significant for each single outcome, all go in the expected direction.

Charlotte Van Cleynenbreugel
Charlotte.VanCleynenbreugel@ped.kuleuven.be

Veerle De Winter

Evelien Buyse

Ferre Laevers
Katholieke Universiteit Leuven, Belgium

NOTES

[ii] http://www.talentenkracht.nl; for English: http://www.talentenkracht.nl/content/files/SITE1765/Brochure_CuriousMinds_eng.pdf

[iii] We made use of factor analysis with Oblimin rotation, while Oberon used Varimax rotation.

[iv] With mean scores of 4.36 and 3.97 for Flemish and Dutch teachers respectively.

[v] With mean scores of 4.44 and 3.97 for Flemish and Dutch teachers respectively.

[vi] With mean scores for learning by design of 4.12 and 3.71 for the oldest and youngest group of teachers respectively, and mean scores for inquiry-based learning of 4.27 and 3.69 for the oldest and youngest group of teachers respectively.

[vii] Due to the small number of male participants, gender differences were not calculated.

[viii] We made use of factor analysis with Oblimin rotation, while Walma van der Molen used Varimax rotation.

[ix] Except for DIFFICULTY and GENDER, for which scales we expected significantly negative t-values, because we expected the Posttest scores to be lower than the Pretest scores on these subscales.

[x] The reported t-statistics in Table 2 will be discussed later.

[xi] Or lower, concerning DIFFICULTY.

[xii] The reported t-statistics in Table 3 will be discussed later.

[xiii] With mean scores for technology of 3.68 and 3.40 for the trajectory and control teachers respectively, and mean scores for learning by design of 3.96 and 3.68 for the trajectory and control teachers respectively.

JOHAN P. DE WILDE, ELLEN SJOER AND MARC J. DE VRIES

11. PRESERVICE PRIMARY TEACHERS' IMAGE OF AND ATTITUDE TOWARDS SCIENCE AND TECHNOLOGY

An exploratory study

INTRODUCTION

An important long-term aim of education about science and technology is to improve the overall image that people have of technology and to help them develop a well-informed attitude. Previous studies have shown that pupils in primary and secondary schools often have distorted and limited views on science and technology and that such views often go hand in hand with a lack of interest in technology or a lack of balance in their appreciation of science and technology (either they are naively positive or negative) (Hodson 2009; De Vries 2005). Information about the image of and attitude towards science and technology is of importance for teaching and learning. If we do not take into account what pupils and students bring in mentally when they enter our classrooms, our education may miss its target. The same holds for the primary teachers that enter the VTB-Pro activities for professional development in teaching about science and technology, and the students in primary teacher training programs. Jarvis (2004) has shown that this also has an impact on the children. The study that is described in this chapter aims at gaining insight into the image of and attitude towards science and technology of pre-service primary teachers. We do not aim at developing new views on the dimensions of their attitudes towards science and technology (as is done, for instance, in the contribution by Asma, Walma van der Molen, and van Aalderen-Smeets in this volume). What we aim at is an exploration of the presence of those dimensions that are common in studies into attitudes towards science and technology in the population of students in primary teacher training. Little has been done to study this population as far as its image of and attitude towards science and technology is concerned. The image part of this study is the most novel part. Thus far, not much research has focused on what primary education student teachers think technology is. Here, our study gives some relatively new insights compared to the existing literature.

Marc J. de Vries, Hanno van Keulen, Sylvia Peters, and Juliette Walma van der Molen (Eds.), Professional development for primary teachers in science and technology. The Dutch VTB-Pro project in an international perspective.

RESEARCH QUESTIONS AND METHODOLOGY

Existing attitude studies for science and technology education

Images of and attitudes towards science and technology are by no means an unexplored domain. In science education, numerous studies have been done and later for technology, too, research studies have been carried out. As far as attitudes towards science and technology are concerned, there is similarity between the various studies in the dimensions that were found: time and again dimensions like interest, relevance and difficulty come up in factor analyses. Instruments for measuring teachers' attitudes towards science and technology usually have a scale for interest, one for relevance or impact on society and humans, one for difficulty or accessibility, and one for gender (Hodson 2009; De Vries 2005, Asma et al. in this book). The existing studies focus on pupils, both in primary and in secondary education, and on teachers. In our case, we dealt with student teachers in primary education teacher training programs. For this group near to nothing is known about their images of and attitude towards science and technology. It can be expected, though, that similar dimensions will be found in their attitudes as in the studies on pupils and teachers. For the image of science and technology, some work has been done (see the contribution by Asma et al. in this volume), but no international 'standard' scales exist, although for science some common features can be recognized in the literature: a scientist is seen by pupils as a male, somewhat elder figure who works in a laboratory with test tubes (Hodson 2009). For technology it is known that pupils in secondary education primarily associate it with artefacts (products) rather than activities (De Vries 2005). The same holds more or less for adults, be it that the range of artefacts they think of is broader than that of pupils. For instance, they tend more to see simple objects like hammers or mugs as technology as well as lasers and robots. In the present study, we used a more open-ended approach, in order to investigate what student teachers come up with spontaneously. Thus, rather than pre-defining artefacts, we investigated how pre-service teachers would define science and technology in an open qualitative manner. In this open-ended approach, however, we searched explicitly for appearance of the known dimensions, as it can be expected that some of those also hold for primary student teachers, while other might be different.

Research questions

The study that is described here has been done in the context of the VTB-Pro professional development activities for science and technology. In our case, the activities were not for primary school teachers, but for student teachers. In Driestar Hoge school, University for Teacher Education, one of the partners in the VTB-Pro project, an activity was implemented that aimed at giving students some science and technology experiences in a context that is recognizable for them (Waterworks; in the Netherlands this is also a socially very relevant topic). This activity was inquiry-based in nature and the responsibility for the content of the activity was shared between teaching staff and students (in the Netherlands this is usually indicated as 'Ontwikkelings Gericht Onderwijs' (OGO): Development-Oriented Education). We wanted to know with what ideas about science and technology the students started with the project. Also, we wanted to have some impressions of how the activities changed the students' perspective on science and technology. This latter part of the study, of course, has limited value, as one can hardly expect dramatic changes to take place due to one project only. But at least it provides a rough suggestion of the impact such a project may have if changes, small as they may be, are in the right direction.

Our research therefore aimed at answering the following questions:

1. What is the image of and attitude towards science and technology of third year students in the Driestar primary teacher education program?
2. How does this image and students' attitude change after these students have gone through one project on science and technology (focused on water works)?

We have separated images and attitudes here, as this is common in earlier studies.

Research methodology

To find answers to the research questions listed above, we have used a combination of qualitative and quantitative methods. Our motive for this is the fact that on the one hand studies among other target groups give clues for dimensions in the image of and attitude towards science and technology, but on the other hand these studies refer to other target groups (pupils and teachers instead of student teachers). As a qualitative instrument for investigating the students' image of science and technology we have used essay assignments. Students were asked to write a one-page essay on what science and technology meant to them. These studies have been read and keywords related to the students' image of science and technology have been tagged and categorised. For their attitudes we have used a quantitative instrument, namely a questionnaire with 5-point Likert scale items. First a pilot version of this instrument was developed, based on the existing literature on attitudes towards science and technology. A slightly revised version of this instrument was used among the larger group of students. The same instruments were used before and after the project.

QUALITATIVE STUDY: ESSAYS

In our study, 19 students in their third year of primary teacher education wrote a one page essay in which they could express their ideas about what science and technology mean to them. We did not ask them to differentiate between science and technology and thus left it to them to decide how to deal with that. After the Waterworks activity we asked them again to write an essay on their ideas about science and technology, now with their recent experiences of that project in mind. The essays contained lots of keywords related to science and technology. By using open ended coding (Strauss and Corbin 1990) we analysed those keywords and categorized them. We took into account some recent insights in the philosophy of science and of technology and thus followed an iterative back-and-forth movement between data and theory. ('sensitizing concepts' suggest direction; see the contribution by Sjoer and Meirink in this book), The coding process was done in a series of sections with four researchers. The coding lead to a consensus on four main categories for the image that the student-teachers have of science and four for the image that these students have of technology. These categories are presented in Table 1 and Table 2. In the Tables we have combined the keywords that we found in the pre and post tests. This was done because the changes in the use of keywords were small. As this was a small-scale qualitative study, it makes no sense presenting the number of times a keyword was used before and after the activity. Evidently, the activity did not have a measurable impact on the students' image of science and technology, probably because the activity was insufficiently focused on doing so. Therefore, the outcomes before as well as after the activity reveal the intuitive image students have about science and technology. In the final part of the chapter, we will reflect on what would have been necessary to cause real changes in the students 'image of technology. In the analysis below, we will mention some minor changes in the use of keywords that we found when comparing the situation before and after the activity.

Table 1. Science

Activities	*Knowledge*	*Disciplines*	*Human/Society*
Reasoning	Theory	Biology	Religion
Inventing	Law	Technology	Non-neutrality
Posing questions	Formula	Physics	Daily life
Hypothesizing	Unspecified	Mathematics	Human being
Data collection		Humanities	Historicity
Explaining			Professions
Publishing			
Unspecified			

Table 2. Technology

Activities	*Artefacts*	*Discipline*	*Human/Society*
Investigating	Mechanical	Theories	Gender
Inventing	Materials	School subject	Professions
Making	Tools		Daily life
Using	Simple		Non-neutrality
Repairing	Complex		Historicity
Unspecified	Electronic		
	System		
	Unspecified		

Some quotes from students' essays can illustrate the meaning of these categories. Let us start with remarks about science. One student wrote:

> Science is rather interesting. Problem solving, posing good questions, making logical connections, posing hypotheses, those are aspects I see as important for good research work.

These show an awareness of the various activities in science.

Another student wrote:

> If you would have to do without all things that contain technology, in fact nothing would be left. And all that has been developed through knowledge development.

This shows an awareness of the social dimension of technology (the daily life omnipresence of technology), but also for science being the development of knowledge.

Some students also mentioned religious considerations:

> I want to know what is behind things, how something works, why it works like that, etcetera. How can it be otherwise, because God gave people capabilities to think and we have to use the talents that God gave us.

Another quote on science:

> Science is rather remote from me. But when I relate it to my own life, I think disciplines like physics, biology, mathematics and chemistry belong to it.

Clearly, this student thinks of science as a set of disciplines.

Now for technology. One student wrote:

> For technology, I tend to think of devices and machines that are very complicated, with gears and engines.

Here we clearly see the Artefacts dimension in the image this student has of technology.

Another student wrote:

> Technology makes me think of high school. Hours of sweating with wood and a blunt saw. With great trouble we made a wooden lamp that was so ugly that it was thrown away immediately.

This student thinks of technology as making activities in the context of a school subject.

A final related to the human/social appreciation of science and technology:

> Science and technology make it possible that constantly new things are discovered that are important for our current society. Science races on and discovers things that are at the edge of what is dignified.

For technology, the four categories derived from the essays fit nicely with the four main areas in the philosophy of technology as identified by Carl Mitcham in his well-known book Thinking Through Technology (1994). This was no co-incidence as we had these in mind when coding the keywords in the essays, but all keywords fitted into one of these categories in a natural way. All keywords were categorized by one of the other researchers and in 90% of the cases the results were the same. The Tables contain the numbers of keywords we found in the essays, both the ones written before the Waterworks activity and after. For science similar categories were found as for technology. The 'Activities' category for technology has a counterpart for science, and so has the 'Human and society' category. The 'Artefact' category for technology has its counterpart for science in the 'Knowledge' category as both are the outcomes of the activities (artefact are the outcomes of technology and knowledge is the outcome of science). Likewise, the 'Technology as a discipline' category for technology and the 'Collection of disciplines' category for science are equivalent.

Although this part of the study is not quantitative in nature, we do want to give an impression of how many keywords were found in the various (sub-)categories. In the science categories, most keywords were in the 'Activities' category and this was even more the case in the post-activity essays than in the pre-activity essays. For technology in the pre-activity essays the emphasis was on the 'Artefact' and 'Discipline' categories, whereby the latter was dominated by the 'school subject' sub-category. In the post-activity essays the emphasis shifted towards the 'Artefacts' category. A small minority of the students did not differentiate between science and technology in their essays (both in the pre-activity and the post-activity essays). Finally, when we compare the total number of keywords in the essays that were associated with science and those with technology, we see that there were approximately twice as many technology-related keywords mentioned in the essays than science-related keywords. This was the case both in the pre- and the post-activity essays.

From this qualitative part of the study we can conclude the following. In the essays the main categories that could be expected based on philosophy of science and technology are present, but they are biased towards certain categories. The effect of the Waterworks activity is not a broadening of the image. No new categories emerge in the students' essays. For science, the students' image focuses on activities. In the pre-activity essays, most of the keywords in this category, though, were rather unspecified (like: "study" or "go more deeply into . . ."). In the post-activity essays we find 'Posing questions', 'Hypothesizing' and 'Data collection' mentioned somewhat more often. Clearly, the Waterworks activity has made students somewhat more aware of what the various activities in science are. It is striking that, contrary to technology, no relation with a primary school subject is made in the essays. Also they do not make much mention of the outcomes of the research activities (knowledge, theories, etcetera). The cause for this may well be the emphasis on the research activities themselves in the project work. Also, we note the absence of the gender issue ("Girls too" for technology). For technology, the effect of the Waterworks activity seems to be a further focus on artefacts. Within the 'Activities' category we see an increase of 'Investigating (artefacts)' being mentioned. The combination of these two observations can probably be accounted for by the fact that a major part of the Waterworks activity was to examine the functioning of artefacts and their history. Finally, from the total numbers of keywords for technology and for science, we conclude that the students find it more difficult to imagine what science is than what technology is.

When we compare these results with studies among primary and secondary school pupils for technology (De Vries 2005), we note that the student-teachers' image was broader. Pupils hardly associate technology with a discipline or knowledge domain, nor do they recognize designing as an important activity in technology. Also we note that the student-teachers are less biased towards electronic devices than the pupils appeared to be. The student-teachers show more awareness of technology also being present in the simple tools they find all around them. For instance, they mention a wooden spoon and a windmill and consider them to be technology equally well as robots or oil refineries.

QUANTITATIVE STUDY: QUESTIONNAIRE

The second part of the research was a quantitative study into the attitude towards and image of science and technology among the pre-service teachers that did the Waterworks project. A questionnaire consisting of 78-point Likert-items was developed on the basis of an existing instrument developed by Walma van der Molen for science and the PATT questionnaire for technology as far as the attitude part was concerned. The attitude part contained statements to which students had to respond by indicating to what extent they agreed or disagreed with them; the image part consisted of topics to which students had to respond by indicating the extent to which they thought they were related to either science or technology (for science and technology two separate groups of topics were presented). This questionnaire was administered to 98 students before the Waterworks activity and 80 students after the activity. Due to time constraints, it was not possible to have the essays written long before administering the questionnaire and therefore the analysis of the qualitative study (the essays) was not used in constructing the image part of the questionnaire. For measuring the image of science and technology, insights from the philosophy of science and technology were used (De Vries 2005. No pilot test of the questionnaire was done because many of the attitude questions were taken from the existing instruments (in particular the Walma van der Molen and PATT questionnaires). A factor analysis was done in order to investigate the main dimensions in the students' attitude towards and image of science and technology[xiv]. Based on that analysis, scales were constructed for each of those dimensions, their homogeneity was tested by calculating Cronbach's alpha, and t-tests were used to investigate differences between the pre-activity and post-activity scores. The number of respondents was very small for a factor analysis. Nevertheless, it resulted mostly in most factors being interpretable without problems, but some not leading to sufficiently reliable scales. Therefore we think that the outcomes of this factor analysis are worth reviewing.

The outcomes were the following. For the students' image of science, four factors were identified: science as explaining, science as experimenting, science as coming up with practical solutions and science as coming up with innovative ideas. These factors show that for many students science has two characteristics that experts probably would associate with technology rather than science (coming up with practical solutions, innovation). This, no doubt, is due to the way science is often presented in newspapers and on television ("scientists have invented a new . . ."). But otherwise the students seem to have a reasonable view on the nature of science. For each of these, it was possible to construct a factor with Cronbach's Alpha higher than 0.70. The factors found here match reasonably well with what we found in the qualitative part of the study (the 'coming up with practical solutions' factor is similar to the 'human-society' dimension in the qualitative part of the study). What we identified as a 'Knowledge' dimension in the qualitative part of the study does not appear as a separate factor in the quantitative part, but merges with other facts. In the attitude towards science the factor analysis revealed the following dimensions: interest in science, social relevance of science, gender and science, difficulty of science, technological relevance of science (the use of science in technology; later, we realized that perhaps this dimension is more part of the students' image of science than of their attitude towards it), and appreciation of science (good or bad). These factors match well with what has been found in studies among primary teachers (see, e.g., the contribution of Asma, Walma van der Molen, and van Aalderen-Smeets in this volume). Scale construction was possible for all factors although the last three factors did not result in scales with Cronbach's alpha's higher than .50.

In the image of technology we found the factors: technology as artifacts, technology as activities, technology as food production and technology as coming up with new ideas and solutions. This matches well with what we found in the qualitative part of this study. It is remarkable that the items related to food technology stand out separately and do not integrate with other items. Apparently, the students see this as a different sort of technology than the more mechanical and electrical items (namely as less associated with technology, as we will see later). But otherwise, the students have a fairly reasonable view on technology, more than secondary school pupils (De Vries 2005; see also the section on the qualitative part of this study on this issue). Except for the last one, all factors could be used to construct scales with Alpha's higher than .70. In the attitude towards technology the following factors were found: interest in technology, appreciation of technology (good or bad) and difficulty of technology. These also match well with what was found for primary teachers (Asma, Walma van der Molen, and van Aalderen-Smeets in this volume), except that surprisingly the gender dimension has disappeared. Items related to gender were included into the other factors. Only the first factor led to a sufficiently reliable Alpha (.86). When comparing the results before and after the Waterworks activity we see no significant differences between any of the scale scores. Also at the level of individual items there were hardly any significant differences, so we can conclude that the activity did not cause any important changes in the students' attitude and image of science and technology.

Let us now look at the scale scores. How do the students think about science and technology? First we will describe their image of and attitude towards science and technology. The students were lukewarm in their interest for science and technology (average scores around the neutral point of the 5-point scale). That was to be expected as they have dropped science and technology in secondary school. They do show an appreciation for the importance of science and technology in society (an average scale score of about 4 on the 5-point scale). The same holds for the issue of science being important for technology. For science, they think this is something in which boys do not perform better than girls. They are also fairly neutral on whether science and technology are difficult. All together we see that the students do not have a very outspoken opinion on science and technology, except that they do recognize the social and technological importance of science. Of all scales, only the ones for the science-technology relation and for the appreciation of science and technology yield scores more than 0.5 from the neutral point of the scale. This confirms the idea that for these students science and technology are not something they feel concerned about and have no outspoken opinions about by lack of insight. As for their image of science, the students give almost equal scores for science as explaining, science as experimenting, science as coming up with innovative ideas (about 4 on the 5-point scale). Science as coming up with practical solutions is somewhat less prominent in their image of science (3.5 on the 5-point scale). As for technology, artifacts are most prominent in their image of what that is (4.1 on the 5-point scale). In that respect they resemble secondary school pupils (De Vries 2005). Then come activities and coming up with new ideas and solutions (3.9 on the 5-point scale). The weakest association is made with food technology, but still it is a positive one (around 3.4 on the 5-point scale). Here their image of technology is richer than that of secondary school pupils.

All together we see that the students do not have very strong attitudes towards science and technology and a reasonable image of science and technology, in which what experts see as essential is somehow present in the students' image. This does not change after having gone through the Waterworks activity.

CONCLUSIONS

The two research questions we have answered in this study, were: What is the image of and attitude towards science and technology of third year students in the Driestar primary teacher education program? How do this image and pre-service teachers' attitude change after these students have gone through one project on science and technology (focused on the water works project)? The image and attitude of students is measured in a qualitative study, using essays, and a quantitative study, using a questionnaire.

The images of technology of teacher students fell under four main categories: technology as an activity, technology as an artifact, technology as a discipline, and technology as a human/society related subject. These categories fit nicely with the four main areas in the philosophy of technology. The effect of the water works project was a further focus on various artifacts where students associate technology with simple tools and systems, such as a spoon and a windmill, not primarily with complex electronic devices. For science similar categories were found. However, students had a better view of technology than of science. Student's images of science were focused on activities, such as explaining, experimenting. The water works project had made students more aware what the various activities in science are, derived from the empirical cycle. Neither a relation with primary school pupils was made, nor with the outcomes of the research activities: knowledge, theories etc.

Based on these findings the entire primary teacher education program was reviewed, not only the Waterworks module, and adjusted to encourage students to develop a good image of science and technology. The limited view on technology should be complemented with images of more complex systems in a global context. The goal of science should be made clear by inviting students to explicate the knowledge that their research has yielded and the contribution to existing theories.

When comparing the results of the questionnaire before and after the Waterworks project we can conclude that the activity did not cause any important changes in the students' attitude towards science and technology. They were lukewarm in their interest for science and technology, they showed an appreciation for the importance of science and technology in society and they were fairly neutral in their thinking about the difficulty of science and technology.

Of course it was disappointing to find that the Waterworks activity did not cause a change in the way students think about science and technology. At the start of the project they felt science and technology were rather unknown for them (though important in life) and after the activity that was still the case. Apparently, the effect of one project is too modest for that. The same was found by Marja-Ilona Koski (see her contribution in this volume) for the effect of an activity that primary teachers did in the context of the VTB-Pro project on their mastery of certain scientific and technological concepts. This is not very surprising. It is well known that attitudes and images are fairly stable and do not change easily over a short time, except perhaps when they are faced with dramatic experiences. For the VTB-Pro project, this means that extension of the activities is needed in order to reach a more long-term effect. In the Waterworks project as it was carried out now, it was hoped that the activities themselves would lead students to changing their image of and attitude towards science and technology. Science and Technology education in primary schools and at many teacher training programs is very action-oriented. The educational concept OGO underlines that orientation. This makes the Waterworks module very appealing for students. What should be added is an active construction of knowledge regarding science and technology. Reflection on the images of students, on the outcome of their research - the results of science - and the connection to the classroom are necessary. The categories, which evidently correspond to the categories in the philosophy of technology, can help the teacher. The development of conceptual 'artifacts' is important, but students should also apply it to their own work, preferably in different contexts. What have I done? How does it relate to existing theories? What do I want to teach children about science and technology? What image do I have of science and technology and what other images are possible? With these additions, the teaching of science and technology may improve and will be much more interesting.

The part of this study that resulted in the most novel insights was the students' images of science and technology. Both the qualitative and the quantitative part resulted in a first impression of the dimensions in this image. It would be worthwhile to extend this part of the study in a more quantitative study with a larger sample in a broader population. Knowing the students' images of science and technology is particularly relevant as changing attitudes can be done best by changing their image of it. Less favorable attitudes are often the result of narrow and biased images of technology. We have seen that most dimensions that experts see as characteristics of science and technology are somehow present in the students' thinking in our study. But the material did not allow us to investigate the weight of the various characteristics in the students' image. We only saw which characteristics were present, not how important they were in the students' image. Knowing that is certainly important if we want to improve their image in order to have an effect on their attitudes.

NOTES

[xiv] We want to thank Dr. Nico Broer, Driestar Hogeschool, for the statistical analysis of the quantitative data.

Johan P. de Wilde
Driestar Hogeschool, University for Teacher Education, The Netherlands
J.P.deWilde@driestar-educatief.nl

Ellen Sjoer
The Hague University, KWT-West, and Delft University of Technology, The Netherlands

Marc J. de Vries
Delft University of Technology, The Netherlands

PART III

PROFESSIONAL DEVELOPMENT FOR PRIMARY SCIENCE AND TECHNOLOGY PEDAGOGY

MARC J. DE VRIES

12. INTRODUCTION TO PART III

Professional development for primary science and technology pedagogy

INTRODUCTION

Role of this Part in the book

In the theoretical framework that was developed for the VTB-Pro project (Walma van der Molen, de Lange and Kok 2009), three main pillars are mentioned as the desired outcomes of professionalization activities. These are: knowledge and skills with respect to science and technology, attitudes towards science and technology, and pedagogical skills for inquiry-based learning and learning by design. Part I and Part II of this book dealt with the first two of these pillars; Part III is related to, but does not fully cover the third pillar. Primary teachers not only need to have knowledge, skills, and attitudes themselves, but they must also have the capabilities to develop educational situations in which the children in their classes also acquire such knowledge, skills, and attitudes in an inquiry-based manner. Part of the VTB-Pro research programme has been dedicated to that. The focus of the studies that are reported in Part III are concerned with the question: what is the pedagogical content knowledge, or PCK as it is nowadays often referred to, that primary teachers need to have in order to be able to deliver good science and technology education? The chapters in this Part are not primarily about the inquiry-based approach, but focus on the content of what is learnt: do teachers have an understanding of what this content is (not in general, as this was the focus of Part I), but at the level at which children in their classes would understand it. Also the studies investigate to what extent the VTB-Pro professional development activities have resulted in primary teachers learning more about how to teach science and technology in their classroom practice.

Topics in professional development for primary science and technology pedagogy

In the International Handbook of Research and Development in Technology Education (Jones and De Vries (Eds.) 2009), the various topics related to science and technology pedagogy that have been researched in the past decades, are discussed. Although the title of the Handbook does not mention science education, it was covered in the book in the many chapters in which the relations between science education and technology education are described. As in the VTB-Pro project science and technology are not

Marc J. de Vries, Hanno van Keulen, Sylvia Peters, and Juliette Walma van der Molen (Eds.), Professional development for primary teachers in science and technology. The Dutch VTB-Pro project in an international perspective.

treated as separate entities, but in their mutual relations, the Handbook is a useful reference for introducing the many topics that can be identified in science and technology pedagogy. Before narrowing the scope to the specific pedagogy-related topics that were covered in the VTB-Pro research programme, I will offer a short survey here of the range of topics that has been studied in other research studies.

Apart from the topics that will be dealt with more extensively later on, topics related to the pedagogy of science and technology, according to the mentioned Handbook are:

- teaching and learning argumentation skills. These are quite important in science and technology, at least when education aims for helping pupils acquire of proper understanding of science and technology. For science, cause-effect reasoning is most important, whereas in technology means-ends reasoning is the primary type of reasoning (De Vries 2005);
- modelling: models are perhaps the most important outcome of science and an important tool in technology. As far as science is concerned, all outcomes of science are descriptions of abstract versions of reality, in which aspects have been left out (abstraction) and irregularities have been 'smoothened' (idealisation) in order to get a simpler description. But often science education does not explicitly deal with the nature of models. In technology education models are used (simple prototypes for instance), but often within a very limited range. In both science and technology education models and their nature certainly do not yet have the place they deserve;
- relations between science and technology and with mathematics: practice has shown that it is by no means easy to find good opportunities to make pupils do projects in which science, math and technology are interwoven in a natural way and still are at their level of understanding. Most 'real' examples are at a higher level. Designing the sort of devices that pupils can design often do not need much science or math. Modelling might be a useful way of connecting math to science and technology education;
- assessment: how to value the rich combination of knowledge and skills that pupils are expected to have in order to complete a research or a design project successfully? The obvious answer is: by using portfolios. But assessing portfolios has shown to be a tricky matter. They contain such a wealth of information that it is hard to select what are truly the indicators of performance;
- dealing with ethical issue in science and technology: everyone agrees that these are important issues, but how to involve pupils in them while avoiding indoctrination? Somehow pupils will have to learn to transform their own general values into opinions about specific ethical questions in science and technology. Here, again, reasoning skills are important.

The issues mentioned here are still under research and debate. Although science education has a much longer tradition in developing pedagogy than technology education has, they both still struggle with such fairly fundamental issues.

Perhaps the most fundamental one is the challenge of doing science and technology education in such a way that pupils really experience what it is like to be a scientist or an engineer. Here, perhaps technology has a better position than science education, in spite of the latter's longer tradition. Still today, science education to a large extent is dedicated to showing pupils the outcomes of science (X's law, Y's formula, etc). The excitement of doing research and reaching an understanding by yourself often is not allowed to pupils. Even when they perform experiments, they know the outcome is already in the book, and the clever pupils will adapt their measurements according to that. Surprises are excluded a priori. No wonder science education in many countries suffers from unpopularity among pupils. There is still a long way to go here, because the more traditional views on science education are still popular among primary teachers (see, for instance, the stdy by Porlan and Del Pozo in Spain). Fortunately the number of projects that focus on inquiry-based learning increases rapidly, and that certainly gives hope for the future.

Focus on concept learning and the role of language

Given the limited resources, the VTB-Pro research programme could not cover all the relevant topics in the domain of science and technology pedagogy. Choices had to be made in order to create a coherent research programme, and those have led to a focus on two issues: concept learning and the role of language. The two are obviously related, because language plays a vital role in learning concepts because concepts to a large extent constitute the language of science. Acquiring this language of science is an important part of being enculturated in science; that is: in becoming a part of the culture of science (that is, as far as a educational setting allows for that; of course the full culture of science as scientists experience it is much more sophisticated). But also the other way round, language is needed in order to develop and learn new concept. Science is and has always been very much dependant on communication, and language is a prerogative for that (symbols are another, and in fact they are a sort of language). There was also a practical reason for having this topic in the VTB-Pro research programme. In primary education policy, there is a tendency to move back to the basics of 'reading, writing and arithmetic'. This is certainly not unique for the Netherlands. A consequence of this tendency is that in order to make science and technology education worthwhile for policy makers to invest in, it is useful to emphasise that science and technology education can contribute to learning these basics.

The limitation to concept learning and the role of language is less severe than it may seem, because in the research programme concept learning is related to several other pedagogical issues. In contemporary thinking about concept learning problem solving and design are seen as appropriate strategies for learning concepts and making them more versatile in the learners' thinking. Problem solving and design then lead to the topic of creativity in a natural way.

CONCEPT LEARNING AND CREATIVIY

Our thinking about learning concepts, about problem solving and design skills and creativity has changed substantially over the past decades. The main lessons we have learnt is that all these are more context-bound than we believed earlier. This has consequences for the way they have to be taught and learnt.

There was a time when there was great enthusiasm about teaching and learning 'general problem solving skills'. The idea was that one could teach and learn how to solve problems by using a set of rule that would apply to any problem. Particularly the 1972 book on Human Problem Solving by Allan Newell and Herbert Simon became a true 'classic' in this field. In a similar way, educators believed in teaching and learning concepts that would apply to a variety of domains. One concept that was particularly popular in science and technology was the concept of systems. This concept had originated in biology (Ludwig von Bertalanffy) and had found widespread acceptance in other domains, among which was the domain of engineering. In that time a vast collection of flowcharts emerged for problem solving and design processes, based on the idea that any problem one wanted to solve or any design to be made in some way or other had the characteristics of a system. A third area in which this generalist thinking emerged was that of creativity. Here, too, a whole range of claimed domain-independent 'creativity stimulating' methods In fact, some of the guidebooks in which such general flowcharts for problem solving, design and creativity methods were published are still reprinted today.

For design, the changed in our thinking have been described nicely by Nigel Cross in his book Developments in Design Methodology (Cross 1984). He showed that the idea that design processes could be prescribed in a domain-independent manner became problematic when practice showed that design problems do differ substantially between domains. Designing a new aircraft is not the same as designing a new house, and for that reason a good strategy for designing that aircraft may not be successful for the case of the house. Simultaneously to that emerging awareness, the idea of the General Problem Solving skills also started waning. (see for instance Lave, Smith and Butler 1988). Also for creativity the domain-independent claims of its nature were more and more criticised (see, for instance, Baer 1998). Finally, the belief in ideal of learning abstract and domain-independent concepts, too, became doubted (Brown, Collins and Duguid 1989).

Gradually the idea of teaching and learning at an abstract level and after that 'applying' what had been learnt to various domains was abandoned and the domain-specific nature or 'situatedness' of these issues was accepted, but now immediately to its full consequences. At first one tried the following approach: teach and learn in a specific application, and then have the learner 'transfer' what has been learnt to another domain. In other words, the initial learning of the general ideas must happen in a specific situation, but what has been learnt still contains the more abstract and general notions and thus can be easily transferred to other domains. There were hopeful examples that this would work even for pre-school children (Brown and Kane 1988). But this hope, too, was abandoned in the end. It became clear that an even more dramatic change in teaching and learning had to be looked for. Perhaps the most outspoken representative of this new approach is the 'concept-context' approach that was developed in the Netherlands, originally for biology education and later extended to other science-related domains such as chemistry and physics (Pilot and Bulte 2006). In this approach the learner goes through a number of different contexts to meet the same concept each time in its domain-specific appearance. The expectation is that (s)he will then gradually develop an awareness of the common features of the concept and thus move to a more generic and abstract level of understanding of the concept.

There is an even more extreme position in which the whole idea of teaching and learning abstract and general concepts is abandoned. I do not believe that is a necessary consequence of the domain-specific nature of concepts. After all, there are studies showing that children do have the capability of learning abstract concepts, even at the primary level (Novak 2005). What seems to be more fruitful is Jonassen's approach. He acknowledges that the idea of concepts having essential properties that can simply be listed and learnt is not appropriate, because it does not help people recognize the varying ways of functioning of these concepts. Concepts ought to help us understand, explain and predict, and learning the essential properties of a concept does not (necessarily) support that. Jonassen therefore suggests that concepts are learnt when learners change their understanding of how concepts are organized in a conceptual framework that relates to personal theories of their world (Jonassen 2006). Such changes occur when they are stimulated to perform actions in that personal world. And that is precisely what the concept-context approach proposes: let learners perform activities in practices that are authentic for them. It also seems quite close to 'activity theory' that claims that learning involves a hierarchy of activities mediated by tools and influenced by objects (Nardi 1996). Other related terms are: situated cognition, cognitive apprenticeship, and distributed cognition. All of these are also related to the earlier term of constructivism that takes into account the finding that learners are not passive receivers of knowledge but construct and reconstruct knowledge based on experiences with their lifeworld.

Underneath these educational issues is the philosophical issue of the relation between reality and science, and this can help us understand why it appeared to be problematic to teach and learn concepts at an abstract level right away. Concepts are by their very nature abstractions from reality. Abstraction means: taking out certain aspects of reality for separate study and leaving out others (about which then, of course, no knowledge claims can be made in that abstraction; this is why natural sciences can never answer questions about the meaning of life, although this is often claimed nowadays by scientists overestimating the scope of their occupation). Concepts isolate characteristics that particulars in reality have in common. The systems concept, for instance, isolates the cooperation between parts in washing machines, cars, mobile phones and what have you, from all the other particularities in those devices. As a consequence, a systems analysis can tell some things about the washing machines, but not everything. Our intuitions by their very nature are based on our experiences with these particulars. In education they are often called 'preconceptions' and in the past were often only seen as 'what is yet to be corrected'. We must not forget, though, that human beings are very good in developing intuitions about reality that work in the sense that for that particular situation they are very helpful. They only become problematic when we follow our need to develop more abstract (scientific) knowledge about reality. But until then they often worked perfectly well to predict what is best to do in a concrete situation. In our enthusiasm for teaching and learning concepts we should never forget the pragmatic value of intuitive daily life notions. A practical example of this could be to take into account indigenous science and technology. That would not only be politically and culturally correct in developing countries, but it can also enrich first world country pupils' view on science and technology (Carter 2007). They will learn to appreciate how much practical knowledge about materials and constructions people in the past had. It will hopefully change their whole idea of 'primitive' people, that were perhaps not that 'primitive' after all.

With respect to our possible over-appreciation of abstract scientific knowledge, the same holds for problem solving and design. Through many experiences we learn to develop ways of manoeuvring through a variety of daily life situations. Like with the concepts, the intuitive strategies we develop have a value in themselves. It is only when we want to develop strategies that are usable in a wider range of situations that they may become problematic. Here too: in our enthusiasm about general problem solving and deign skills, we should not forget the value of more specific problem solving and design strategies. The history of teaching and learning as sketched above has shown us that ignoring these does not lead to a very effective educational strategy.

THE ROLE OF LANGUAGE

In the previous section I made a plea for showing the processes of science and technology (and the excitement that comes with those). To do that will contribute substantially to the pupils' language development. Pupils will have to interact and communicate in science and technology education when practiced in that way. Language in science and technology education is a theme of its own right. So far only modest attention has been paid to it, when compared to, e.g., concept learning and problem solving and design processes. Yet it is an important issue. I will summarise briefly what literature says about it so far[xv].

As stated before, language is part and parcel of doing science, and also for teaching and learning science. The language that is used is by no means 'neutral' in the sense that language is not just a vehicle for conveying information, but it contains certain values by itself. That can either help or hamper, depending on whether or not these values match with the values that learners hold. Several studies in this domain have shown that the way science is conducted in the western industrialized world implicitly carries a way of looking at the world around us that is very much dominated by ratio. Of course this is not surprising, as rationality plays an important role in science, but there is more to science as the philosophy of science and the philosophy of technology have shown (Lee 2005; De Vries 2005). Non-rational elements such as power, money, interests, reputation, emotions, all play a role in science too. Language can either hide or reflect that. The latter of course gives a more realistic image of science and technology. This image might also be more gender neutral, as particularly feminine values (that may be held by men as well as by women) then also feature in the picture we draw of science and technology. This can go all the way down to the simplistic suggestion to use 'she' as well as 'he' in textbooks when referring to a 'scientist' or 'engineer'. But it can also be practiced by using terms that reflect emotions, such as 'an impressive phenomenon' or' a exciting result', or 'a pleasingly simple formula'. All this has to do with the way language represents the real nature of science and technology.

A second area of research in this domain is the way language plays a role in teaching and learning science and technology. When science education and technology education are purely a matter of a teacher transferring knowledge to children, they themselves will only learn both the content and the language in a passive way. As we know, this is not the most effective way of learning. Making them use the language themselves while doing active project work is a much better way of learning both the content and the language. In this way, learning science and technology and language acquisition go hand in hand and two goals are served at the same time. Literature surveys (Lee 2005, Carter 2007) provide evidence for that. It is known that in some cultures questioning and inquiry are not encouraged (Lee 2005), and for children from those cultures the use of language will have to be chosen carefully in order to avoid the impression that science is an impolite matter.

CHAPTERS IN THIS PART

The chapter by Koski and De Vries deals with primary teachers' understanding of some basic concepts in physics as they were taught and learnt in the professional development activities in the VTB-Pro project. The results were rather disappointing. It appeared that teachers themselves lack an understanding of these concepts, even at the level of activities for children. This is, of course, rather alarming, even more because it also appeared that the limited amount of time spent in the professional development activities was not sufficient to bring them to a level of understanding that would enable them to implement the learning of these concepts with children. In some cases the activity even enhanced conceptual confusions. Apparently, more than a short course consisting of less than ten days of immersion in this content is needed to make a real change. The chapter by Slangen, Van Keulen and Gravemeijer deals with a similar situation. It studies the effect of a relatively short course on primary teachers' understanding of certain concepts. In this case these concepts were not physics concepts, but concepts related to robotics. In addition, these authors also studied the extent to which the activities helped these teachers to acquire skills in dealing with these concepts in classroom situations. Here the results were more positive. The teachers appeared not only to have acquired a basic understanding of the concepts on their own level, but also the some ability in 'translating' these concepts to classroom level. Comparing these two chapters immediately raises the question of why it is that the second chapter shows such better outcomes of the professional development efforts than in the first chapter. This is probably related to the level of expertise of the trainer. Slangen, who conducted the training activity himself, has an extensive experience in this type of activities and is well acquainted with the technological content. The trainers in the Koski/De Vries chapter had sufficient understanding of the content to do the training, but in informal conversations with the researcher overtly admitted that physics for them, too, was not easy. This also means that they were not as a aware as Slangen was of the conceptual difficulties in learning this content. His suggests that the expertise of the trainer is an important factor in the learning that takes place (or fails to take place) in the professional development activities.

These two chapters both deal with concept learning, the first area mentioned in the earlier sections of this Introduction. The next two chapters deal with the second area: the role of language in learning about science and technology. The chapter by Damhuis and De Blauw professional development activities can successfully train primary teachers in acquiring capabilities for incorporating high quality interactions in science and technology classroom practice. Their findings confirm that short-term activities are not sufficient for making real changes. The authors suggest that the teachers should be supported during their teaching practice and not in activities that take them out of their practice. Henrichs, Leseman, Broekhof and Cohen de Lara have analysed the natural discourses of teachers and kindergartners during science-related lessons. They showed that teachers' being comfortable with the science content of an activity is an important factor in the quality of the conversations that occur during the lessons. The authors claim that relatively light interventions can raise the quality of conversations in the teachers' lessons considerably. This seems to contradict what Koski/De Vries and Damhuis/De Blauw suggest, as according to them the effect of short interventions is limited. We must realize, though, that two different effects are at stake here. Henrichs et al. refer to reasoning mainly (in particular predicting and explaining). These are general capabilities that teachers may already have before becoming involved in science activities. Koski/De Vries and Damhuis/De Blauw refer to more science-specific content that was entirely new to teachers. For learning that, more time may be needed.

NOTES

[xv] I will make use of a survey presented by Maaike Hajer (Hogeschool Utrecht, the Netherlands), presented at the Onderwijs Research Dagen 2010 (not published in English).

Marc J. de Vries
Delft University of Technology
Delft, the Netherlands

MARJA-ILONA KOSKI AND MARC J. DE VRIES

13. CONCEPT LEARNING IN PROFESSIONAL DEVELOPMENT

INTRODUCTION

In this chapter an exploratory, qualitative study into primary school teachers' preconceptions in science and technology is presented. The purpose of this study was to get a first impression of the effect professional development activities in the VTB-Pro project had on primary teachers' knowledge of concepts. Learning concepts is one of the aims of the professional development programs, next to attitude development and the development of pedagogical skills in teaching science and technology. To be able to assess the effects of these programs, it is necessary to know the ideas with which the teachers enter the program, also in terms of their scientific and technological preconceptions. Investigating such preconceptions is by no means a new research field and for that reason it was possible to use existing research methods for that. New in this study is that it is done for primary school teachers. Existing studies allow us to compare with other target groups such as primary school pupils and secondary school teachers. First, we will briefly sketch the activities in which the teachers took part in order to show the background against which our study should be read. Then we will present what has been found in previous studies and from that we derive our own research questions and methodology. Next, we will describe the collection and analysis of the data and finally draw conclusions about the primary school teachers' preconceptions and first impressions as to what effect the professional development activities had and can have.

CONTEXT OF THE STUDIES INTO CONCEPT LEARNING

The training, referred in this article, for the primary school teachers was organized by Kenniscentrum Wetenschap & Techniek West (KWT-West; Expertise Centre for Science and Technology, region West). The six sessions of this training, each one afternoon long, are presented by a trainer, who was trained also by KWT-West beforehand. The number of participants, their age and in-service experience in the groups differed and each of these groups met roughly once a month. The training sessions were aimed at providing more knowledge about science and technology as well as solutions and examples of how to deal with the topics in classrooms. The six topics covered in the sessions were:

1. Flying
2. Survival
3. You, your class and science
4. Developing living quarters for sharks and corals using the expert method

Marc J. de Vries, Hanno van Keulen, Sylvia Peters, and Juliette Walma van der Molen (Eds.), Professional development for primary teachers in science and technology. The Dutch VTB-Pro project in an international perspective.

5. Your own theme (content of this session teachers can choose) and
6. Learning, living and working with water.

Each of these sessions provided background theory of science and technology needed, in addition to hands-on experiments. Teachers were also encouraged to do homework: trying out the learned methods in their own teaching. After completing the sessions, teachers got a diploma stating what they have done as a conclusion of the training.

Activities related to concept learning

Conceptual change does not happen easily and trainers need to be provided information on where to pay ample attention and how to improve knowledge delivery. The method, used at the moment, does not give enough time for teachers to understand the concepts being taught, as we will show. Based on the results of this study, we claim that making the relation between theory and practice is one of the problems. Teachers have knowledge that covers only the surface level of a certain theory and this hampers them in fully exploring the possibilities in practice. Because what was learned in training is not consolidated, it is not sure that the concepts were really learned or applied in practice. Therefore research and attention on how training teaches the concepts is essential. As a general suggestion, we argue that instead of putting the emphasis on providing the material to work with the concepts in classroom, the focus should be more on describing the concepts so that a good understanding emerges with the teachers.

THE STUDY INTO CONCEPT LEARNING

If we want to see a different approach towards science in the classroom we need to include activities that change teachers' ideas enabling them to deepen their every day science practices. However, a minority of primary school teachers has fairly good knowledge of science content and besides this they do not have the confidence in teaching science (Appleton 2003). Unfortunately, the choice of not teaching this topic is no longer available and teachers can not avoid teaching science even if their confidence about it is low (Harlen and Holroyd 1997). Avoiding science conflicts with a desire expressed by scientists from university that more experience in performing science research, as well as developing their own critical thinking skills should be practiced to support their profession (Taylor et. al 2008).

Previous research includes investigation regarding student confusion about the concepts being taught and failure of recognizing contradicting answers (Loverude et al. 2003). Kruger et al. (1990) found that both teachers and primary school students have the same wrong interpretations about science concepts when testing primary school teachers' knowledge about force, gravity and friction. As well as the teachers and the students, teacher trainees hold the same wrong ideas about air and air pressure (Rollnick and Rutherford 1990). She (2002) discovered when testing the conceptual change of air pressure and buoyancy that students are more reluctant towards the conceptual change if the underlying concepts are not comprehensible.

Research questions

Compared to existing literature this research examines in-service teachers and what are their underlying concepts or mental sets. More precisely, how these practices and models need to be taken into account for an effective change to happen so that science practices do not only exist in the classroom but have a broader usage as well. Therefore if primary school students and teachers share the same misconceptions before developing classroom activities, the possible factors stopping the transform process in a teacher need to be examined. The aim of this article is to make the difficulties in theories visible and to see what types of problems in teacher's knowledge level this brings out.

DESCRIPTION OF THE CHOSEN RESEARCH METHOD

With the decision to do a research about concept learning, a reasonable start for the project was to test the knowledge of the concepts before the training takes place. Lewis (1999) states that understanding the conceptions and misconceptions is an important prerequisite for better teaching and improved learning. Based on this idea about students' knowledge before teaching, we apply the same into the teacher training as well. Better perceptions of the needs of the training help to adjust the training and develop it further.

The data was collected by means of a questionnaire. First model for a suitable approach was examples in a book of Unesco Source Book for Science. In this book possible practical approaches to physics concepts are introduced for teachers to use them in their classroom. An idea what to really look for with the questionnaire came from the article of "Force Concept Inventory" by Hestenes et al. (1992). They have created questions for students to test what are the common sense beliefs and misconceptions of the physical world. Based on this article it became comprehensible to what extend it is possible to find results with this method and what is worth looking for.

The design of each individual question is either based on our ideas or it relies on the article "Diagnosing and Dealing with Student Misconceptions: Floating and Sinking" written by Yin et al. (2008). Even though the article is designed to test student misconceptions, it provides not only ideas but usable tests to see the misconceptions of the teachers in the training. Questions presented in this article were altered keeping in mind the suggestions that theory of Andragogy (Knowles 1980) states how adults learn and how teaching adults should be designed.

Data collection

The data was collected during the spring 2010, between the months of March and June. Concepts in two themes were tested; theme water and air. The collection of the data was regulated by the training schedule and content. Due to the problems in practical arrangement, concepts in the theme air were tested only once. Concepts in the theme water were tested three times, twice with teachers before they had their training and once again with a group of teachers after the training. Analyzed data consists of 24 answers. 13 teachers gave an answer to the questions about water, and as stated above two of the teachers have answered twice. 11 answers were given to the questions about air.

The final versions of the questionnaire include seven questions about water and eight questions about air. In both questionnaires first question is more of an introductory question to the topic. In a case of water law of Archimedes was tested as well as concept of density. Also teachers' ideas about what determines whether an object sinks or floats were asked in different ways. In the air questionnaire teachers answered to questions concerned with density, weight and force of air. Questionnaire also included more theoretical questions about Bernoulli's principle.

DATA ANALYSIS

Data analysis is divided in two parts. First are introduced three issues that need to be further examined to overcome problems in teacher training as well as an experimental question about teachers' abilities of system thinking. Second part introduces more detailed analysis about the answers of teachers that gave replies to both before and after training questionnaire. This second part serves as a support and confirmation to the findings introduced in the first part of the analysis.

When teachers were asked to describe what water is, majority answered with terms such as: *"Water is liquid"*, *"It is necessary for life"*, *"We use it daily"* and *"We use it in everywhere"*. Air was perceived in a same way: *"Not visible"* or *"Not touchable"*, *"You can feel it in the wind"* and *"You can not live without it"*. These types of answers indicate practical, representational, down to earth qualities to the science and technology concepts. Therefore a supporting approach to the themes would be to give examples to the teachers that are sensible and closely related to everyday life. It looks like teachers have the same tendency as students to link their concepts of matter to tangible properties (Davis et al. 2002).

Universality of Theories

The questionnaire continued inquiring teachers to compare the weight of two glass bottles, one filled with air, the other one vacuum. In the answers more than half of the teachers stated that adding or removing air from the bottle does not change the weight. However two of the teachers replied correctly that bottle with air weights more but just as well, two teachers replied that the vacuum bottle weights more. Considering that more than half gave a reply that air has no weight, it looks like air is considered as something without a quality of a mass. Questions about ideas on air and air pressure have been asked from African teacher trainees (Rollnick and Rutherford 1990), however the questions were about air's existence, and does it occupy space or exert pressure.

To examine this further, a question about does air have a force was included to the research. It was asked if it is possible to pump air into a swimming ring that has five books on top of it. Eight teachers thought it is possible (one answer was just positive without explanation). Replies included answers such as:

"Yes, the air goes in to the ring and everything is lifted",

"Yes, the band becomes firm",

"Yes, the band expands and the books will be lifted",

"The books will rise if you put enough air in. The mass of the air is more than the books" and

"This can be. The air pressure in the ring is on the given moment greater than the weight (pressure of the book towards the ring) *of the books and that's why the books can be lifted* (whether they stay on top of the ring is another question)*"*.

Contrary to the previous question, here teachers could imagine the conditions or they even have experienced similar situation. This appeared to trigger them to reason in a way that air has a mass that changes conditions. Three teachers explained the situation with changed condition of the swimming ring (band becomes firm, expands). Four teachers explained the phenomenon with using terms such as the increase of mass or the weight of the air. The mass of the air or the weight was bigger than the mass of the books.

However, three teachers doubted this because of the alignment problem of the books.

"It depends on how and where the books lie",

"It is possible but if the books are not well arranged, they will fall" and

"One book can well be lifted by one ring. But a stack of 5 books is heavy. If you pump the ring that has 5 books on top of it, succeeds it only partially. The pile will slide off and fall over. After that you can continue pumping the ring".

Findings contradict with Rollnick's (1990) research about explanations of air and its effects on other objects, where in the answers it has been stated that air is invisible and has power. Here it is not clear whether it is understood that air has a mass, but even if it is a known fact, the knowledge does not have a universal usage. Theory might be well learned but to apply it and recognize the situation where it could be used seems to be difficult. When asked in a way that teachers can picture the situation, the results are better but this does not necessarily ensure that the concept is understood. A situation where teachers have to purely rely on theory and imagine the situation demonstrates the difficulty of applying knowledge whereas a situation that is more concrete results to better explanations and shows understanding.

It appears that the knowledge level may not be solid or stable enough, when talking about what is triggering scientific thinking or more precisely what is stopping it. The assumption that is made among children that learned theories lack universality (Yin et al. 2008) can be detected in teachers' answers as well. Therefore to promote universal use of knowledge presenting several examples, even ones that do not work, might help to clear the ideas about the concepts. Either approaching the concept through many different examples or focusing on explaining one example with enough time could be valuable notions for the trainer in helping the teachers grasp the concept. However, despite the method, trainer should pay more attention on checking whether a concept is learned and even more importantly how it is learned.

Relationship of Theory and Practice

The training introduces practices, which should encourage more scientist-like behavior in the teachers. One trainer approached her in-service teachers by saying *"What we do in here is as much as science as anything"*. Despite the effort, something is blocking the teachers to behave more like scientists. Relationship that the scientists have with theory and practice, the open and personal relationship where both parts support one another, is more difficult to achieve among the teachers.

As a concept, lighter than air is generally well understood and it was the most popular answer to a question what makes the other balloon rise and the other one stay put (Figure 1.).

Figure 1.Two differently behaving balloons.

"The red balloon is filled with heavier gas than the yellow. The gas in the red balloon is also heavier than air around it",

"In the yellow balloon there is gas that is lighter than air around it. In the red balloon the gas is not lighter because the red balloon does not sink down or rise upwards",

"If there is helium in, is this gas lighter than air and is pushed upwards" and

"In the yellow there is helium and in the red one there is air that is blown. Helium is lighter than air".

However the answers also included replies like:

"There is probably gas in it",
"Due to the gas rises the balloon" or
"The other balloon is filled with gas".

When analyzing these answers it seems that also other characteristics of air have a tendency to be understood incorrectly. In this case most of the answers were correct, but for the research and development purposes the incorrect ones are more interesting at this moment. The incorrect replies give an impression that air is not seen as a gas and it necessarily does not have anything to do with whether a balloon rises or stays put. Contrary to this, it has been reported, when asked from teacher trainees about the nature of air that air consists of a variety of gases (Rollnick and Rutherford 1990). On a foundation of this research it seems that air is perceived as something sort of a "zero state" and not as mix of gases. Balloons that rise are those once from fairs and they are filled with gas and that makes them rise. Hence there is knowledge from the training which is not applied to this situation because knowledge from a fair situation is more suitable for on this occasion.

When asked about Bernoulli's principle, almost half of the teachers have the idea that because there is a longer distance for the air to travel on the top of the wing, the air floats slower and due to this high pressure is created. The reason why lift is created is because air moves faster on top of the wing creating a lower pressure to the top part than below the wing. Lift is about high pressure but it should be below the wing. It appears that there is an idea where the top part of the wing operates separately from the below part of the wing. So a longer distance means slower travel even thought the plane pushes through the air with both parts of the wings at the same time.

Above two answers about the characteristics of air show the gap between theory and practice. The theory stays as a theory, learned for a reason, and the practice is another world. Certain investigation process is missing, which would make the creation of the relationship between theories and practice more visible and present. Since both of these topics are included to the training, the connection between difficult theories should be made even more tangible than before. It should not be like it is at the moment that the theories are used to present scientific concepts and a different knowledge of the concept is used when it is applied to practice. Most likely it is not enough to explain a concept once but it needs to be repeated. Trainer needs to come back to it and to make a connection to another practice as well. From this the process of not being able to explain a concept with a misconcept starts to happen.

Learning how to freeze thinking process

Floating of a big, iron boat was explained correctly by terms of up- and downwards forces:

"It has to do with the up- and downwards forces",

"Due to the heaviness of the boat, it has to do with the upwards pressure of the water" and

"Because the upwards pressure of water is bigger then the downwards force of air".

Here the answers leave space for doubt but on a whole it seems that the underlying theory is understood. But answers also included replies in which just a term is mentioned or a bigger confusion is observable:

"Buoyancy of the water",

"Surface tension, buoyancy",

"The form of the boat determines that it floats. The hollow shape determines" and

"Iron doesn't absorb water".

Mentioning just a term is a sign of knowledge gap but declaring surface tension as a reason for a boat to float, describes confusion. However signs of similar confusion as among students, when they have stated that the amount of surface contact with water determines sinking or floating (She 2002), can be detected in the answers. Here the form and shape were also reasons for the boat to stay on the surface. In general, these types of answers show that teachers share the same problem with their students of replying just with terms without stopping to think. The confusion can be seen as follow-up of this type of action and as a lack of fully understanding the concept and its use in practice. Replies give a reason to doubt the fact that re-thinking about what has been experienced in the experiments does not happen after the training. Doing something should be stopped at one point to really discuss why something like that can happen and what the reasons are for it.

Another misconception, also found by Yin et al. (2008) among students, was when teachers were asked to compare same block placed in the water differently (Figure 2). First it was put into water vertically and then it would sink. Second time it was put in horizontally and it was asked whether it sinks or floats.

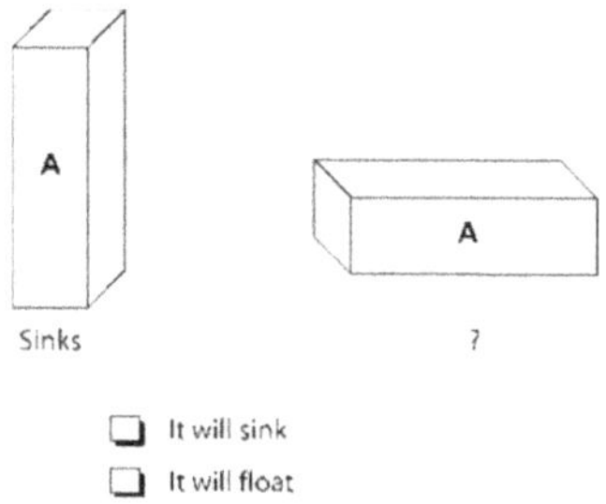

Figure 2. Block placed into water in two different ways (Yin et al. 2008).

Yin et al. (2008) reports that children state that horizontal objects float. In this case half of the teachers stated correctly that on the second time block will sink as well. However the other half, six teachers, answered that it will float. Thus the same misconception can be observed among in-service teachers as what students have.

Answers such as above imply gaps in the knowledge in terms of explaining an exact theory. For the future relevance, it is important for the introduced theories to not to stay on the abstract level or remain hanging in the air. The provided training material attempts to improve this by instructing teachers to answer to a question "What happens?" Based on the results it looks like this is not enough. Teachers also need to learn to think in a way that they can answer to a question "Why this happens?" Whilst answering to a why-question, teacher needs to apply the knowledge to be able to explain. Learning to stop and re-evaluate the situation helps to avoid simplest mistakes and encourages questioning the thinking path. Additionally this acquires extra attention from the trainer not to let the teachers just observe the experiments or run through them too quickly, but actively involve them during the experiment and encourage them to think and make assumptions.

Thinking in Systems

As an experimental part of the research thirteen teachers were given a question about system and system thinking. Setting for the question was much influenced by the ideas of a system from the field of computer science but for general definition, an Oxford dictionary definition of the word system, was used:

> A set of things working together as parts of a mechanism or an interconnecting network; a complex whole and a set of principles or procedures according to which something is done; an organized scheme or method.

Idea of a system was presented with a picture of a coffee machine accompanied with a description from its user manual of how to make coffee. Question was divided into three sub questions and the first one asked about which elements in the coffee machine have an affect on each other. Most of the answers were like the following three:

"On/off–button -> (red) lamp on",
"Filter -> coffee quality" or
"Aroma control button -> stronger or milder coffee".

Even though it was given in the question that the elements have to have an affect on each other, most of the teachers replied in a straight forward way where one component influences another and the state of the later one changes. One teacher replied after listing components *"... coffee jug and drop lock in the filter holder"*. Even though this answer does not explain the relation between these two components, the relation is clear to everyone who has used coffee machine. When coffee jug is placed under the filter it releases the drop lock allowing the coffee to drip into the jug. If the jug is not under the filter, the lock is active and if the case is that the machine is on, only the filter is being filled. The dependency and interaction is made visible between the two elements and certain thinking "backwards" which is common is system thinking, appears in the answer unlike in the three first ones.

The second question asked about what types of influences can come from outside of the coffee machine and how do these affect the system. Mostly word influence was understood as a synonym to malfunction. Only one teacher replied that an influence could be human action: *"put on* (machine) *and put coffee* (in)*"*. Here the influence is stated clearly but the affect on the system is missing.

"Power supply -> no power -> no coffee" and
"Disturbance in electricity -> machine stops".

Above answers can be interpreted in a way that parts are working together and when (negative) influence affects the system, its functioning is not normal. These answers give the source of the influence, the influence and the effects. In the following answers this type of reasoning is not visible. They describe the effect but the source of the influence or the type of influence is missing.

"Jug out of its place -> overflow" or

"No coffee powder -> only warm water".

In the end most of the answers described states of the machine or the end product.

"Filter isn't placed correctly",

"No water in the water tank" or

"Coffee type -> taste",

"Water -> no coffee".

These answers are end results produced by the system when something has gone wrong. They do not state the influence to the system or the effect on the system itself, only how the end product is like.

The answers on the third part of this question describe the possible situations or changes occurring within the system in the following way:

"Water to steam" or

"Cold plate to warm plate" and

"Solid (coffee powder) to liquid (ready coffee)".

First two answers are not explaining the situation changes within the system but the last one is a result of processes, changes, happening within the coffee machine. Eight teachers explain the following situation in a more or less detailed way:

"From water to coffee" or

"Water becomes warm, sets off, and mixes itself with the coffee powder -> water gets color and taste" or even more detailed

"Water becomes warm -> evaporates and rises up through the pipe, cools there and condensates, runs through the filter and mixes there with the coffee powder and falls to jug. Jug is warmed by the plate under it".

In system thinking, elements are interdependent and cycles appear within procedures. Answers present a linear way of thinking about processes and systems. Overall structures, patterns and cycles are difficult to find in the answers. Even though there are signs of system thinking the idea of network and complex whole is missing. Mostly independent parts are described or behavior of water or coffee powder is followed.

Concept Understanding Before and After the Training

With two teachers it was possible to test the preconceptions and then the knowledge of the same concepts after the training. Here are presented three questions from both tests.

The training material explains the law of Archimedes on a relatively detailed level. In the questionnaire following scenario is given to the teachers: a child plays in a bath tub with a toy boat that is filled with toy blocks. Teachers are asked to tell what happens to the boat and to the water level when the child starts putting the blocks into the water.

Teacher A: *"Boat rises in the water"* and *"Water level will be lower"*

Teacher B: *"The boat sinks (depending on the buoyancy and weight) further to the bottom or sinks"* and *"The water will rise as much as the content of the blocks"* then the teacher also continues commenting and the water level of the bath tub *"Depending on the content and how far the bath is filled with water, it (water) will flood or not"*

Teacher A starts correctly explaining the situation. The reasoning becomes incorrect when it is stated that the water level goes down, although this could be related to the rising of the boat, meaning water level is compared to the position of the boat. In the case of the other teacher, a clear misconception of the Archimedean law is seen in the part where teacher claims that the water level will rise as much as the content of the blocks is.

In the after test teachers were asked to simply explain the basic idea of Archimedean law. The obscurity of the theory becomes clear with the following answers. Teacher A replied with one word that it is about the gravity while the other teacher replied by writing: *"Apple falls from the tree and this action provides the needed reaction"*. Findings of these to questions support earlier stated difficulty of applying theory into practice.

Concept of density was asked by giving a picture of two glasses, filled half way with liquid (Figure 3). In the other one an ice cube is floating and in the other one it is at the bottom. Both ice cubes are identical in means of weight and size. Teachers were asked to explain why.

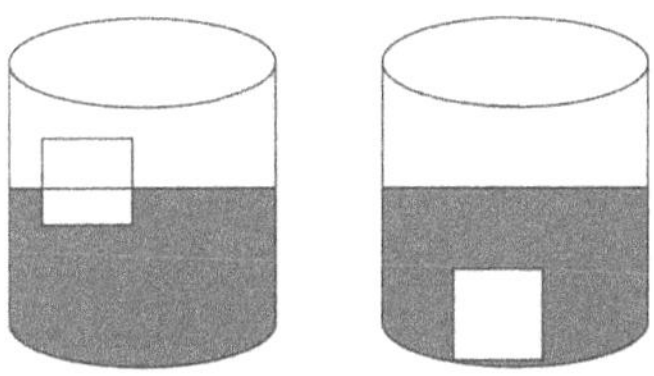

Figure 3. Two ice cubes in different liquid.

Teacher A: *"In the other cube there is more air?? That's why it floats"*

Teacher B: *"The liquid is a different type (e.g. salt or soap)"*

On the second time the question was described by words without the picture. However, this time it was given that the other glass contains water and the other one alcohol. Teachers replied the following:

Teacher A: *"The density of alcohol is bigger than water's"*

Teacher B: *"It is a matter of another density, where the molecules in the ice are closer to each other than in the water around them. In alcohol this is not the case"*

The answer of teacher A to the first question shows the misconception that objects with air in them float. This similar result was reported among students by Yin et al. (2008). Second time the term density appears but what it really means is wrongly understood. Correct order is that density of water is greater than the density of an ice cube or alcohol. Contrary to this in Loverude et al. (2003) students mostly stated it correct that behavior of a floating object is due to lower density than what the density of water is.

The first answer of teacher B is in a sense correct (liquids are indeed different and that's why the scenario is like presented), but when compared to the second one, it shows the same problem than what the teacher A has. Density as a term is used but it is understood the other way around.

Density as a term is known after the training but how it works is vaguely understood. It seems that greater density of a liquid makes an object sink. Teacher A states this clearly but also teacher B ends up to the same conclusion. The first sentence of teacher B is incorrect but the second one, if you assume that the density of ice cube is all the time the same, makes sense. Indeed the molecules in the ice cube are closer to each other than the alcohol around them. But now the densities are other way around than what they should be. Again according to this, alcohol has the greatest density, then ice cube and then water. In this case teachers are able to explain the theory which implies that the theory is learned. However the theory is wrongly understood and this could be due to the lack of right type of explanation and connection to an experiment.

Another question about floating and sinking was presented as a picture again (Figure 4) and this question was not altered between the tests. The given answers support the findings of the previous question. The balls in the picture look the same from the outside, they have the same mass and volume, but the ball B one is hollow from the inside and the balls are made from different materials. It is said in the text that the ball A will sink and the question is will the ball B float or sink. The teacher A replied in both tests incorrectly that it will float and the teacher B stated correctly in the after test that it will sink as well. In the Pretest teacher B replied: "*It depends on the material they are made. If made from same, it will sink*".

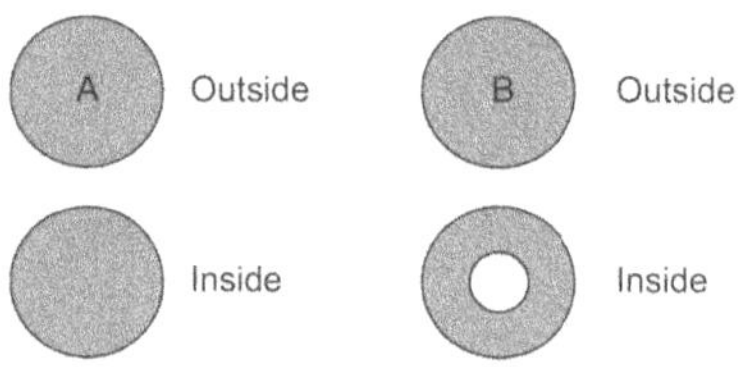

Figure 4. Comparison of two balls (Yin et al. 2008).

Here the same misconception of hollow objects floating repeats itself in a case of one teacher, similar to results of Yin et al. (2008) among students. The other one first relates it to be dependent on the material meaning in a Pretest teacher B also had incorrect idea about it.

CONCLUSION

The used method does not give enough time for teachers to understand the concepts being taught. Similar results have been reported among students (Vosniadou and Ioannides 1998) and these applied practices do not promote students to construct their own knowledge (Bencze 2001) and this seems to be a problem among teachers in training as well. Conceptual change is not happening easily and trainers need to be provided information on where to pay more attention and how to improve knowledge delivery.

Based on the results of the analysis making the relation between theory and practice is one of the problems. Teachers have knowledge that covers the surface level of the theories and this stops them to fully explore the possibilities in practice. Fear of failure is too big for to teachers to try out the learned method. The values that scientists state that are any significance are relatively far to reach for a teacher. However this could be overcome by giving the theory and doing enough time individually and together as well as repeating it several times. In this fashion, the training shows teachers how science works but also gives them time and frequency of tries to do science themselves; as it should be done.

Also learning to ask questions in a more suitable way seems to have an importance. Not only to ask “what happens” but also “why it happens”. It is vital to know when to freeze the thinking process. Based on the results, the provided training does not focus or give enough support to this type of teaching. However, this could also be due to the type of training that the trainer in this case provided. The trainings were more focused on showing examples and providing material for the classroom experiments. Teachers in this training did the experiments alone without a feedback and they were never challenged to explain why the experiment did not succeed nor to explain why it works the way it works. Therefore the knowledge that was offered stayed on a superficial level without changing teachers’ knowledge of theories or the way to apply the exact knowledge into practice. She (2002) argues that students are not willing to change their ideas if the underlying theories are not properly understood. It is reasonable not to expect anything else from the teachers as well. It is challenging to change a working concept into something that can not be related to. Thus research and attention to how training teaches the concepts is essential as well as moving the focus more on describing the concepts than on providing the material to work with the concepts. Because what was learned in training was never checked, it is not sure that the concepts were learned or even applied in practice. Showing as many experiments as possible has a good reason underneath it but just running through them has almost the same effect as not showing them at all.

The observed problem that students often leave the classroom without changing their misconceptions (Hand and Treagust 1988) is visible among teachers in this research as well. Teachers share the same misconceptions than their students and changing these concepts is a complex process, and this change seems not to be in a close reach with the current methods. It is a reasonable result not to obtain big changes with training that has just six sessions, especially when teachers in a more than a year long in-service training still hold on to their old concepts (Jarvis and Pell 2004). Even university students may fall into using similar misconceptions but they have years to change their ideas and they are constantly exposed to the scientifically accepted point of view. However, in-service training improves the attitude and confidence about science teaching (Jarvis and Pell 2004). This could also be observed in the followed training. Teachers were enthusiastic and more aware about science and technology but most likely did not have a clear idea of what they knew, just like Jarvis and Pell (2004) have reported.

Teachers have their ideas and theories how things work and on this relatively short period of training, these ideas and theories do not conflict enough with the scientific theories to make the wanted change (Hand and Treagust 1988). These theories could be relatively easily overcome if something better was made available to replace them (Hestenes et al. 1992), but this does not happen with the training at the moment. It is logical to hang on to these misconceptions because they work in everyday life. In everyday life situations it is acceptable to have more than one theory about a concept and these can conflict with each other (Posner et al. 1982). It is only a problem in a scientific world when one universal law can not be applied into all its occasions.

FINAL REMARKS

Our study shows that primary teachers have various preconceptions that do not match the concepts that are used and taught in science and technology. As stated in the introductory chapter to Part II, often preconceptions do have a value in dealing with practical situations. That should make us a bit cautious in using the term 'misconceptions' in situations like this. Often preconceptions help us deal with practical situations in a fairly successful way. But they remain limited to that particular situation and do not provide a broader view on reality by connecting different situations and phenomena. For that reason, we do value the learning of the 'proper' concepts, that is, the way they are used and taught in 'real' science and technology. Our study indicates that this is not an easy matter in the case of primary school teachers. Their lack of background in science and technology is not easily compensated by a limited number of activities directed at learning certain scientific and technological concepts. Although our study is too small to provide a sound basis for general conclusions, it does suggest that the activities have not made a great impact on the way the teachers conceptualized certain phenomena. In the VTB-Pro project, time for activities focused on concept learning was limited. As a consequence expectations for change must be modest. Our study has shown that substantially more is needed in order to realize real changes in the teachers' thinking. Hopefully, there will be opportunities after the VTB-Pro project has ended, to enforce what has been initiated now.

Marja-Ilona Koski
m.i.koski@tudelft.nl

Marc J. de Vries
m.j.devries@tudelft.nl
Delft University of Technology, the Netherlands

LOU SLANGEN, HANNO VAN KEULEN AND KOENO GRAVEMEIJER

14. PREPARING TEACHERS TO TEACH ROBOTICS IN PRIMARY SCHOOLS

INTRODUCTION

Science and technology play an important part in present-day society. People need some degree of scientific and technological literacy to cope with the challenges and possibilities of the many products, devices and processes available. Robotics is a good example of this. For many people it is a black box. The way it functions is hidden under its superficial appearance, which may be that of a coffee machine, a squeaking baby doll, or a thermostat. People may learn to push the right buttons but remain unaware of what goes on underneath. Suchignorance may alienate people from technology. Primary schools can play a decisive role in preparing children for their future, yet science and technology do not have a strong focusin primary education. Examination of learning and teaching robotics illuminates the problems encountered in improving technological literacy through primary education.

We are in the process of preparing a teaching experiment on robotics in primary schools. In relation to this, we investigate in this paper whether it is possible to prepare teachersadequatelyto implement the intended pedagogy with the help of an in-service teacher education course thatwe developed. In view of the forthcoming teaching experiment,we were especially interested in the content and character of the knowledge, insights and attitudes of the teachers. We therefore capitalized on qualitative measures.We reporthow teachers developed the required knowledge and skills in three domains, i.e. subject matter knowledge, pedagogical knowledge, and pedagogical content knowledge. We examined whether teachers were able to accommodate the content knowledge, concepts and approaches we proposed, whether they developed a personalized version of pedagogical content knowledge, and whether they increased their pedagogic ability with respect to scaffolding pupils' learning of robotics. We draw some conclusions with regard to the professional development of primary school teachers in areas of science and technology with whichthey in generalare unfamiliar.

Marc J. de Vries, Hanno van Keulen, Sylvia Peters, and Juliette Walma van der Molen (Eds.),
Professional development for primary teachers in science and technology.
The Dutch VTB-Pro project in an international perspective.

THEORETICAL FRAMEWORK

Learning and teaching robotics

Several studies show that children in primary education can learn to open black boxes and develop technological literacy with respect to robotics and automated systems (Krumholtz, 1998; Levy & Mioduser, 2008; Mioduser et al., 1996; Nourbakhsh et al., 2006; Petre & Price, 2004; Resnick & Martin, 1991).

Robotic direct manipulation environments (DMEs)can activate pupils' higher-order thinking (Slangen et al., 2008) and conceptual development (Slangen et al., 2010). A robotic DME such as Lego® Mindstorms® NXT(Astolfo et al., 2007)is a toolkit with which pupils can build and programme robots. Its effect largely derives from the stimulus to facilitate reciprocal interaction between pupils and between pupil(s) and teacher when discussing the manipulative materials. Therefore, pupils may learn robotics best when working on realistic robotic problems including designing, constructing, programming, testing and optimizing, and at the same time having a discourse in cooperative learner-learner or learner(s)-teacher situations. Teachers best support such a learning process by means of scaffolding and dialogic teaching (Alexander, 2010; Lepper et al., 1997; Wyeth et al., 2004; Xun & Land, 2004). For that reason, teachers themselves need knowledge ofwhat pupils have to learn about robotics and how pupils learn this.

Consequently, one may distinguish between subject matter knowledge (SMK), pedagogical knowledge (PK), and pedagogical content knowledge (PCK) (Carlsen, 1999; Park & Oliver, 2008; Rohaan, 2009; Zeidler, 2002). Teachers' SMK is shaped through the quality and quantity of information, conceptualizations, and constructs of the particular domain(Zeidler, 2002). Teachers' PK is the understanding of generic instructional variables such as classroom management, pacing, questioning strategies, et cetera(Zeidler, 2002). PCK is defined by Shulman (1987) as anamalgam of SMK and PK. The core of PCK is the teachers' understanding of the way pupils best learn the concepts of a certain domain and how theyovercome learning difficulties (Rohaan, 2009).

Teaching conceptual knowledge in a constructionist context of robotics

Social constructivism postulates that knowledge development is the (re)construction of personal mental models and concepts under the influence of experience and discourse (Bodner, 1986). Concepts and conceptual change derive from the use of analogies, imagistic representations, thought and material experiments and (mathematical) analysis and reasoning (Nersessian, 2008).Pupils develop shared conceptions to make sense of the surrounding world and to communicate successfully. Experience and active manipulation of tangible objects are presumed to be helpful (Papert, 1993), as are teachers' questioning and inquiry strategies(Xun & Land, 2004). 'Children don't get ideas; they make ideas'(Kafai & Resnick, 1996). The robotic DME encourages pupils to construct and reconstruct concepts through manipulating the robotic material environment. Pupils appeal to their conceptual knowledge to find solutions, in the meantime changing or refining these concepts as a result of experience and reflection (Norman, 1993). Therefore, concepts are tools to tackle problems as well as learning outcomes of that same process.

Conceptual learning is reinforced when an empathic and well-informed teacher engages in dialogue with the learners, focuses pupils' attention on important phenomena and concepts, and helps to make tacit understanding explicit. Teaching for conceptual development is a much more difficult and subtle process than delivering information through direct instruction. Learning to teach for conceptual development can also be seen as a constructivist endeavour that can be supported by engaging teachers in meaningful experiences and reflective discourse.

Subject matter knowledge in robotics

Proficient teachers know why and which conceptual subject matter knowledge is of importance to pupils learning robotics. In a previous study, we showed that solving problems with robotic DMEs involves pupils' understanding of the core concepts 'Robot', 'Function', 'System', 'Control' and 'Sense-Reason-Act Loop' (Slangen et al., 2010).

Robot. A robot is a material construction of sensors, processors, actuators and algorithms that performs predefined tasks in interaction with an ever changing outer environment (Wisse, 2008). Robots are sophisticated technical systems that function autonomously or by remote control. Teachers need to recognize that pupils often tend to approach robots as animated entities with human or animal characteristics such as volition, consciousness, intention, emotion or reflexes, and consequently this may hinder their comprehension of robotics (Ackermann, 1991, 2000). Teachers can help pupils to develop from a more psychological conceptualization towards a more technological conceptualization.

Function. From a technological perspective, function is 'the action or purpose for which something has been designed, or that users ascribe to it' (Hacker et al., 2009). Pupils use 'function' with different connotations and teachers need to be aware of this. Function in robotics can refer to (1) underlying processes that make up the internal activity of a robot, (2) external activities or roles of the robot, (3) the major objective as the sum of all internal and external functions, (4) the contribution to a larger system, (5) a feature to adapt or reproduce itself (Mahner & Bunge, 2001). Having a well-developed function concept helps teachers to support pupils to analyse the actions a device (robot) or sub-device has to fulfil in order to serve its purpose.

System. Anderson and Johnson (1997)describe a system as a group of interacting, interrelated, or interdependent components that form a complex and unified whole. A robot is a system constituted of tangible, interrelated and interdependent components as well as intangible processes, interactions, relations and information flows. Robotic DME problems can confront pupils with phenomena that help them to develop insight in the goals or functions, the order within and between, the fundamental structures, the information flow and relations between (sub) systems, and the system feedback processes. In this way, pupils understand that systems have input, processes and output, and that a system is a dynamic structure in which actions are the results of its design. Teachers can help pupils to explore and analyse phenomena that relate to system effects and discover and recognize patterns.

Control. The concept of control is fundamental in understanding the specific nature of automated or robotic systems (Mioduser et al., 1996). Firstly, control refers to a process: the power a person or a device has to influence the actions of a system, its components or related systems. Secondly, it refers to the device that is designed to regulate a system, e.g. a programmable logical controller (PLC). Control mechanisms regulate the state of a system by comparing the value of preset variables with the actual input values and executing predefined algorithms that generate output. Understanding the concept of control means being able to translate an intended functionality into a programmable rule or algorithm. Even if the icon-based programming language of Lego Mindstorms in terms of its 'grammar' and 'syntax' is quite simple, teachers may challenge and support pupils' in-depth reasoning.

Sense-Reason-Act loop. The Sense-Reason-Act loop is the most defining robotic concept. A robot is a system based on capabilities of sensing (S), reasoning (R) and acting (A), which repeat in succession and form the so-called S-R-A loop (van Lith, 2006). This mechanism represents the interaction an artefact like a robot has with a changing or (partly) unknown environment. The human method of influencing the environment by means of perceiving, reasoning and acting is by proxy conducted by a technological device, the robot. The S-R-A loop implies that sensing continuously generates new information that is fed into the 'reasoning' facility, which in turn enables consequential actions. There are, however, fundamental differences between the S-R-A process of a robot (which always follows an algorithm) and humans who use consciousness and will to interfere with the environment. Teachers can help pupils to understand this difference by reflecting on their experiences with the robotic DME.

Pedagogical knowledge and scaffolding

In a previous study, we showed that pupils can design, build and programme robots that solve a problem in reality, and that discourse with other pupils and a proficient teacher leads to conceptual development with regard to the concepts described above (Slangen et al., 2010). To achieve such outcomes in a design- and inquiry-based setting, teachers need pedagogical knowledge and scaffolding skills (Lepper et al., 1997; Wyeth et al., 2004; Xun & Land, 2004). Learning objectives that focus on pupils' conceptual knowledge development through inquiry and design require teachers who stimulate and (verbally) scaffold exploration and explication (Barnes & Todd, 1995; Vosniadou et al., 2001; Wells, 2002). Scaffolding techniques build on (1) intelligent and informed tutoring, (2) nurturing an affective relation, (3) Socratic dialogues (4) raising pupils' attention and awareness of relevant phenomena, (5) open, non-instructive communication, (6) reflection on results and processes, and (7) encouraging and motivating pupils (Lepper et al., 1997).

Pedagogical content knowledge

We define pedagogical content knowledge (PCK) as personal knowledge of teachers formed by the fusion of subject matter knowledge, pedagogical knowledge and knowledge of the context. An example of PCK from the domain of robotics is as follows. Novices often solve robotic problems by programming long linear structures ('do this, then do that, then ...') (Slangen et al., 2010; Wyeth et al., 2004). Even when pupils know about the availability of iterative or conditional program codes (often using 'if this, then that' types of reasoning) they donot use them, even though they may shorten or improve the program. Apparently, pupils still lack such structuring capacities. Teachers should be aware that pupils tend to solve problems by programming action after action. Experts solve such problems by searching for repetitive and conditional structures and by developing these into subsystems that each has its own individual input and output. Teachers can help pupils to overcome their difficulties with iteration by asking questions such as 'Does the robot have to do things that happen more than once?', 'What must the robot do if it hits the wall?'. Teachers who know this type of learning problem can recognize it in practice and stimulate conceptual development, if they possess relevant PCK. For this study, PCK is required that enables teachers to help their pupils elaborate their imprecise or anthropomorphic previous knowledge into conceptions that rely on function, system, control and sense-reason-act looping, in a context that is defined by an inquiry- and design-based pedagogical approach.

Professional development in relation to robotics

In this study, we presume that developing scientific and technological literacy with regard to robotics requires teachers who have the appropriate subject matter knowledge, pedagogical knowledge and pedagogical content knowledge outlined above. Such teachers and corresponding teaching practices are not, however, available in abundance. On the contrary, although robotics may be important for society, it is not widely taught in primary schools in the Netherlands and most teachers have little or no experience of teaching robotics. Currently, it is difficult if not impossible to investigate how teachers develop pupils' understanding of robotics in day-to-day classroom settings.

Teachers normally generate pedagogical content knowledge from their own teaching practice but it can also be acquired through professional development

(Smith, 1999). Intensive and comprehensive professional development programs can be effective in transforming teachers' ideas about teaching and learning as well as their teaching practice. In particular, teachers can be assisted to shift from a teacher-centred approach relying on instruction to a more learner-centred approach, such as inquiry- and design-based learning, and to align all the elements of the teaching situation in order to achieve positive learner outcomes (Hackling, 2007; Prebble et al., 2004; Stes et al., 2007). Kirkpatrick (1996) suggests dividing the impact of professional development into how teachers react, what they learn, what they do, and what the results are.

RESEARCH QUESTIONS AND METHODOLOGY

In this chapter we report on a qualitative study that explores how experienced teachers learn to teach robotics. The main research question is: 'How do primary school teachers develop the ability to support pupils' inquiry- and design-based learning of robotics?'.

We assume that a proficient teacher requires subject matter knowledge, pedagogical knowledge and pedagogical content knowledge to be outlined in the theoretical framework. Learning to teach robotics implies improving these capacities, and a professional development approach may be instrumental in achieving this aim. Regarding the intended teaching experiment we decided to develop and execute a course for in-service teachers that should help us answer the following research questions:

1. How do teachers acquire the subject matter knowledge (SMK) they need?
2. How do teachers elaborate their existing pedagogical knowledge (PK) into an approach that suits the requirements of inquiry- and design-based teaching, i.e. a scaffolding approach?
3. How do teachers develop pedagogical content knowledge (PCK) with respect to function, system, control and sense-reason-act?
4. How does the professional development trajectory influence teachers' attitudes and self-efficacy with regard to teaching robotics in class?

From a research point of view, data gathered during this course should contribute to educational theory on how teachers in primary education can develop pedagogical content knowledge in the domain of science and technology. To meet this objective, we capitalized on qualitative parameters that informed us about the three processes that are especially relevant. Firstly, we had to provide teachers with information on robotic concepts and on pupils' conceptual development in solving robotic problems in an inquiry- and design-based setting. Secondly, we had to activate and elaborate teachers' scaffolding knowledge and skills, and thirdly we had to confront teachers with robotic problems in order to help them recognize and tackle the conceptual problems that pupils may have in such situations, and help them construct relevant personal pedagogical content knowledge.

Kirkpatrick (1996)developed a four-level model for the impact of professional development and our course focused on the first two levels, that is,(1) on how teachers reacted to the course and its content and (2) on what they learned from it. In a subsequent study, we will focus on the two next levels, (3) behaviour in the classroom and (4) student learning outcomes.

Instruction can be an effective strategy to provide teachers with the information they require, but it may fail when teachers have to construct new concepts. Just like pupils, teachers learn more effectively when they are able to explore their own questions, design solutions to problems, and are stimulated to explicate their ideas.

Furthermore if the processes mentioned above were well chosen and well executed, then the behaviour, discourse and opinions of the participants would help to answer our research questions. Each process had its own presuppositions on effectiveness and a consequent approach, outlined below.

Acquisition of knowledge

In order to be able to teach robotics in a design- and inquiry-based setting in primary schools, teachers should have subject matter knowledge of robotics and knowledge of pupils' conceptual development. In the course, we provided much of this information through reading assignments, short presentations and video fragments showing pupils solving robotic problems in interaction with each other and an experienced teacher (the first author). Our conjecture was that primary school teachers are able to understand the information provided, and that they are able to enlarge their knowledge of the concepts during subsequent problem-solving activities.

Activating scaffolding skills

Experienced teachers have a large repertoire of teaching skills. We assumed that the participants would be familiar with scaffolding in general, although they might have relied on direct instruction methods in the context of teaching science and technology and never tutored pupils' learning processes in robotics. We conjectured that they would adapt and elaborate their pre-existing generic skills to robotics when we showed examples of successful scaffolding on video and practised scaffolding techniques ourselves during the assignments with the teachers.

Development of pedagogical content knowledge

The participants were informed about the conceptual difficulties pupils experience when solving robotic problems. We conjectured, however, that this would not be sufficient to enable them to recognize and adequately react to the conceptual problems that might occur in the teachers' own teaching practice. For that reason, we confronted participants with typical conceptual problems in robotics and engaged them in reflective discourse, in order to help them develop personal pedagogical content knowledge as a versatile tool for use in teaching practice.

Course construction

In line with our theoretical framework and the assumptions mentioned above, we developed a course on teaching robotics that consisted of four three-hour sessions and various assignments to help teachers construct relevant PCK on the basis of the information given, experience, reflection and discourse when necessary. We made ample use of the insights gathered and the video recordings produced on pupils' conceptual development from previous research (reported in: Slangen et al., 2010). The contents of the course are outlined in figure 1.

First session

The objective of the first session was to draw teachers' attention to the necessity to develop SMK, PK and PCK appropriate to the subject, i.e. robotics.

Information on types of teacher knowledge was provided. To expand teachers' SMK and PCK on the material components, the computer programming features and the concept of control, tangible Lego Mindstorms materials were provided and analysed in a combination of instruction and exploration. Teachers, analysing a problem in pairs, solved simple reason-act programming tasks, wrote and tested the program, and reflected on the effects.

Second session

The first objective of this session was to acquaint teachers with the characteristics of programming (instructions and syntax) and with conceptual thinking (function, system, control, and S-R-A). The second aim was to improve teachers' sensitivity to pupils' approaches, difficulties and (mis) conceptions and enhance teachers' repertoire and role in the dialogical teaching process.

Teachers in pairs practised programming, focusing on the idea that concepts are pieces of scientific and technological knowledge as well as mind tools to be used in teaching and learning. The pairs solved robotic programming problems, such as analysing a given robot and a desired functionality to convert their findings into an appropriate program, or predicting and testing a robot's operation from a program provided on the computer. The teachers were shown a video recording of pupils solving a robotic problem with interactive teacher support. Through manipulating tangible automatons and robots the teachers elaborated their overarching concept of 'robot' and the derived concepts (function, system, control, and S-R-A). All activities were discussed, reflected on and, through interventions of the trainer, explicitly related to conceptual learning.

Third session

The main objective of this session was to develop teachers' PCK, enabling them to use the concepts to analyse robotic problems, recognize possible difficulties pupils can have, and experience scaffolding techniques that may help pupils to overcome difficulties.

More detailed information (SMK) on the concepts and on pupils' conceptual development through design- and inquiry-based assignments was provided. Information on scaffolding techniques (PK) was also presented. Video recordings of interaction between a teacher and pupils were shown and analysed. Two robotic design problems (a soft drink vending machine and a cat rescue robot) were conducted and discussed. Additional programming instructions were offered.

Fourth session

The main objective was to improve teachers' confidence and self-efficacy with respect to the ability to develop and execute robotic problem-solving tasks in their own classes.

Teachers presented and discussed problems they themselves had designed. Some problems were selected to be solved and discussed within the group. Problem, assignment, execution and solution were reflected upon.

Figure 1 Outline of the course

Course delivery

The first author acted as the teacher educator. The course was delivered to two comparable groups of in-service teachers, a total of fifteen experienced primary school teachers (seven male and eight female, with an average of 17 years of teaching experience ranging from one to 36 years), recruited from schools in the southern province of Limburg in the Netherlands. Most of the teachers did not have experience of teaching robotics. Motives for participation bythe majority of the teachers were to increase personal knowledge and skills with regard to robotics and learn how to teach robotics in school.

Data gathering and analysis

Data were gathered in the form of answers to semi-structured questionnaire items, mind maps and notes of the participation, and video and audio recordings of all meetings. These materials were analysed (a) to enable qualitative retrospective analysis (Cobb et al., 2003) of the participants' subject matter knowledge, pedagogical knowledge (i.e. scaffolding skills) and pedagogical content knowledge, (b) to understand the mechanisms for development of these forms of knowledge, and (c) to estimate the possibilities and limitations of teaching robotics by these teachers in their primary schools.

Questionnaires were used at the beginning and the end of the programme. Initial answers (Q1) were compared with the final answers (Q2) against the background of the conceptual framework of robotics as presented in the theory section, in order to detect change and development with regard to knowledge of robotics, knowledge of pupils' conceptions and ideas on effective teaching.

Mind maps drawn at the start of the first meeting (MM1) and redrawn in the last meeting (MM2) provided additional information on teacher knowledge and how teachers related various aspects of robotics to each other.

Video recordings (VR) allowed for analysis of teachers' activities and spontaneous reactions to various interventions, such as assignments and questions. Teachers' behaviour was also compared with the way pupils reacted to similar assignments in a previous study (Slangen et al., 2010).

Audio recordings (AR) allowed for discourse analysis. This enabled investigation of teacher understanding of the concepts and derivation of the mechanisms and characteristics of conceptual development in the context of the interventions.

Notes (N) taken by the participants were analysed for additional cues on concept development.

We used Atlas-ti to tag items in the questionnaire, notes, and audio fragments. Frequencies within the labels were used as a guide towards patterns and interpretations.

In general, we constantly compared teachers' ideas and actions with the theoretical framework. Did teachers take the relevant subject matter knowledge into account in their answers to questions, in their utterances, in their decisions? Did teachers mention or apply scaffolding techniques whenever adequate? Labels (e.g. 'uses the systems concept') were developed and applied to data units by the first author. The second author checked the interpretations. In the case of disagreement, the original data were reinterpreted until agreement occurred.

Cues on attitudes and self-efficacy were mainly derived from the answers to the questionnaires and the discourse. We elaborated our data into preliminary texts for a member check (Silverman, 2006). Ten of the fifteen participants reacted and agreed with the texts. The main reason for non-response was lack of time, not disagreement.

RESULTS

In this section, we report on the patterns in the teachers' actions, reactions and utterances resulting from the instruction, assignments, and discussions during the professional development programme. We illustrate these patterns with representative examples. The abbreviations mentioned before refer to the different data sources.

Acquisition of subject matter knowledge

In the first session the teachers drew a mind map to visualize their knowledge about robotics and in the questionnaire teachers described what particularly characterized robotics. We analysed the mindmaps and the outcome of the corresponding questions by means of classifying remarks referring to the conceptual labels (function, control, system, and S-R-A). From the mind map and the outcomes of the questionnaire we concluded that teachers' initially predominantly characterize robotics using rather generic concepts that denote 'function' (38 scores) and 'control' (15 scores). The function seems to be to help humans and robots do this efficiently and accurately:

> *(Q1): To save costs in the industry. Labour that is boring doesn't have to be done by humans anymore. Higher production with fewer people.*

Teachers, however, did not explicitly focus on the necessity to perform a functional analysis of problem and context before designing, constructing, and programming a robot.

Although the term control is explicitly mentioned in four mind maps only, the mind maps and answers on the questionnaire showed that most teachers regard robots as programmed, that is, controlled machines:

> *(Q1): A programmed machine or automaton that, with a simple operation, executes a task for someone.*

Most teachers (12) showed in one or more of their answers on the questionnaire that learning to programme is part of what pupils have to learn in robotics. One teacher thought that programming was too difficult for children to learn.

The concept of systems thinking was never explicitly mentioned in the initial mind maps and questionnaire. Although a few teachers hinted at a relation between causal loop thinking and programming, just one of the teachers knew that a robot was an autonomous interactive machine controlled by sense-reason-act loops.

From the comparison of the mind maps MM1 and MM2 we saw that most teachers (11) integrated several new concepts like function, system, control, and S-R-A. Moreover, they (5) stressed the systematic problem-solving approach. From analysing fragments of the problem-solving discourse in the third session we concluded that most teachers actively used conceptual knowledge to find a solution to the given robotic problem. The concepts function and system were used more frequently than control and S-R-A. Gradually, teachers refined and elaborated on their knowledge. From the answers in the questionnaires, we can deduce that the information we intended to convey was received and elaborated upon. Most teachers (11) indicated they possessed more knowledge about the use of the concepts in relation to a robotic problem-solving process and were able to define robotics using the concepts of function (12), system (9), control (12) and sense-reason-act looping (12):

> *(Q2): In reality [robots] are machines that can observe, reason and act. [I learned] process thinking, problem analysis, and dividing a problem in modules. [I know about] concepts: function, system, control, and coding.*

Some teachers referred to S-R-A looping by means of mentioning the causality in this process:

> *(Q2):[I learned about] thinking in loops, 'if-then' reasoning, actuators, sensors and what they can do.*

Acquisition of knowledge concerning pupils' conceptual development

During the course we discussed the way in which pupils can acquire an understanding of robotics in terms of the four concepts, and how they encounter and solve problems in an inquiry- and design-based setting. Several teachers (eight) initially conjectured that pupils would have difficulty with systematic problem analysis and with programming the robot:

> *(Q1): [Pupils and teachers will have difficulties] investigating and finding solutions themselves. Teachers and pupils are used to giving or receiving answers quickly.*

Gradually, as a result of the activities and discourse during the meetings, the teachers (10) became more aware of the possibilities of using the robotic concepts as pedagogical tools to teach pupils to analyse robotic problems systematically:

> *(N): The concepts are a kind of stepping-stone for teacher and pupil. Without using the definition of the concepts, pupils develop a kind of awareness of the existence of these concepts. This can be done by discourse while building and programming.*

Activating and elaborating pedagogical knowledge and scaffolding skills in an inquiry- and design-based setting

By the end of the course most teachers (13) said in their notes that scaffolding pupils' thinking in their attempts to design, investigate and solve problems could be important for achieving conceptual understanding:

> *(N): Scaffold pupils by asking questioning, reasoning together, redirect processes, give solutions, et cetera.*

A few teachers stressed that pupils' self-directed learning needed to be supported:

> *(Q2): The teacher is supposed to support pupils with their learning process. They should not find out everything all by themselves. The learning process is determined by discoveries of the pupils but also by support of the teacher in instances where this is necessary.*

About half the teachers initially preferred self-directed inquiry learning, whereas the other teachers preferred mixtures of instructive introductory activities with inquiry-based learning. Although the importance of supporting this learning was stressed by most teachers (10), only a few teachers specified a generic intention to supply verbal scaffolds right from the start:

> *(Q1): Stress discovery learning. What is robotics? Build, test, evaluate, improve, et cetera. Take a role as a coach, stay as much as possible out of sight and where needed stimulate, help, challenge.*

From the audio recordings, teachers' notes, and answers to the questionnaire, we deduced that, when commenting on video fragments, all teachers became aware of the importance of dialogic teaching and verbal scaffolding to help pupils construct new and refined knowledge:

> *(AR): Teacher: The pupils, by their way of thinking, by combining, approach a higher level. Researcher: What do you mean by that? Teacher: Two know more than one, however, you have to communicate with each other in a special way. It means listening to the other person, which stimulates the other person's thinking.*
>
> Most teachers recognized the features and possibilities of teaching robotics by verbal scaffolding and were able to denote characteristics that were consistent with theory:
>
> *(AR): Summarize, challenge to reflect, paraphrase pupils' thinking, give information, ask guiding and open questions, stimulate thinking, stay in interaction, et cetera.*
>
> *(N): Ask questions, reason together, redirect thinking, give solutions, et cetera.*
>
> *(Q2): A way of asking questions that helps the pupils to clarify or solve the problem.*

A few teachers (four) also recognized that scaffolding challenged and motivated pupils. Three teachers reflected on the discourse on scaffolding conceptual learning as a something that made them aware and focused their attention:

(Q2): Especially the awakening process. It is all very abstract. Discussing this approach makes you aware.

Development of pedagogical content knowledge

At the start we did not find many indications of teachers' PCK. For instance, the notion that pupils should learn to understand the robot as a functional system and that they might have difficulties with analysing functions and with system thinking was completely absent (MM; AR). Initially, teachers seemed to be unfamiliar with the idea of teaching for conceptual development as opposed to teaching for the correct answer or the solution of the problem, that is, building and programming a robot. Moreover, several (12) statements in the questionnaire indicated that they saw learning robotics as a generic problem-solving or discovery process involving learning to think and not as a process influenced by content-specific elements:

(Q1): Learn problem-solving thinking.

(Q1): Learn analytical thinking.

(Q1): Learn a higher level of thinking.

Confronting the teachers with exemplarily robotics problems that could be analysed and solved by using specific concepts (i.e. function, system, control and sense-reason-act) was intended to overcome this and help teachers develop flexible and detailed pedagogical content knowledge. The assignments led to better insights into the four concepts themselves.

Most teachers (Q2) showed that they knew that the concept *function* referred to actions or objectives that a robot could achieve. They showed in their handling of the assignments on the automatic drinks machine and the cat rescue robot that they were able to analyse a problem using functions and sub-functions:

(AR): Teacher1: At least [the drinking machine] has to deliver drinks. Teacher 2: It should serve drinks fully automatically. Teacher 1: Therefore, it has to be able to perceive sound. Teacher 2: So a microphone has to be built in. Teacher 1: Yes. Teacher 2: It must react to sound. Teacher 1: After delivering the drinks it should supplement the empty spaces.

The function analysis, however, remained unsystematic and shallow in most cases. Participants tended to jump to the next stage (designing or programming the robot) too early and consequently were confronted with unanticipated functionality problems:

(AR): Teacher 1: We did not take notice of [...] when the touch sensor hits the floor the robot should start riding [...] and our robot cannot ride.

> *Teacher 2: Who says that the robot cannot ride? Teacher 1: We did not define that function.*

Grasping the features of the *systems* concept was more difficult. At the end of the course fewer than half of the teachers characterized a system as a group of interacting, interrelated, or interdependent components that form a complex and unified whole:

> *(Q2): A system is the whole (in this case the robot) that is able to execute several functions. A system consists of more systems and is related with other systems.*

> *(Q2): [A system] can come out of very different components. Within a system is some coherence and predictability.*

When executing a robotic problem-solving task some teachers explicitly defined subsystems and others recognized relations between parts of a system:

> *(AR) Teacher 1: Then we have to go to the system. Teacher 2: Transporting system. Teacher 1: Yes. Teacher 2: And the motion sensor. Teacher 1: Perception. How we do it doesn't matter at this point; we just need a perception system.*

> *(AR) Teacher 1: Yes, but movement is also detected when the front legs pass the sensor. Teacher 2: Yes. Teacher 1: The cage should not close at that moment because the back part of the cat is still outside the cage. Teacher 2: Yes, then you rather put the bait and the sensor at the back of the cage.*

Most teachers reflected on the educational possibilities of the systems concept to support pupils' learning. It helps to map the complex reality in identifiable entities (part-whole thinking). Although teachers themselves experienced the importance of understanding and recognizing the input and output relation between entities, none of them expressed this in the questionnaire:

> *(N): Thinking in systems and subsystems is difficult. Pupils are superficial and expect immediate results. The teacher has to develop a structure, present small parts, give pupils enough time to understand the structure.*

The essence of the *control* concept was not difficult for teachers. They were aware that all robotic actions are based on algorithmic structures that are programmed by humans into the controlling device.

To control a robot one has to decide which contextual information the robot should take into account and how the internal reasoning process should be executed to generate relevant output in terms of actions. When programming and explaining some simple programs, most teachers recognized and were able to practise basic programming principles such as linear and parallel structures, iteration, causal loops, parameters, and input and output instructions to communicate with sensors and actuators:

> *(AR): Teacher 1: If temperature.... Teacher 2: Warm, high. Teacher 1: Then catch. Teacher 2: If temperature is high then catch ... Teacher 1: The robot descends. Ride and search, that is the most difficult, ride search, ride search. It should search systematically.*

The execution of the assignments revealed that most teachers relatively easily manipulated the software and designed simple programs. As with pupils, errors or ambiguities resulting from unfamiliarity with programming instructions and syntax were resolved by testing the program:

> *(AR): Teacher 1: Click on 'unlimited', there. Teacher 2: Degrees, revolutions, seconds. If we do not know how large the area is we do not know the number of seconds. Teacher 1: But last time something was said about 'unlimited'. You have to interpret that in another way. It is not the same as 'infinite'. Teacher 2: Length of time .... Teacher 1: Yes, change that in ... Teacher 2: In 20 seconds. Teacher 1: Let's see if that has an effect. Is that all right? Teacher 2: That is all right. 30 seconds. Teacher 1: Yes. Teacher 2: Yes, but then I must change everything. That one [a 'move' instruction] too.*

The *sense-reason-act* concept was initially unfamiliar. Teachers' problem-solving activities (VR) and reactions in the questionnaire showed that all of them arrived at an explicit understanding of the concept:

> *(Q2): I learned to think in sequences, the if-then way to solve problems. This stimulates problem solving and analysing pupils' thinking. I learned anticipating and thinking in loops.*

The teacher pairs actually used S-R-A constructions when solving problems:

> *(AR): Teacher 1: But a touch sensor is also possible. Teacher 2: Yes. Teacher 1: As soon as it touches the box then the touch sensor is pushed in .... and it has a choice.... Teacher 2: Touch sensor, is that the name? Teacher 1: If the cat touches the box... Teacher 2:... Then the switch is activated and the box moves. Teacher 1: And if the touch sensor is hit. So we need a touch sensor. The touch sensor gives a signal to the trap. Teacher 2: The trap closes.*

We noticed that sometimes the teachers used the same strategy to circumvent problems as the pupils did. For example, they used predetermined information instead of dynamic environmental information that had to come from sensor data:

> *(AR): The pillar... I can imagine the robot is walking against it, a hard object. Then, if you know the circumference of the pillar, you can programme in such a manner that it goes around it.*

The teachers also explicitly indicated that the concepts had become tools for teaching and learning. This we can interpret as signs of developing pedagogical content knowledge:

(Q2): Understanding the concepts enables the pupils to analyse and solve problems. It helps pupils to understand whole-part relations. This helps to split up the problem in modules.

(Q2): This is the main subject of problem-solving thinking. Starting from the problem to think about what is expected (functions), what is needed to execute that (system) and how this can result in good control.

(Q2): Because these concepts can help the teacher to let pupils themselves develop. Tutoring the processes becomes clearer.

(Q2): By means of these four concepts someone can analyse the problem well. So you prevent missing some thinking steps when resolving the problem.

Self-efficacy

From the answers on the questionnaire (Q2) we deduced that, at the end of the course, most teachers (11) also intended to use robotics in their school. Most teachers (10) clearly felt competent with regard to teaching robotics in their own classrooms in the near future. This is best expressed by the following quote:

Q2: I now know more about robotics and I think I am capable of coaching and supporting the pupils.

Several teachers (six) were concerned about not having enough knowledge and fluency in programming to support pupils' programming:

(Q2): I developed a good basis to work with pupils in school. I am able to solve parts of a problem by analysing it together with the pupils. A stumbling block could be the coding [programming] of the robot.

Two teachers, evidently not feeling confident, expressed some reserve about bringing robotics into school at this stage. They wanted to develop more knowledge and skills first. The need to develop more knowledge and skills, especially programming skills, is also mentioned by other teachers (six) despite their self-confidence.

Four participants suggested that learning robotics is worthwhile in itself but is also a means to achieve more generic scientific and technological goals:

Q2: Robots are a means to increase pupils' problem-solving capacity.

Sometimes, however, teachers doubted their ability to use this open approach in normal classroom practice, when they have to take care of twenty to thirty pupils. Problems were expected in quickly and correctly recognizing and understanding pupils' problems and providing just-in-time and adequate scaffolds:

(AR): Working with small groups and having enough SMK and PCK makes immediate and adequate interventions easier. Delay in teachers' reactions or interventions can frustrate pupils' learning.

Member check

All ten respondents who reacted to the member check agreed on the preliminary text and conclusions. One respondent added that in her opinion several participants found certain topics in the course to be very difficult. Other respondents (three) also pointed to this, especially with regard to programming. Most respondents (eight) said they planned to start using robotics in school or had already started. Most teachers (nine) expressed higher self-efficacy with regard to teaching robotics.

CONCLUSIONS AND DISCUSSION

In this study, a professional development course was used to prepare primary school teachers to teach robotics in an inquiry- and design-based setting. We investigated their learning processes using data from questionnaires, mind maps, discussions and observations of teachers solving robotic problems, and used qualitative measures to indicate trends. We checked our interpretations with the participants and conclude that many of the intended outcomes have been achieved and that the results can be summarized as follows.

(1) Through a professional development approach, teachers with little or no experience of robotics can be convinced that robotics is a suitable topic for primary science and technology education. (2) Teachers have learned subject matter knowledge and acquired knowledge of how pupils learn to solve robotic problems in an inquiry- and design-based setting. (3) Teachers have acquired pedagogical knowledge to monitor pupils' progress and affect this through scaffolding. (4) Teachers have acquired experience in using robotic concepts in problem-solving contexts to anticipate pupils' learning difficulties in order to teach the pupils to solve the problem themselves. (5) In terms of the Kirkpatrick levels, teachers have learned the prerequisites for pedagogical content knowledge in classroom practice. (6) Teachers recognize that robotics contributes to technological and scientific literacy in a generic sense, in that it provides a context for problem solving and the development of higher-order cognitive skills.

We did not anticipate the anxiety with which teachers focused on programming. Apparently, for many teachers, getting stuck or making a mistake in the program is more dramatic than a flawed functional analysis or a clumsy sense-reason-act loop. This may be interpreted from the perspective of traditional instructive approaches, in which teachers are supposed to know all the answers and never make mistakes themselves. A minor flaw in the program can lead to complete failure of the robot and this may be seen as an immediate threat to teacher authority. In this respect, we did not completely succeed in preparing teachers for a design- and inquiry-based approach in which the process and the resulting conceptual development have more value than correct solutions to capstone problems.

Our claims are limited by the characteristics of our interventions. In the first place, although we claim that teachers are able to develop subject matter knowledge and pedagogical content knowledge in a professional development program, there may be other ways to achieve this thatwe did not investigate, e.g. through direct instruction or reading.

A second limitation is that our claim that teachers can acquire PCK is necessarily based on interpretations of actions and utterances, since PCK cannot be measured directly. We know that teachers possess the prerequisite subject matter knowledge and also possess sufficient knowledge and understanding of scaffolding (pedagogical knowledge). We also know that they are acquainted with learning problems of pupils and with interventions that might help pupils to overcome these problems. What we do not yet know is whether they put this into action when it really matters: in normal classroom practice. That will be the subject of a subsequent study.

Lou A. M. P. Slangen
Fontys University of Applied Sciences, the Netherlands
l.slangen@fontys.nl

J. (Hanno) van Keulen
Fontys University of Applied Sciences & Utrecht University, the Netherlands
j.vankeulen@uu.nl

Koeno Gravemeijer
Eindhoven University of Technology, the Netherlands
koeno.gravemeijer@esoe.nl

RESI DAMHUIS AND AKKE DE BLAUW

15. HIGH QUALITY INTERACTION IN SCIENCE AND TECHNOLOGY EDUCATION

How teachers link cognitive and linguistic development

INTRODUCTION

This chapter describes the approach of incorporating high quality interaction in science and technology education, employing interaction as indispensible tool and strategic purpose simultaneously.

First we argue why high quality interaction is vital in science and technology education (S & T education). We describe the concrete features that mark high quality and demonstrate why high quality interaction needs to be incorporated in teacher professionalization. Next we focus on research findings that show how teachers are actually able to learn to realise the desired interaction in S & T education. The findings result from small scale quantitative studies and from a qualitative description of an S & T lesson.

INTERACTION IN SCIENCE AND TECHNOLOGY: HANDS ON, MINDS ON, TALK IT OVER

In science and technology education the popular adagio is *hands on, minds on* (introduced by Driver 1983). We will argue here, that this adagio lacks a third component. Although 'minds on' implies to 'think aloud together', our work with practitioners has made clear the need for an explicit statement. Children need ample and high quality opportunities to talk to others, i.e. peers and teachers, about discoveries, ideas and solutions. Without such opportunities science and technology education will not reach its goals. So we propose an extended adagio: *hands on, minds on, talk it over*. This adagio could also help to avoid the trap of the so called 'pseudo-inquiry' (see Harlen & Léna, chapter 1): plenty of practical activity, but a lack of involvement of the children *in making sense* of phenomena or events in the natural world (italics by the authors). The proposed extension is grounded in two major theories: socio cultural learning theory and language acquisition. These can be fruitfully linked: "the very same conversations that provide the opportunity for the child to learn language also provide the opportunity to learn *through* language." (Wells, 1999, p.51). Interaction in S & T education needs to be thought- and talk-provoking.

Marc J. de Vries, Hanno van Keulen, Sylvia Peters, and Juliette Walma van der Molen (Eds.),
Professional development for primary teachers in science and technology.
The Dutch VTB-Pro project in an international perspective.

Indispensible tool: active learning

Here we probe learning theories of science & technology education for the role they assign to dialogue[xvi]. The kind of science and technology that children are to learn does not merely consist of a wide array of facts that can be transmitted to the children by a teacher or a textbook. The main objectives of S & T-education in primary school are (Van Graft & Kemmers 2007):

- children get familiar with the process of 'doing science'
- children acquire knowledge: not mere facts, but concepts in their context
- children develop an inquisitive attitude

These aims can only be reached fully, if interaction plays a prominent role. This role is eminent for three reasons:

(1) learning originates from language in dialogue
(2) reasoning processes are acquired through dialogue
(3) learning to engage in dialogue creates an open mind for new ideas

Each of these reasons is explained here.

(1) learning originates from language in dialogue. When learners are engaged in problem-solving activities with peers and teacher, they learn through language. The aims of S & T education are in accordance with the current view on learning as a socio cultural process (Vygotsky, 1978; Leontiev, 1981). Learning is not a matter of transmission from an expert (teacher) to novices (the pupils), but consists essentially of transformation or co-construction by active learners (the pupils) with the support of a facilitator (teacher). In the social exchange children make meaning of the world surrounding them. In such transactions classroom discourse functions as a thinking device (Wertsch & Toma 1995). "… language is the essential condition of knowing, the process by which experience **becomes** knowledge" (Halliday 1993, p.94). In interaction children acquire new concepts and interconnect concepts. Moreover, it is *oral* interaction that permits them to function at their highest cognitive level. The abstractions needed to cope with written language restrict the level of cognitive operations a child can manage with as much as three years (Hammond 1990, cf Halliday 1993, p.110; Snow & Kurland 1996; Dickinson & Tabors, 2001)

(2) reasoning processes are acquired through dialogue. Learning through dialogue is expanded in approaches such as dialogic inquiry (Wells 1999), inquiry learning (van der Linden & Renshaw 2004; Flick & Lederman 2006), inquiry-based science education (Harlen & Allende 2009), and dialogic teaching (Alexander, 2004; Mercer & Littleton 2007). Dialogue is also part in the Dutch version of the content-based approach: 'Taalgericht vakonderwijs' (Content based language education) (Hajer & Meestringa 2009). Through interaction children think things through, construct representations, and reflect on solutions and explanations. For instance, when making a prediction before executing an experiment "one is involved in a form of theorizing, as one examines one's beliefs about the phenomenon in question and relates them to any other knowledge one has that is relevant to the possible outcomes of the experiment. As important as the actual predictions that students make, therefore, are the *reasoning processes* that

lead to them." (Wells, 1999, 215; italics by authors). Thinking and exchange of thinking constitute integral elements of science activities. Children's 'small ideas of science' gradually are linked together towards broader understandings, the 'big ideas' of science (see Harlen & Léna, chapter 1). It is this process that is brought about by interaction. The central task of science education is to scaffold student engagement in such discourse (Metz 2006).

(3) learning to engage in dialogue creates an open mind for new ideas. Inquisitiveness and wonderment lead to thinking and talking. So far, we viewed dialogue as a means to an end, i.e. the construction of knowledge. However, dialogue as an end-in-itself has been discussed recently as a broad educational aim - in the challenging form of a written dialogue (Wegerif et al. 2009): "to be more dialogic (…) is to be more open to other voices, more able to question and to listen and so more able to allow new unanticipated meanings to emerge" (op. cit. p.185). It is the *creative space* of dialogue (op. cit. 197) that is emphasized here.

Strategic purpose: language acquisition

Within the field of linguistics interaction is since long accepted as a major source for language acquisition (Gass & Mackey 2006), whether first language (since Bates 1976) or second (since Hatch 1978). Here we outline only the basic idea. Details concerning the required quality of interaction are the focus of the rest of this chapter. Learners need to receive comprehensible input that provides the model for the target language (Krashen 1980). Such input must be complemented by pushed, comprehensible output (Swain 1985, 1995, 2005). By speaking the learner goes from semantic processing merely focussed on comprehension to complete linguistic processing needed to construct new and accurate linguistic knowledge. In sum, learners must be challenged to produce language in order to learn (see overview in Damhuis 2008). Interaction brings about language proficiency in a broad sense, and in more specific sense communicative competence. Both are important objectives in language education.

This central role of interaction in language education can be used as a strategic purpose for S & T education. In the Dutch curriculum only a limited portion of time is allotted to S & T. Therefore alliance with other subjects is a good strategy to secure more learning time spent on S & T topics. Linking S & T topics with language education combines mutual objectives while each subjects maintains its own merits. Moreover, as reported earlier, dialogue can be seen as a broad educational aim for any subject. So, interlinking language education, S & T and other subjects has a great advantage.

HIGH QUALITY INTERACTION MADE CONCRETE

A considerable body of research is dedicated to identifying features of interaction that enhances linguistic and cognitive development (e.g. Halliday 1993; Wells 1999; Nystrand et al., 2003; Alexander, 2004; Mercer & Littleton, 2007; Schwarz et al. 2009). The educational context varies from language arts to history to science and mathematics (Boersma et al. 2005, Lampert 1990, Lemke 2001, Mesa & Chang 2010, Sharpe 2008, Zion & Slezak 2005). The level of concreteness of descriptions varies strongly. General terms related to 'thinking', 'reasoning processes' and 'developing understanding' are made more concrete in the following:

- "questions are structured so as to provoke thoughtful answers (…);
- answers provoke further questions and are seen as the building blocks of dialogue rather than its terminal point (Alexander, 2004 p.32).

Working with teachers made us aware of their need for even more concrete descriptions of what 'exactly' they could do in their everyday practice. Based on the arguments for interaction as an indispensable tool and a strategic purpose presented in earlier sections, we focus here on three characteristics that are vital specifically for S & T education:

(1) Rich content in combination with extensive output
(2) A coherent and productive line of enquiry
(3) Deepening feedback

Table 1. Features of high quality interaction that are essential for S & T education (in brackets the reference to the item on the LIST Checklist, Damhuis, de Blauw & Brandenbarg 2004, Damhuis & De Blauw 2008)

Child	*Teacher*
1. Rich content combined with extensive output	
takes turns on his/her own and expands on his/her turns (3c)	refrains from asking questions continuously (3c)
answers at length in response to open questions (3d)	asks when necessary open and inviting questions (3d)
responds at length on his/her own to teacher's statement (3e)	makes a thought provoking statement occasionally (3e)
uses and expresses a higher level of thinking (complex cognitive language functions, such as comparing, reasoning, making conclusions) (4c)	encourages to use and verbalize a higher level of thinking (complex cognitive language functions, such as comparing, reasoning, making conclusions) (4c)
2. Coherent and productive line of enquiry	
continues expressing and verbalizing communicative intentions (4a)	connects to the content of child's conversation (contingent discourse) (4a)
expands on content (4d)	builds on the content of what the child says (4d)
uses my support to express his/her meaning (negotiation of meaning) (4b)	supports the child in clarifying his/her meaning (negotiation of meaning) (4b)
3. Deepening feedback	
contributes in a well structured manner (5b)	structures the contribution of the child and summarizes when needed (5b)
accepts translation and makes use of it (at a later moment) (5c*)	translates the child's contribution into more appropriate language and encourages the child to respond (re-voicing) (5c*)

These characteristics were dissected into concrete child behaviour that shows opportunities for linguistic and cognitive development, and corresponding teacher strategies to foster such behaviour: table 1. Together with the more general characteristics of communication prerequisites and language input they form the LIST Checklist for talk- and thought-provoking interaction[xvii] (Damhuis, de Blauw & Brandenbarg 2004, Damhuis & de Blauw 2008). LIST is the acronym for Language acquisition through Interaction Strategies for Teachers. Thus, we identified concrete strategies a teacher could use in real life, presented and used within the framework of language- and thought-provoking interaction.

(1) Rich content in combination with extensive output. This combination is required for thought- and talk-provoking interaction, a refined construct of important aspects of exploratory talk (Mercer & Littleton 2007) and dialogic spells (Nystrand et al. 2003). Interaction only qualifies as thought- and language provoking if it contains features of extensive output and rich content in combination (De Blauw et al. 2010). Within the perspective of language and cognitive development form and function together determine the quality of a response.

Example 1: Teacher: *How could they solve this problem?* Child: *They have to take out the dead fish, because they poison the water.* This thought provoking question has an open form, allowing for an extensive answer, and thus is. simultaneously talk-provoking.

Example 2: Teacher: *What is poisoning the water?* Child: *The dead fish.* Although still thought provoking, the question form is closed, allowing only a limited output: not talk-provoking.

(2). Coherent and productive line of enquiry. High quality interaction is not a series of autonomous question-answer exchanges. But answers are followed up by the teacher and used to build a coherent line of thinking and reasoning. This category builds on the concepts of exploratory talk (Mercer & Littleton 2007), coherent line of enquiry (Alexander 2004), 'uptake' (Nystrand et al 2003), and productive dialogue (Wegerif et al. 2009).

(3) Deepening feedback. This encompasses feedback that leads to more in-depth thinking and talking. One of the actual strategies concerns *structuring and summarizing* children's contributions. This makes the scientific process explicit, and draws attention to important content of children's' contributions to support acquisition of knowledge (objective 1 and 2 of S & T education). Moreover, it contributes to the coherence of the line of enquiry.

Revoicing. Children express new ideas, suggestions and reasoning often in everyday language, hesitatingly, in search for words, and support the conveyance of their intentions by gestures. For cognitive and linguistic development this is the optimal moment to revoice this contribution: in her feedback the teacher (a) values the contribution, (b) offers the proper wording in scientific and technical terms, and (c) invites the child to consider approval (O'Connor & Michaels 1993, 1996; Damhuis 2008).

INTERACTION COMPONENT OF TEACHER PROFESSIONALIZATION IN S & T

How can we contribute to teacher professionalization in S & T education? In order to realise a rich learning environment for science and technology education, VTB-Pro builds her program for teacher professionalization on three pillars (Kuijpers & Walma van der Molen, 2007):

(1) knowledge of scientific and technological concepts and the skills necessary for the scientific thinking process,
(2) attitude towards science and technology,
(3) pedagogic-didactic skills, especially with respect to inquiry learning.

Obviously, the realisation of high quality interaction constitutes a major didactic skill (pillar 3), and hence has to be included in professionalization. In addition, thought- and talk-provoking interaction helps teachers in their own S & T development (pillar 1). It should be employed to make explicit the scientific thinking processes at their own cognitive level, feeding their insight in children's thinking processes and how to foster those: positive impact on didactic skills (pillar 3). Also, when teachers themselves get involved in proper thought- and talk-provoking interaction, it fosters their own inquisitiveness, contributing to a positive attitude towards S & T (pillar 2). It may even help them to overcome their perceived inadequacies in S & T knowledge. Naturally, teachers need a basic knowledge to ensure rich content and the deepening of insights by pupils in the phenomena at hand (see pillar 1). But with the proper interaction skills in their repertoire, they are able to function as facilitator of the discovery process of their pupils and won't feel uncomfortable when they admit not having all knowledge at hand. Their interaction skills relieve their anxiety about unpredictable pupil contributions to the creative dialogue. In sum, an important part of teachers' professionalization in S & T education must consist of learning how to realise thought- and talk-provoking interaction.

IS HIGH QUALITY INTERACTION IN S & T EDUCATION LEARNABLE?

We argue that learning how to realise high quality interaction must be a part of teacher professionalization in S & T education. This only makes sense, if the way in which a teacher interacts with her students is viable to change. We show that teachers are actually capable of realising different types of interaction describing a training course we developed, and presenting findings concerning the efficacy of this course.

The core of the professional learning course we developed for (student) teachers in primary education (De Blauw & Damhuis 2006) is the classroom interaction checklist described previously. The course is especially structured in order to yield actual changes in classroom conversations. Major elements are: the use of video footage of teachers' own classroom conversations, and team meetings combined with individually oriented teacher guidance in the classroom (Damhuis & De Blauw 2008). During the training, participants choose their own learning objectives from the checklist and practice these chosen strategies in their classroom interactions. Videotaped interaction in the participant's own classroom forms the base for each coaching session with the participant.

This interaction course was experimentally integrated with enquiry learning in S & T (Van Graft & Kemmers 2007) to develop a *Language, Science &Technology teacher training*[xviii].

From eight different schools in The Netherlands teachers and their pupils in Dutch grades 3 and 4 (age 6-8) participated. In the team meetings the teachers experienced S & T activities and learned about S & T education and the role of high quality interaction. Most importantly, co-operatively they designed S & T activities for their classes and practiced their strategies for realising thought- and talk-provoking interaction. Two experimental themes were developed and conducted in the classroom. Simultaneously, the teachers practiced the learning objectives from the LIST Checklist, and were coached on these points.

Can this specialised approach be successful? First we present results of small-scale quantitative studies on learn ability of interaction strategies of the LIST Checklist. These studies use data from training courses where the link between language and thought was realised within a broad range of school subject areas, not specifically S & T. Next the focus is on S & T. A qualitative analysis of a lesson in the theme *Sound* is offered. This detailed analysis demonstrates how teachers may realise thought- and talk-provoking interaction in S & T education.

SPEAKING TIME AND QUESTION BEHAVIOUR

To what extent are the concrete strategies from the LIST Checklist learnable in the real life practice of education? Here we present findings on two aspects: general speaking opportunity for children, and question behaviour of teachers. Although we work with distinct, actual strategies, we do not consider them as ends-in-itself (a risk signalled by Mercer and Littleton, 2007, p.35), but within the framework of thought- and talk-provoking interaction (see earlier section).

Research questions. Innovations for early S & T education should start both at in-service level with teacher and at pre-service level .with student-teachers. The first research question originates from the general aim of the training course: (1) Do (student) teachers create more speaking time for children after the course? Within this broad aim, participants choose their specific learning objectives. Two important strategies led to these specific research questions: (2) Do (student) teachers ask less questions after the course? (3) Do they pose relatively more open questions than closed questions after the course?

Method. Conversations between (student) teacher and pupils were videotaped before and after the course. A 10-minute episode of each video was analysed.

Speaking time for teacher and for children was measured in seconds, using a stopwatch while watching the video recording.

Questions were transcribed from the video-recording, and categorized for type with respect to the extent of appropriate answer. Besides open and closed questions (example 1 and 2 in earlier section) a what-action question was distinguished:

Example 3: Teacher: *What should we do now?* Child: *Close this pipe.* This type may create more opportunity for active participation than the closed question, but generally still less than the open question.

Inter-rater reliability established for three juries of two persons each reached an average Cohen's kappa of 0.86 (Damhuis et al. 2010).

We expected proportion of speaking time by the teacher, total number of questions by the teacher and proportion of closed questions by the teacher to decrease. An increase was expected for proportion of speaking time by the children and proportion of open questions by the teacher.

This study presents data from three teachers who participated in the LIST course as an in-service training and 8 student teachers following the course within the 4th year of the initial curriculum. They all had chosen one or both question strategies as learning objectives.

Results. All three teachers and five out of eight student teachers succeeded in providing children more speaking time after the training: figure 1. They reached a far better level than the 60% teacher proportion in traditional classroom discourse (e.g. Cazden 1988).

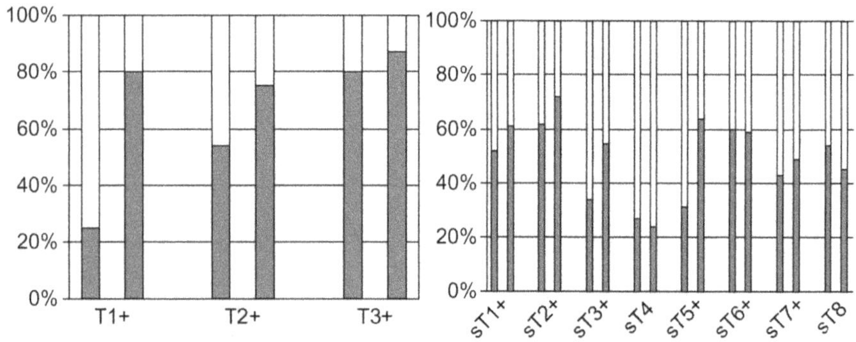

Figure 1: Speaking time by Teachers (T) (white) or student Teachers (sT) (white) and children (grey) before the course (left column of pair) and after the course (right column of pair); (+ = expected direction of change).

With respect to the total number of questions all three teachers changed towards the expected direction, but only three student teachers: figure 2. The expected relative increase of open questions was realised by two teachers and six student teachers: figure 3. The third teacher showed a decrease of open questions, but he already used the lowest total number of questions (figure 2).

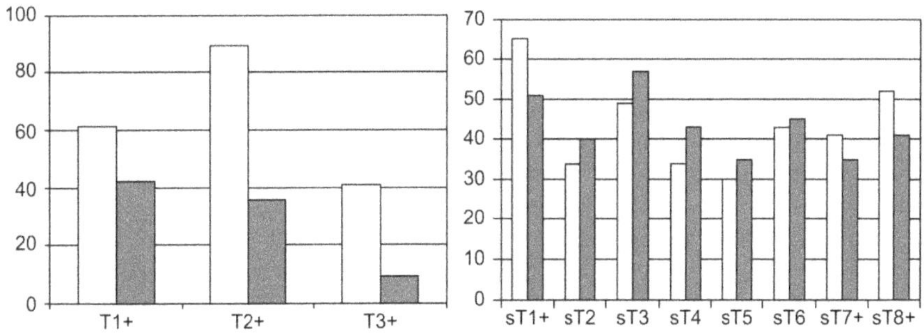

Figure 2: Total number of questions by Teachers (T) and student Teachers (sT) before (white) and after (grey) the course (+ = expected direction of change)

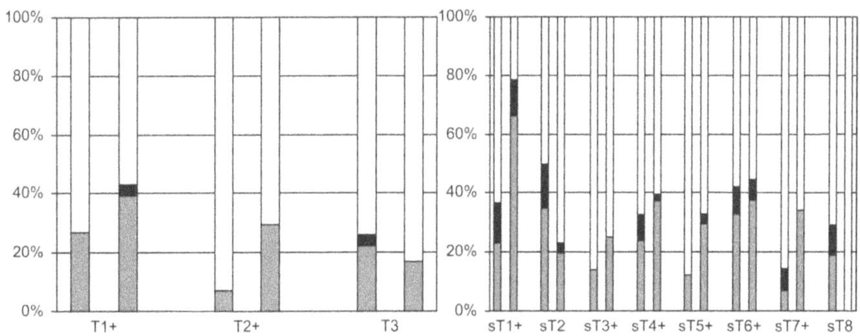

Figure 3: Percentage of type of question on total number of questions by Teachers before the course (left column of pair) and after the course (right column of pair); open questions (grey), what-action questions (black) and closed questions (white); (+ = expected direction of change)

Discussion and conclusion. Change is possible for (student) teachers, although not yet for all of them. The teachers actually changed their interaction in general as well as with respect to the chosen strategies of question use. Most student teachers improved with respect to type of question. Noteworthy is that these (student) teachers started out at quite different levels, but each was able to improve. The student teachers seem to have ignored the conditional part of the strategy of open questions: if necessary. This may require more emphasis during the course.

It requires further exploration of internal and external factors to explain differences and to find ways of optimizing the professionalization course. Future analysis will involve more participants in order to draw conclusions on the statistical significance of changes in the interaction. In addition, features of the quality of content will be investigated (see De Blauw et al. 2010).

PRACTICING HIGH QUALITY INTERACTION: A LESSON ON *SOUND*

What does high quality interaction in S & T education in primary schools actually look like? It is time to dive into reality.

Here we present Wendy and her class of 20 children aged 7-8 (Dutch grade 4) as a good practice example of what high quality interaction is about and how it looks like on a daily basis. Wendy participated in the collaborative project on developing interaction strategies in S & T education. Until this project Wendy, an experienced teacher, worked on S & T guided solely by a regular, frequently used Dutch textbook.

The class is involved in the theme 'sound'. Normally Wendy follows *Leefwereld* (*Lifeworld*). As part of the special course on interaction and S & T education, she now has enriched the theme with thought- and talk-provoking dialogues. Today the children participate in a search for characteristics of 'sound'. Sound is everywhere, you can hear it, see it and feel it. The teacher's goal is to let the children discover aspects of the concept 'sound'. We will focus on the specific role of interaction strategies used by the teacher, describing the lesson according to the steps of the so-called Inquiry Learning Cycle for S & T education developed by Van Graft & Kemmers (2007) and Kemmers et al. (2007).

Step 1: The problem

By way of orientation the children close their eyes for half a minute and have to describe what they hear. This prepares children for thinking about *sound.*

Would it be possible to hear nothing at all? This is the problem teacher Wendy brings into focus. The issue is explored jointly in a whole group conversation. Children put forward several suggestions for places where you hear nothing.

Teacher: *On the planet Venus? Have you been there? Then how do you know?*
Dennis: *Because there are no people and all that. And also not wind.*
Teacher: *Is there no wind?*
Dennis: *I think (smiles)*

Sounds are everywhere: on an uninhabited island, in an empty building, in a shed in the middle of a meadow. To reach complete silence a lot of things have to be not there. There are extremely few places where it is totally soundless, they conclude.

Strategies. What strategies does the teacher use? Teacher Wendy raises amazement and inquisitiveness. Nonverbally her face invites children to explain their ideas. The teacher keeps silent, gives listening responses, occasionally asks an open and inviting question, and makes use of summarizing to combine building blocks into the line of enquiry (Alexander 2004), thus deepening the content.

Coaching. What are important coaching objectives? In our in-service and pre-service courses for high quality interaction teachers learn that the-whole-group setting is not ideal for children to express ideas. A regular switch between whole-group and peer exchange in pairs is more effective in providing opportunities for children to produce extensive and rich output around a thought provoking issue. Wendy will try this later.

Step 2: Exploration

Now the theme is established, teacher Wendy introduces a second problem: *How does my sound come to you?* First responses are *with my ears* and *because you talk and then it goes a bit further*. Wendy follows up those ideas: *But how does that happen?*

Two children, Wesley and Mike, are each supplying pieces of the puzzle, at different moments in the line of enquiry. They are deeply thinking, as can be seen by their hesitating way of talking and the way they use gestures to convey their intentions. Teacher Wendy scaffolds how to put these pieces together, combining the building blocks offered by different children.

Wesley: *That's when you talk then eh there comes always, well look, for instance when I blow then it also comes further out of my mouth* (hand indicates movement away from mouth). *And when you talk then it is I think the same.*

Teacher: *That is has to do with air then?*

Wesley: *Yes, that it flies a bit further like that. All very quickly.*

Teacher: *Okay, so it has to do with air.*

(Then Diana contributes spontaneously, but on the earlier topic of 'where is no sound at all'. Teacher accepts and then allocates the turn to Mike)

Teacher: *Mike, you want to say something on how sound comes to your ears?*

Mike: *Yes, I think it just goes to your ears* (supports his words with a gesture) *then to your brains and then your brains think "Oh somebody is talking to me, then I have to talk back".*

Teacher: *Okay. So the sound comes from my mouth and then what happens next Wesley?* (combines now) *The sound comes from my mouth* (indicates movement with finger).

Wesley: *and then uh* (points)

Teacher: *then you said it went through...?*

Wesley: *the air and then it comes slowly, then it goes very fast further to Mike*

Teacher: *then it comes into Mike his...?*

Wesley: *his ear and that really goes so fast, you just can, yes, that is just, well that is awfully fast.*

Teacher: *Sound goes very fast. And Mike and then you say then it comes into my ear and then what is happening to it?*

Mike: *And then it goes very fast and then my brains think someone is talking to me and then I talk back.*

Teacher: *Okay. I think that's a very clever idea.*

Strategies. What are the strategies the teacher uses in the example above? The teacher offers ample opportunity for the children to contribute ideas. She encourages them to use complex cognitive language functions. She re-voices Wesley's expressions into words belonging to science- and technology concepts: *it has to do with 'air'* (O'Connor & Michaels 1996). By interlinking children's ideas she builds up a coherent and productive line of enquiry (Alexander 2004, Wegerif et al. 2009). She designates the children's answers as building blocks and involves the children themselves in actually performing the building. It is by the interaction strategies of the teacher in this classroom example, that the children are actively and jointly involved in expressing the reasoning process.

Coaching. On which points teacher coaching has to focus? There is a fine balance between providing opportunities to all children to actively participate and sticking to the conversation topic. The guideline we suggest for teachers is to focus on the line of enquiry and opportunities for building blocks, by taking up the contribution of a child in the ensuing turn (Nystrand et al. 2003) .

Step 3: Setting up the experiment or plan

Next Wendy demonstrates two experiments. Experiment 1 makes use of a washtub with water on a table in the middle of the circle. The children are invited to predict what will happen when she would throw in two marbles. *Sink* is the first reaction. Wendy continues: *And what else?*

Theodore:	*Well for instance when you throw a very big stone in the water, then it goes into the water indeed but then also water comes out again. Because there is no air in that stone. And when, when throw in water for instance, then it just goes fuller.*
Teacher:	*Then it gets fuller yes.*
	… (a bit further in the conversation) …
Diana:	*Well, look, then you suddenly hear 'ploomp' and then it splashes a bit upwards.*
Teacher:	*And then what is happening with the water?*
Diana:	*Well, that moves very fiercely to and fro and all that.*

Strategies. The teacher models *hands on, minds on, talk it over.* Prompting predictions is thought provoking (Wells 1999). Wendy creates the opportunity for active, extensive output. The children utilize the opportunity to put into words what they expect to happen. They use complex cognitive language functions such as reasoning, concluding, cause-and-effect. Thus, the teacher realises the combination of being thought ánd talk provoking that is required for high quality interaction. Some ideas already prelude on 'waves' and sound: this strand becomes more explicit in the next step.

Coaching. Special attention is needed for re-voicing, translating the child's contribution into more scientific and technological terms, is difficult for teachers. In this classroom dialogue Wendy actually did not reply with extensive feedback to Theodore's contribution, but she could have re-voiced as follows: *Right, so the level of the water rises, because of the volume of the stone. That's is good prediction.* She then could have built on: *And so, if we would drop in a very tiny stone, we would see nothing.* Acting that way a teacher can deepen the child's idea, and provoke talk as well as thought by using a thought provoking statement in stead of a question: rich content and extensive output.

Step 4: Conducting the experiment

After all these predictions it is time now to demonstrate what happens throwing the marbles into the water.

Teacher: *Manja you said it will make waves and that is also like what happens with sound. When I produce sound with my voice, then it goes a bit in waves through the air, then it reaches you* (supports with hand- and arm gestures)

Next, the teacher performs another experiment, going through step 3 and 4 once more: children predict and observe. On the table there is another tub. This one is covered with plastic foil and some icing sugar. By beating a drum Wendy demonstrates that 'by sound things can be moved'. One of the children adds that she sometimes feels her father's drumming in her stomach.

Strategies. One of the strategies the teacher uses here is that she revoices the children's observations into scientific language. She also explicitly draws a conclusion, taking up children's contributions (Nystrand et al. 2003), modelling step 5 of the Inquiry Learning Cycle. Acting in this way the reasoning process becomes audible for everybody (Wells 1999).

Coaching. Coaching draws the teacher's attention to the fact they have to assess which level of insight in the phenomena their pupils can handle. In this first orientation on the theme children get a bit closer to understanding 'sound', although not yet the full scientific picture of 'sound makes particles move in waves'. Teachers search for a balance, and a gradual deepening of insight.

Step 5: Drawing conclusions

After these experiments Wendy draws a main conclusion about what they all have experienced now: sound is something one can hear, feel, and see. When children came up with a counter argument, the experiment was repeated and talked over until everything was clear.

Strategies. By using several interaction strategies the teacher has integrated high quality interaction in steps 3, 4 and 5. She discusses the observations and lets children reflect on how phenomena might be connected. So they are learning through language (Halliday 1993). Wendy also models how they can go about investigating their ideas. She creates a spontaneous bridge between children's amazement and the process of doing science, objectives 1 and 3 of S & T education. With these child contributions and her own feedback she enables children to acquire knowledge of the features of 'sound', objective 2.

Step 4: Constructing an object and Step 5: Testing the object

The whole group activity has ended now. In couples the children go working on assignments that concern 'sound'. In various ways children go through steps 3, 4 and 5 of the Inquiry Learning Cycle. For instance the hands-on activity of constructing a telephone from two plastic cups and some string.

Strategies. What and how do the children learn? They discover while doing and discussing how things work and don't work. The teacher frequently looks into their activities, having them express verbally what they see, predict, compare and are trying: the reasoning process (Wells 1999).

Coaching. Here coaching focuses on requirements of small-group work. It is essential in these assignments that children not just follow a step by step instruction. They need to consider several possibilities, reason about expected success, and share their ideas with others. This creates the opportunity for building up new ideas, understandings, concepts, even though they are still pre-concepts instead of full scientific concepts. In step 6 we will see that Mike has picked up the concept of 'vibration'. The teacher is needed to deepen the exchanges. She may do this by following up (Alexander 2004, Nystrand et al. 2003) on what the children are saying and doing when she joins a pair or group. This is also her role when the pairs reconvene in the whole group.

Step 6: Presenting

After each couple of pupils has fabricated and tested their telephone, a whole group discussion follows. The children tell how things went and what they discovered. The teacher offers feedback. At the end of the presentations the teacher summarizes the findings.

Here we see how two children, Diana and Mike express their findings from their pair work.

Diana: *Yeah look, when you talk through it just very softly you can still hear it quite well. And very softly from the big distance then you just cannot hear it* (means 'without the phone').

Teacher: *Okay.*

Diana: *And I also thought it very funny, for then we were going to construct it, but then we did not know that you had to hold taut the string, and then we went very loosely, Tonia stood eh like there and me here. And then I heard like: Tonia can you hear me. And then she couldn't hear me every time and I also not. And then we went into the hallway and then we held it taut for we wanted to do it very far away, far from each other, and then we did hear it. And it did not matter whether there was a knot in, or not.*

Mike: *Well, I think that when you talk through the cup* (accompanies his words with gestures and sounds). *Look, I think it goes zoof zoof, the sound goes through the little hole, that air. Then the vibration goes there through the wire, swoosh, and then it arrives at the other cup.*

Teacher: *Okay, so your sound, what you say, goes through that wire to the other side, into the ear.*

Mike: *Yes.*

Teacher: *You have discovered many things.* (Then addressing the whole group) *What do you think I mean. What have they discovered?*

Children repeat the discoveries, using their own words. Wesley repeats, Diana adds to it. The teacher emphasizes the important elements.

Strategies. What strategies are central here? Wendy creates great opportunities for the children to express verbally their process of discovery. She requests clarification when the verbalization is not completely comprehensible or not specific enough. So she elicits pushed output (Swain 2005). Together they will negotiate meaning. Wendy's feedback is focused on S & T content, and more specifically on the thinking processes (Wells 1999). She draws the attention of the speaking child and all peers to justified conclusions that bear on the scientific content.

Step 7: Deepening and broadening

In the discussion of the results of their telephone-construction and testing, it is Theodore who presents a counter-experience. He himself could **not** hear Mike that well. He suggests an explanation. Wendy replies by advising him to study this more thoroughly.

Theodore: *Well, I discovered that I could not hear Mike so clearly. For with Mike there* (stammers) *was, wered, were, look* (rises and walks towards Mike's cup on the other table), *very many holes in there.*

Teacher: *Okay, that might have to do with it, yes.*

Theodore: *Then the air escapes from it!*

Teacher: *To investigate it again, to investigate whether it makes a difference, you should construct a new one and then you should use each of them. Hey, do I hear it better now with the cup with the hole or do I hear it better with the cup without the hole in it? You can investigate that.*

Theodore: (nods)

Strategies. The teacher uses an important strategy: she takes up Theodore's contribution (Nystrand et al. 2003) and uses it to stimulate further research on the telephone. She models possible research questions. In addition, with this step the teacher will ensure to repeat and focus on all things the children have done, have seen, and have concluded. She structures, summarizes and re-voices in appropriate language, providing correct content in her feedback.

Coaching. Coaching should concern several ways of deepening and broadening. It may be difficult for teachers to extend the findings to new situations. An example in this theme could be the step to cell phones. This interlinks the concepts of sound as 'waves through a wire' and 'sound as waves through the air'.

High quality interaction in S & T education: conclusions

The example of the classroom interaction of Wendy and her pupils demonstrates how it is possible to integrate high quality interaction into S & T education with 7-8 year old children. This ensures that the main objectives of S & T education in Dutch primary school as quoted earlier (Van Graft & Kemmers, 2007) are met.

Wendy teaches according to the Inquiry Learning Cycle for S & T education enriched with high quality interaction. In this way she shows that:

- her children get familiar with the process of 'doing science' by exploring issues about *sound*, discussing problems with *sound* and solutions, and constructing telephones.
- they acquire knowledge: not mere facts, but concepts in their context: in this lesson they are beginning to form concepts of sound, waves, air, vibration.
- they develop an inquisitive attitude: children contribute ideas moving from own experiences with sound to wondering about the phenomenon itself; they use complex cognitive language functions, such as reasoning, cause-effect and conditions; children themselves are building up the line of enquiry and are actively and jointly involved in expressing the reasoning process.

That children are involved in such a language- and thought provoking way of learning is possible because their teacher demonstrates the features of high quality interaction as shown in Table 1. (1) she offers a rich content combined with extensive output, (2) she offers a coherent and productive line of enquiry, and (3) she offers deepening feedback.

Teachers and student teachers can acquire these classroom interaction competencies by courses especially focused on high quality interaction in S & T lessons.

CONCLUSIONS

In this chapter we have demonstrated that high quality interaction is vital for S & T education. So we propose an extended adagio for S & T education: *hands on, minds on, talk it over*. High quality interaction is thought- as well as talk-provoking. The characteristics are (1) extensive output in combination with rich content, (2) coherent and productive lines of enquiry and (3) deepening feedback. To become applicable in every day practice by teachers, these characteristics are made feasible and workable in the form of the desired child behaviour and complementing teacher strategies.

We found that teachers and student teachers are indeed capable of incorporating high quality interaction in their S & T lessons, when they participate in specially designed courses. Therefore, in order to realise the objectives of S & T education, courses like this must be included in pre-service and in-service professionalization of teachers in primary education. One must bear in mind here, that a single workshop or lecture does not suffice. To acquire strategies for high quality interaction teachers as well as student teachers must participate in team meetings, practice in their own classes over a longer period of time and be coached during their practice. Investing in primary school teachers this way is a prerequisite and a guarantee for successful early S & T education.

NOTES

[xvi] For general aspects of learning theory and S & T, other chapters in this volume offer more comprehensive discussions.

[xvii] In the LIST Checklist the characteristics are arranged in five main categories: prerequisites, language input, active participation, quality of the content, and feedback.

[xviii] This design research was conducted by Resi Damhuis & Akke de Blauw for the National Centre of Language Education (Nijmegen), and Marja van Graft, Pierre Kemmers and Tjalling Brouwer for the SLO, the Netherlands institute for curriculum development (Enschede).

Resi Damhuis
Marnix Academie, ,Utrecht, The Netherlands
r.damhuis@hsmarnix.nl

Akke de Blauw,
Marnix Academie, Utrecht and Universiteit van Amsterdam, the Netherlands

LOTTE F. HENRICHS, PAUL P.M. LESEMAN, KEES BROEKHOF AND HANS COHEN DE LARA

16. KINDERGARTEN TALK ABOUT SCIENCE AND TECHNOLOGY

The situation preceding a teacher-directed intervention

In their daily lives, young children encounter numerous phenomena that are part of scientific domains, such as mathematics, physics and biology. By nature, children are curious and motivated to learn about the world and have strong intuitions about the nature of the phenomena they encounter. From infancy, children form visuo-spatial and sensorimotor representations of their personal experiences, and create mental ‘scripts’ or ‘models’ that include information about the temporal and causal structure of actions and events, its obligatory and optional components, and the associated roles (French, 2004; Nelson & et. al, 1986). This has most extensively been investigated in the realm of arithmetic and mathematics, where it has been shown for instance that 9-month-old infants possess a number sense, that is, an ability to discriminate numerosities, and to add and subtract, that develops and refines steadily over the early childhood years (McCrink & Wynn, 2004; Xu & Spelke, 2000). A major step in number sense development, as related to early math, is the integration of nonverbal ‘approximate’ intuitions with ‘exact’ language representations, starting with mapping count words onto nonverbal number representations (Ansari, 2009; Zur & Gelman, 2004). In the present study, we focus on children’s early understanding of physics, in particular the phenomena light reflection and air pressure. We presuppose that children as young as 5 years of age already have an intuitive understanding of these phenomena based in sensorimotor representations of previous experiences with these and similar phenomena. However, they still need to learn to map particular scientific words and expressions onto these representations in order to be able to articulate their insights in a conventional way and to share them with others. Sharing insights is thought to promote the development of more accurate, comprehensive and abstract scientific knowledge through processes of instruction and (peer) co-construction. Several studies in number sense and early math development indicate that teacher-guided age appropriate counting and math activities, in which language representations are systematically related to particular observable events, actions and outcomes, is essential for early math development (Klibanoff et al., 2006; Siegler & Ramani, 2008; Zur & Gelman, 2004). For instance, the amount of math talk in preschool classrooms strongly predicts growth in children’s mathematical ability (Klibanoff et al., 2006). We assume that for

Marc J. de Vries, Hanno van Keulen, Sylvia Peters, and Juliette Walma van der Molen (Eds.),
Professional development for primary teachers in science and technology.
The Dutch VTB-Pro project in an international perspective.

promoting young children's science understanding similar verbal support is needed. In this chapter, we will describe what conversational elements can be identified in science-related conversations between kindergarten teachers and a small group of 5-year-olds. We will discuss to what extent these conversations provide children with the opportunity to practice the verbalization of the sensorimotor intuitions they possess.

Cognitively challenging speech during elementary science interactions

Verbalisation of intuitive knowledge, thus, is considered important for further development of scientific knowledge and reasoning in young children. In turn, science activities in the classroom may stimulate a particular language use that matches the cognitive challenges of these activities. Gelman and Brenneman (Gelman & Brenneman, 2004) state that "science should be considered content for mathematics and literacy experiences" (p.159). In line with this view, we propose that science-related conversations provide excellent opportunities for children to practice the skills they need to use language as is expected of them in school, often referred to as 'academic language'. Academic language can be described as the kind of language use that is functional for displaying knowledge and to efficiently convey a cognitively complex message. Such displays of knowledge often concern topics beyond the here-and-now and lack the possibility of relying on shared situational context or material environment (Henrichs, 2010; Scheele et al., in press; Schleppegrell, 2004; Snow & Uccelli, 2009; Wong Fillmore & Snow, 2002). While 'doing science' (French, 2004; Yoon & Ariri Onchwari, 2006) children can still partly rely on the material environment, but are simultaneously challenged to find accurate wording for the phenomena they observe, which may not necessarily be directly perceivable (e.g. the effects of air pressure or reflection) (Spycher, 2009).

Greenfield et al. (Greenfield et al., 2009) distinguished eight 'process skills' that children are expected to learn before entering grade school. This distinction was based on an extensive review of 29 pre-kindergarten and kindergarten science standards and 10 early childhood curricula in the United States (c.f Greenfield et. al. p. 240). The resulting eight process skills were *observing*, *describing*, *comparing*, *questioning*, *predicting*, *experimenting*, *reflecting* and *cooperating*. Closely related to these process skills, is what Tenenbaum and Leaper (Tenenbaum & Leaper, 2003) call 'cognitively demanding speech'. Tenenbaum and Leaper, focus on three speech forms that they consider cognitively challenging: causal explanations, conceptual questions, and scientific vocabulary. The first two speech forms, causal explanations and conceptual questions, map perfectly on the categories as proposed by Greenfield et al. The final speech form, however, scientific vocabulary, is a new addition to the aforementioned categorization.

A relevant vocabulary classification is the three-tier concept proposed by Beck, McKeown and Kucan (Beck, Mckeown, & Kucan, 2002). In this categorization, words are distinguished based on their academic utility. High frequent, everyday words are referred to as tier-one words (e.g. table, sit, go). These words will be 'picked up' easily from the daily language environment. Tier-two words are high frequent for mature language users, and are generally more sophisticated than tier-one words. Tier-two words are typically useful across disciplinary areas (e.g. *result*, *experiment*, *describe*) and can be described as 'general school words'. These words are more likely to require explicit instruction in order to be learnt because they occur primarily in written language as opposed to the daily language environment. Finally, tier-three words are domain specific words that carry content meaning (e.g. *force*, *transparent*, *resistance*). As the term suggests, they are primarily relevant to a specific field or topic.

RESEARCH PURPOSE AND RESEARCH QUESTIONS

The data presented in the current chapter, form part of a larger randomized intervention study. In this intervention study, teachers will be trained in academic language and domain-specific technical vocabulary regarding reflection and air pressure. They will be taught about the theoretical background of academic language, and they will receive training on the possibilities of promoting academic language in their own students while they are involved in scientific reasoning.

In this chapter, we focus on the data collection wave preceding the training. It is considered important to obtain a view of teachers' natural way of interacting with their students during science-related activities, before providing them with training. The research purpose of the larger study is to ascertain whether teacher training can promote young children's use of academic language with science activities and whether more sophisticated language use with science activities results in deeper exploration and understanding of the phenomena at stake. In this first measurement, we will investigate the extent to which three different science-related activities elicit scientific reasoning skills, based on the conceptual frameworks of Tenenbaum and Leaper (2003) and Gelman and Brenneman (2004). The three science-related activities in focus involved familiar class room material on the one hand and new material brought along by the researcher on the other hand. The research questions we pose are: *a*) What is the distribution of scientific reasoning elements during science-related conversations, for both teachers and children? *b*) Can we identify differences in this distribution between an activity with familiar science-related material as opposed to activities with new science-related material?

METHOD

Participants

Fifty nine teachers participate in the over-all intervention study. A random sample of 30 teachers will be reported on in this chapter. On average, the teachers had 12.72 years of experience as a kindergarten[4] teacher (SD 10.99), ranging from 1 to 35 years. Ninety percent of the teachers taught a combination class of the

[4] The term kindergarten is officially no longer used in the Netherlands since 1986. Grade 1 and grade 2 of primary education correspond to the kindergarten years (age 4 – 6).

kindergarten grades 1 and 2 of primary school (with children in the 4 to 6 years age range). Ten percent of the teachers had a 2nd grade only class (with 5- to 6-year-olds). Most teachers worked part-time, with a mean of 3 days.

We contacted 90 schools in the area of Utrecht, the Netherlands, to recruit teachers. The student population of the schools included a maximum of 30 % children from ethnic minority background, to preclude effects of different conversation strategies to children with different language backgrounds (Aarts, Demir, & Henrichs, 2008). Thirty-one schools agreed to participate in the study (positive response rate 33%). The teachers in the current sample (N=30) teach at 17 different schools.

Procedure

Research assistants made video recordings of the teachers, while they interacted with a small group of four 5-year-old children during three science-related activities. The choice for the particular children was left at the teachers' discretion, the only criterion being that four children of approximately 5 years of age participated, with a minimum age of 4;6 and a maximum of 5;6. The teachers had received the materials needed for the activities well in advance of the day of the observation, accompanied by a basic explanation. They were allowed to keep the materials for their class rooms. We asked the teachers to conduct the activities as they would normally do during similar activities. Explicit instruction on how to conduct the activity or how to talk about what the children saw was not provided.

Science-related activities

The three science-related activities consisted of two new activities[5] and one control activity. The materials for the two new activities were new, whereas the material for the control activity was known and already present in the classroom. Below we will elaborate on the three activities.

Air pressure The first activity concerned a basic understanding of air pressure. Two syringes –without the needles– were joined together with a flexible rubber tube. Pushing one of the plungers down, results in the other plunger moving upwards, due to the air pressure. The second element of this task was a plastic frog, attached with a rubber tube to an air-filled plastic bulb. Squeezing the bulb makes the rubber legs of the frog unfold and jump.

Mirrors The second science task concerned reflection (mirroring). The teacher and the children were provided with a simple plastic periscope, and with two upstanding mirrors and a puppet. During the task the participants could find out that by means of the reflection in mirrors you can see what would otherwise be out of sight.

[5] The two new science related activities were inspired by the research program 'Curious Minds'. See http:www.talentenkracht.nl.

Control task: construction material At the very beginning of the research project, we surveyed the kinds of science practices that were common in the teachers' classrooms. We did this, in order to make an informed choice for a control task, relevant to the science domain, not introduced by us. The outcome of the survey suggested that construction material was a typical activity that teachers associated with science. The teachers pointed out that their pupils would typically build objects with moving parts, which reminded them of the technical domain. Consequently, we asked the teachers to guide the children in building an object of Knexx© (a construction material ubiquitous in Dutch kindergarten classrooms) that is able to 'drive and fly'. We provided this standard assignment in order to attain a minimal degree of comparability.

Data preparation and data analysis

The conversations during the three science-related activities were verbatim transcribed, according to the conventions posed by CHILDES (MacWhinney, 2000). First, we identified the 'on-task' utterances. Utterances that were considered procedural were excluded from further analyses. To this category belonged utterances in which turns were divided (e.g. *now Tom can try*) and utterances that were not related to the activity, for instance because they were addressed to another child or utterances serving to keep the order. The on-task utterances were subsequently coded for scientific reasoning, following a coding scheme. The coding scheme is presented in Table 1.

We distinguished between nine categories of conversational elements. Four of these categories were considered categories of academic scientific reasoning: *predicting*, *comparing*, *explaining* and *generalizing* (Greenfield et al., 2009). Within the categories, we distinguished between 'ask' and 'provide', e.g. whether an explanation was asked for or whether an explanation was provided. The data were analysed by means of ANOVA repeated measures, with 'activity' as the within-subjects variable, and 'scientific reasoning' as the dependent variable. We conducted a post-hoc procedure to identify the differences between the three activities with regard to scientific reasoning.

Table 1: Coding categories for on-task utterances

	Category		*Example*
Scientific reasoning	Naming	ask	Does anyone know what this is called?
		provide	This is called a syringe.
	Observing	ask	What happens?
		provide	The other one goes up!
	Predicting	ask	What will happen if I squeeze the bulb?
		provide	It will start walking.
	Comparing	ask	So, does this work in the same way as the periscope?
		provide	The frog moves because of the air, just like the syringes

Table 1 (continuing): Coding categories for on-task utterances

Category		*Example*
Explaining	ask	How is this possible?
	provide	Because the air pushes the plunger upwards.
Generalizing	ask	Can you grab air?
	provide	Air is transparent
Other (minors, feedback, repetitions)		Very good; I know; This is funny!

Secondly, we investigated the words that were used by both teachers and children to express their ideas during the science activities. First, we made a list of all words used in all transcripts. Subsequently, from this word list, tier-two and tier-three words were identified. The tier criteria were applied broadly, because of the very young age of the children. As a result, words such as *air*, *push*, and *press* were considered domain-specific technical words, thus belonging to tier-three, whereas in the original classification by Beck et al. these words might not have made it to the tier-three category. For our research purposes however, given we are working with such a young group of participants, relevance to the domain of the activity was the main criterion. Finally, we counted the occurrence of the identified tier-two and tier-three words in each individual transcript.

RESULTS

Scientific reasoning during the science-related tasks

Table 2. presents the descriptive statistics of the total number of utterances, the procedural utterances, and the resulting number of on-task utterances for both teachers and students. As can be seen, the rate of on-task utterances varied between 74.60% for the air pressure activity and 84.62% for the mirrors activity. Table 2 and Table 4 present the rates of scientific reasoning (predicting, comparing, explaining and generalizing) and the other on-task conversation elements (naming, observing and 'other') during the three science-related activities.

Table 2: Means and (standard deviations) for the number of on-task utterances (N=30)

	Utterances			
Teachers	*Total*	*Procedural*	*On-task*	*% On-task*
Construction	109.10 (37.04)	19.73 (11.44)	89.37	81.92 %
Air pressure	107.50 (38.24)	27.30 (14.20)	80.20	74.60 %
Mirrors	130.00 (38.11)	20.00 (12.66)	110.00	84.62 %
Students	*Total*	*Procedural*	*On-task*	*% On-task*
Construction	104.73 (57.60)	17.17(14.04)	87.57	83.61 %
Air pressure	80.50 (31.98)	18.60 (11.37)	61.90	76.89 %
Mirrors	110.97 (44.74)	15.07 (10.81)	95.90	86.42 %

A very large part of the conversations comprised utterances that belong to the category 'other' because they were literal repetitions, minors (e.g. 'yes', 'no', 'don't know'), short items of feedback, or utterances that otherwise carried little content. Indeed, the teachers very often repeated children's comments to make them accessible to the whole group. Furthermore, for all three activities, the element 'observation' occurred most frequently. It can thus be concluded, that for a large part, both the teachers and the students talk about the here-and-now and about the ongoing events they observe while being involved in the activity.

Table 3: 'Scientific reasoning' percentages for teachers (N=30)

Teachers	*Construction*	*Air pressure*	*Mirrors*
Naming	6.23%	7.87%	8.22%
ask	4.67%	6.67%	4.55%
provide	1.56%	1.20%	3.67%
Observe	19.64%	23.07%	34.94%
ask	12.06%	16.40%	26.61%
provide	7.58%	6.67%	8.34%
Predict	8.98%	11.17%	6.87%
ask	6.24%	9.81%	5.39%
provide	2.75%	1.36%	1.48%
Compare	0.52%	2.44%	2.37%
ask	0.29%	0.86%	0.76%
provide	0.22%	1.58%	1.61%
Explain	9.35%	16.40%	6.90%
ask	8.03%	12.29%	5.39%
provide	1.32%	4.10%	1.51%
Generalize	2.33%	1.89%	0.55%
ask	1.27%	0.93%	0.06%
provide	1.06%	0.96%	0.48%
Other [a]	52.69%	37.32%	62.19%

Note. [a] The category 'other' includes repetitions, feedback, minors etc.

It can be seen from Table 3 and Table 4 that for the mirrors activity in particular, the speech form *observe* was highly frequent. When children were looking through the periscope and looking in the mirrors, indeed they often kept repeating what they could see, as they found the images surprising. It appeared to be difficult for the teachers to scaffold the children towards reasoning *about* what they saw, for example by means of predicting, comparing or explaining. They were more likely to continue asking the children questions such as '*what do you see?*' thus promoting answers about observation.

Table 4: 'Scientific reasoning' percentages for students (N=30)

Students	Construction	Air pressure	Mirrors
Naming	9.29%	12.18%	6.96%
ask	1.30%	2.38%	0.83%
provide	7.99%	9.80%	6.12%
Observe	29.14%	23.37%	45.82%
ask	3.58%	2.02%	3.41%
provide	25.55%	21.36%	42.41%
Predict	17.56%	12.47%	4.75%
ask	2.06%	0.48%	0.22%
provide	15.50%	11.99%	4.52%
Compare	1.00%	2.78%	0.67%
ask	0.07%	0.52%	0.03%
provide	0.92%	2.26%	0.64%
Explain	7.50%	17.54%	4.81%
ask	2.66%	0.46%	0.45%
provide	4.84%	17.08%	4.36%
Generalize	3.49%	1.03%	0.22%
ask	0.36%	0.08%	0.00%
provide	3.13%	0.95%	0.22%
Other [a]	33.16%	30.41%	36.79%

Note. [a] *The category 'other' includes repetitions, feedback, minors etc.*

In general, teachers were more likely to *ask* for contributions belonging to the scientific reasoning category than to *provide* these. With the exception of the speech form comparing, teachers asked more often than they provided. In contrast, as can be expected, children were more likely to provide scientific reasoning in the form of answers than to ask for these.

A composite measure 'scientific reasoning' was created by pooling the elements 'predicting', 'comparing', 'explaining' and 'generalizing', relative to all on-task utterances (i.e. including the category 'other' but not the procedural utterances). For the measure 'scientific reasoning', we did not distinguish between 'ask' and 'provide'. Table 5 presents the means and standard deviations for this measure.

Table 5: Means and (standard deviations) for scientific reasoning by teachers and students

Scientific reasoning	*Construction*	*Air pressure*	*Mirrors*	
Teachers	21% (.08)	32% (.10)	14% (.06)	M < C < A
Children	30% (.10)	34% (.10)	10% (.06)	M < C = A

Note. C = Construction activity, A = Air pressure activity, R = Mirrors activity

We conducted ANOVA repeated measures analyses to investigate whether the three different activities yielded different rates of scientific reasoning. For the teachers, the results of the analysis revealed a main effect of task ($F(2,58) = 40.74$, $p < .001$, η^2_p=.584). Post hoc analyses showed that different rates of scientific reasoning occurred during the three activities. All these differences were significant. The percentage of scientific reasoning elements was lowest during the mirrors activity and occurred most often during the air pressure task.

For children, also a large main effect of task was found ($F(2,58) = 60.30$, $p < .001$, $\eta^2_p = .675$) The post hoc analyses revealed that this effect was mainly caused by the very low percentage of scientific reasoning during the mirrors activity. The difference between the construction material activity and the air pressure activity was not significant.

General school words and domain specific words during science talk

We explored the extent to which the teachers and the children used tier-two words and tier-three words while they talked about the science activities (Beck et al., 2002). In Table 6 the occurrence of these words is presented.

Table 6: Means (and standard deviations) of tier-two and tier-three words

	Construction	*Air pressure*	*Mirrors*
Teachers			
Tier 2 (general school words)	5.00 (4.38)	3.80 (3.38)	5.50 (3.17)
Tier 3 (domain-specific words)	33.73 (14.59)	26.47 (12.77)	49.10 (20.09)
Students			
Tier 2 (general school words)	1.60 (1.61)	0.67 (1.41)	1.23 (1.89)
Tier 3 (domain-specific words)	30.50 (18.81)	15.80 (6.41)	20.63 (11.45)

Note. The figures in this table are absolute numbers.

It is noteworthy that both teachers and children use very few tier-two words (general school words) when they talk about the science activities. Words from this list that emerged relatively frequently include *explain*, *idea*, and *example*. It is striking though, that words that seem so relevant to the acts of observing, predicting and explaining, such as *result*, *experiment*, *effect*, or *difference*, occur in so very few conversations.

Tier-three words (i.e. domain specific words) occurred more often, relative to the tier-two words. However, the number of tier-three words, presented in Table 6, correspond with approximately 4% of all words uttered by the teachers and 2% of all words uttered by the children. Table 7 presents some examples of tier-three words for the three different domains.

Table 7: Tier-three word examples

Construction	*Air pressure*	*Mirrors*
Air screw	Air	Binoculars
Drive	Blow	Direction
Engine	Distance	Mirror
Helicopter	Press	Opposite
Plane	Push	Periscope
Transport	Syringe	Slant

It should be noted that during the air pressure activity, the word *air* was highly frequent, though the word *air pressure* did not occur once. Similarly, during the mirrors activity, the word *mirror* not surprisingly was highly frequent, while for instance the word *reflection*, which would very accurately describe the phenomenon, occurred in only a very few occasions. The choice for assigning the tier-three label to the words *air* and *mirror* could be the topic of debate. As was mentioned earlier, we did choose to categorize these words as tier-three words for our research purposes, because we are dealing with such a young participant group.

CONCLUSION AND DISCUSSION

The data presented in the current chapter shed light on the natural discourse practices by teachers and kindergartners during science-related activities. By natural, we mean that to the best of our knowledge, we captured the conversations as they normally take place, without directing teachers towards particular language use.

The data show that science-related activities do elicit speech forms that are part of scientific reasoning, including observing, predicting, comparing, explaining and generalizing. The data also show, however, that the kinds of science-related material can make a difference for the degree of challenge during the activity. We asked whether new material would elicit more scientific reasoning that familiar material did. Our results show, that it is not the newness that made the difference. One of the new activities, the mirrors activity, yielded fewest scientific reasoning occurrences. The familiar material yielded slightly higher rates of scientific reasoning, and highest rates were found in the air pressure activity.

The participating teachers mentioned that they found the mirrors activity complicated. They felt insecure about their own knowledge of the task and therefore hesitated to involve the children in exploring the technology behind the activity. Similar findings have been reported in other studies as well (Lena, 2006; Peters & DeVries, July 2010). Teachers' insecurity seems to be an important factor in the quality of the conversation that evolves during science related activities. The other new activity, the air pressure activity, elicited significantly higher rates of scientific reasoning. Teachers felt more at ease with this activity, knew better what they wanted to hear from the children, and were therefore better able to scaffold the children towards high quality reasoning. Even though the construction material activity yielded a higher rate of scientific reasoning, it should be noted that the nature of these speech forms was, in general, considerably different from the scientific reasoning speech forms during the air pressure activity. This is due to the coding categories. During the construction material activity, the scientific reasoning speech forms *predicting* and *explaining* were the most frequent for both teachers and children. For this activity, utterances where teachers and children talked about the conditions that were necessary for the object to drive and fly were coded as predictions. This was plausible, because similar to making a prediction, one talks about a hypothetical issue, not present in the here-and-now. In the air pressure task and the mirrors task, the category of prediction was more straightforward (e.g. *what do you think will happen if I press down this plunger?*).

Similarly, utterances in which the teacher asked the children how they were going to proceed to build the object, were coded as explanation. We did this, because in this way children were prompted to account for their actions and verbalize why a particular action was called for. Even though we argue that the coding as we applied it, is plausible, a problem of comparability did arise. However, holding on to our research purpose of sketching the situation as it is in Dutch class rooms, we needed to include a science-related activity familiar to all teachers and all children. In the survey preceding the study, some teachers filled out that they engaged in 'true' scientific activities every now-and-then, such as discussing the theme of 'float and sink' or assembling an 'autumn table', but if these topics occurred at all, it was only occasionally and it was always within the frame of a particular classroom theme. Thus, our choice for construction material was the most logical, though it did not result in optimal comparability between the activities.

Our results show, that when teachers talk with kindergartners about science-related activities, they do so in relatively general terms. General school words do not occur frequently, and the range of domain specific words is small (the same words are repeated very often). It thus seems that there is ample room for improvement of the conversations being held.

The teachers and the children thoroughly enjoyed working with the material. We suggest capitalizing on this enthusiasm by explicating the affordances of science-related for academic language learning. We argue that with a relatively light intervention, the quality of the conversations could be boosted considerably. We expect that by providing teachers with some basic knowledge of academic language and the affordances of science-related activities to promote academic language skills (of both the teachers themselves and children) the activities will become much more challenging and fruitful. In addition, teachers should be provided with some content knowledge of the scientific concepts involved as well, to overcome the insecurity they experience in the technology realm.

Lotte Henrichs
L.F.Henrichs@uu.nl

Paul Leseman
P.P.M.Leseman@uu.nl
Utrecht University, Department of Education

Kees Broekhof
Hans Cohen de Lara
Sardes, Utrecht

PART IV

IMPACTS OF PROFESSIONAL DEVELOPMENT

SYLVIA A. F. PETERS

17. INTRODUCTION TO PART IV

Impacts of professional development

INTRODUCTION

Role of this part in the book

Primary science education should be an essential aspect of development of education, like reading, writing and mathematics were in the last two centuries (Lena, 2006, 2009). This view has led to new developments worldwide, based on inquiry pedagogy, for example in large EU programmes in Europe but also in China, Brazil and the United States (Lena, 2006, 2009). The results of these new approaches also led to new teacher training concepts and insights. In this section an introduction of factors with an impact on professional development is presented starting with general findings on professional development and reform movements and followed by a focus on findings of research on science and technology teacher education.

Teachers are central to school improvement efforts. They are the most significant and costly resource in schools. Considering the changes and reform movements in science and technology teaching during the last decades, the pace of change has been such that developing and learning science and technology teaching has become a major issue internationally (Harlen, 2009). The question is how to engage teachers in all new developments in the context of continuing professional development. And which types of professional development activities engaged in are effective especially with respect to science & technology? It is generally known that the way science & technology is taught in schools is not effective enough. It is necessary to focus on building greater capacity and stability (Lena, 2006). Therefore most European countries promote a renewal of science education through 'inquiry based' methods and teachers being part of a network (EC, 2007).

FACTORS CONTRIBUTING TO EFFECTIVE PROFESSIONAL DEVELOPMENT

Professional development is defined as those activities that develop teachers' skills, knowledge, expertise and other characteristics (OECD, 2009). This development can range from formal to informal, in the form of courses, workshops or formal qualification programmes through collaboration in networks across

Marc J. de Vries, Hanno van Keulen, Sylvia Peters, and Juliette Walma van der Molen (Eds.), Professional development for primary teachers in science and technology. The Dutch VTB-Pro project in an international perspective.

schools or within schools, sharing their good practices. Successful programmes are programmes focused on the development of teachers' learning communities. Developing schools as learning organizations and ways to share expertise and experience systematically seem to be effective too (OECD, 2005).

Teacher quality

Fullan (1993) states that quality of professional development is the key for educational improvement. The OECD emphasizes the need for high quality teaching and the ability for all students to have access to high quality of their teaching (OECD 2005; Education Council of the Netherlands, 2008). Teacher quality is related to teaching experience, qualifications, experience and indicators of academic ability or subject-matter knowledge. A large programme focused on the content students were to learn, active learning and follow-up to enhance teacher quality had an impact on content focus, active learning, and follow-up and follow-up on knowledge and professional community (Ingvarson, Meiers & Beavis, 2005).Top performing nations even have a strategy with highly selective training programmes so they recruit top talent in its teaching profession (McKinsey & Company, 2010). Furthermore, research shows convincingly that reaching teaching quality improves learning of students (OECD, 2005; 2009; Dagevos, Gijsberts & Van Praag, 2003; McKinsey & Company, 2007).

Teachers' general perceptions on professional development

Almost all teachers engage in informal dialogue with others to improve their teaching and the vast majority reads professional literature. However, in general about 11% of teachers does not take part in any of the structured forms of professional development for 18 months (OECD, 2009). In general, teachers report a moderate or high impact for the types of professional development they had undertaken. The most effective types of development were *Individual and collaborative research*, *informal dialogue to improve teaching* and *qualification programmes.* It is striking that up to 90% of the teachers report a moderate tot large impact of these types on their development as a teacher(OECD, 2009). As teachers' perceptions and views are important factors believed to influence their behaviour this view has to be involved in the context of developing continuing professional development. Furthermore, regarding the activities having the highest impact in the view of teachers, it has been found that paying for activities and their investment in time enlarges the impact of professional development activities (OECD, 2009).They also need extra individual support. Strengthening the system of teacher appraisal and feedback can develop subsequent changes in their teaching within schools, according to teachers reports. Mostly teachers have few incentives to improve their teaching. Professional development, especially in the case of Science and Technology, might be effective when they share stable professional relationships with other teachers, such as networks for teacher development and mentoring (OECD, 2009). Networks can be used as effective components of teachers' professional development compared with more traditional forms of in-service teacher training and stimulate morale and motivation (Walma van der Molen, De Lange & Kok, (2008).

The role of leadership

Instructional leadership is essential for school reform, with leaders focusing on improving instruction (Fullan, 2010; McKinsey & Company, 2007; OECD, 2009).The quality of professional development is determined significantly by strong leadership (OECD, 2009). Leaders facilitate teachers for training and influence teachers and their work by appraisal and feedback. This may have a strong positive influence on teachers.

Online professional development

To build capacity for teacher change, online teacher learning can be promising, especially in the field of science and technology. However, little is known about best practices for the design and implementation of these models. Evidence on teacher development, student outcomes and long term effects are often lacking (Dede, Ketelhut, Whitehouse, Breit & McCloskey, 2009). Also, teacher purposes and needs are needed to develop effective programmes. A mentoring programme for science teachers, using scientists and experienced science teachers to provide mentoring services in an online environment, showed that teachers mainly used the programme mentors for advice and resources, not to increase their content knowledge or teaching skills (Jaffe, Moir, Swanson, & Wheeler, 2006). One of the successful approaches for enhancing teacher training is a case-based learning environment (Kinzer, Cammack, Labbo, Teale & Sanny, 2006; Verhoeven & Kinzer, 2008). It presents authentic multi-media lessons in the classroom presenting a rich environment to learn compared with print-bases cases. Other examples of online teaching are OPD workshops having a strong impact on instructional practices (O'Dwyer& Masters et al. (2010). One of the effective aspects is the flexibility of the digital curriculum (such as ability to adjust text, navigation, features, colors, pace, order, etc).

FACTORSCONTRIBUTING TO EFFECTIVE PROFESSIONAL DEVELOPMENT FOR PRIMARY SCIENCE AND TECHNOLOGY EDUCATIONSPECIFICALLY

EC (2007) stressed the widespread idea that there is a need for professional development in teaching science leading to a reform movement in science teaching. Research suggests strongly that there is a connection between attitudes towards science and technology and the quality of science and technology teaching as is already noticed earlier in this book (EC, 2007). Teachers at primary schools often lack effective skills and self-confidence and motivation (Harlen & Holroyd, 1997, Lena, 2006). Furthermore, science and technology subjects are often taught in a formal and traditional, much too abstract way which can have a negative impact on the attitude towards science and technology (Jarvis & Pell, 2004). As Harlen & Lena state in the introduction of this book, teachers also seem to have less understanding of natural sciences and they view science teaching as ensuring students to know and memorize these facts. By changing the science-teaching pedagogy, a reversal of school science-teaching was introduced to increase interest in science and technology by using inquiry-bases methods (EC, 2007). The pedagogical practice of inquiry based methods are proved to be more effective in increasing the interest and attainments of students, from the weakest tot the most able. This also leads to an increase in teacher motivation. Being part of a network allows them to improve the quality of their teaching.

Pedagogical content knowledge and attitude

The teaching profession especially in the context of science and technology is a learning profession. Over the past 10 years teacher confidence developed increased reasonable in primary science in the UK, but it still is critical (Murphy, Neil & Beggs, 2007). Teachers also report the ability to teach science as a major issue of concern in primary science. One of the key aspects of professional development activities focused on science and technology is aiming at building confidence and modifying teachers' view (Murphy, Neil &Beggs, 2007, Introduction of Harlen & Lena in this publication, 2011).Teacher training focused on skills and knowledge of science and technology teaching has positive effects on science- practice at schools and on the outcome and attitude of students (Osborne & Dillon, 2008). Science knowledge is a very significant factor that influences primary teachers'confidence in teaching science (Harlen & Holroyd, 1997).The study of Supovitz & Turner (2000) on professional development based upon intensive and sustained training around concrete tasks indicates that inquiry-based teaching has a powerful effect on both inquiry-based practice and investigative classroom culture. Their study focused on subject-matter knowledge, connected to specific standards for student performance, and embedded in a systemic context. The Primary Connections approach, including inquiry and investigative approach, also showed improvements in teachers'confidence, self-efficacy and practice, students' outcome and the status of science within schools (Hackling, Peers & Prain, 2007). Jarvis & Pell (2004) studied the attitudes and cognition of an in-service programme and their effect on pupils. It showed that in-service professional development activities had positive outcomes with regard to teachers'confidence, self-perception of ability, attitudes to managing science in the classroom and science understanding. There were gains in science understanding of concepts. The study indicated a relation between change in attitudes, skills and knowledge of the teachers and the student attitude and attainments.

CPD and co-teaching

The effect of CPD (continuous professional development) on teachers'classroom activities in science was studied by Bennet, Braund, & Lubben, (2010). They found that classroom impact is affected by teacher characteristics as professional enthusiasm to change. The impact increases when resources, ideas, strategies and policies promoted at the course link closely to practice in the school. Effective CPD integrates small changes into existing practice. Co-teaching is two or more teachers teaching together. In recent years co-teaching has become an important element of science teaching. Teachers share the responsibility for planning, teaching and evaluating lessons together, working as collaborators in classroom practice. Research indicates that it can be very effective for both teachers and students (Murphy & Scantlebury, 2010). Murphy and Scantlebury (2010) studied a blended CPD approach and co-teaching with teachers and student teachers having equal roles in science teaching. By teaching together, teachers and student teachers shared the CPD. It was found that co-teaching science in primary schools, enables student teachers to bring scientific expertise to the classroom, which can be shared with the pedagogical expertise of the classroom teacher to improve classroom practice. It can also improve the confidence of science teaching. The findings showed CPD to be effective and sustainable.

Pre-service training and primary science and technology

Higher education institutions need to enhance the preparation of new primary teachers to ensure that they are all confident and effective teachers of science. They could also increase their partnership work with schools and other CPD providers in relation to primary science (Murphy and Scantlebury, 2010). In the Netherlands pre-service training of primary science and technology teaching has been developed under the influence of the Lisbon Strategy and was emphasized under the influence of the programme VTB-PRO during the last four years. In a relatively short time the colleges implementing a science and technology inquiry-based programme in their curriculum showed a clear policy for science and technology and a continuing professional development programme through all years of pre-service training. But there is still concern about the confidence, the interest and effective teaching primary science and technology of the pre-service teachers. The schools of teacher training have the ambition to further increase their regional partnerships with schools, science clubs and industries and technical higher institutions and universities (Kat & Van der Neut, (2010).

CHAPTERS IN THIS PART

This part IV provides insight into the impact of the design of professional development and focuses on essential issues of professional development in science and technology. The research of Uum & Gravemeijer presents the results of programme development through active involvement of teachers and experience in assignments based to develop pedagogical content knowledge and attitude. They suggest that it is important for teachers to experience changes in their thinking, but also to change their teaching practice accordingly and to engage in practical activities.The study of Sjoer & Meirink goes into an analysis of teachers'exchanges of their experiences with science and technology education. The experiment is focused upon the collaboration between teachers developing their own science and technology curriculum for instructional classroom practice.The study of Segers, Peeters, Strating & Verhoeven is focused on a digital learning environment to enhance knowledge and attitude of student teachers. As this has a large impact on the attitude of students is shows, with other research, that it is very promising to integrate a digital learning environment in the curriculum of teacher trainingstudents. Their perceived competence improves.

In summary, the studies in this part describe the impact of different factors contributing to the professional development of primary science and technology education. It can be concluded from these studies that science and technology training has to be based upon practical experiences and the exchange of these experiences in the context of a theoretical framework. Furthermore, a digital environment with different role models and situations can enhance their knowledge of science and technology and their teaching attitude and. The practical implications that can be derived from the studies are that informal exchange, involvement in practical activities and the availability of role models in a digital environment as part of teacher training areeffective factor tochange the teacher practice of teachers.

Sylvia Peters
Kennisnet, the Netherlands
s.peters@kennisnet.nl

MARTINA VAN UUM AND KOENO GRAVEMEIJER

18. CHARACTERISTICS OF EFFECTIVE PROFESSIONAL DEVELOPMENT FOR PRIMARY SCIENCE AND TECHNOLOGY

ABSTRACT

The professionalization program VTB-Pro (elaborating technology in primary education, professional) was founded in the year 2007 and enabled 5000 teachers and 5000 student teachers to professionalize themselves in the field of science and technology. This chapter describes the investigation of three professionalization programs developed by the knowledge centre for science and technology in the southern part of the Netherlands. The participants report that the investigated professional development programs for primary school teachers contributed to positive changes, in particularly in pedagogical content knowledge and attitude. The element that emerges from different investigation methods as the one that contributes the most to these changes is the participants' own activity in assignments that took place during the course meetings. Program developers are advised to stay close to the teaching practice of participants and let them engage in practical experiences as much as possible. Key words: Science and technology, professional development teachers, professional learning teachers, teachers primary school

INTRODUCTION

Because of expected shortages of scientists the European Council aims to achieve 15% more student outflow in higher scientific and technological education in the year 2010 (Council for education, Youth matter and Culture, 2007). Initiatives regarding this aim in the Netherlands are provided for by the Platform Bèta Techniek, founded in 2004 (Deltaplan Bèta/Techniek, 2003a). Within the platform different programs have been launched including the professionalization program VTB-Pro. This program started in 2007 to achieve professionalization in science and technology of 5000 teachers and 5000 student teachers. Within regional science and technology knowledge centres professionalization programs for (student) teachers have been developed that are carried out by elementary teacher education colleges (pabo's) (Kuijpers & Walma van der Molen, 2008).

The professionalization programs of the science and technology knowledge centre in the southern part of the Netherlands are offered in three regions, in this chapter indicated by A, B and C. In each region an in-service teacher education course, that was part of the professionalization program, was attended and

Marc J. de Vries, Hanno van Keulen, Sylvia Peters, and Juliette Walma van der Molen (Eds.),
Professional development for primary teachers in science and technology.
The Dutch VTB-Pro project in an international perspective.

investigated. Each course encompassed six day parts and a similar amount of study time. The objective of the courses was to foster the teachers subject matter knowledge, attitude, and pedagogical content knowledge in the field of science and technology (S & T).

CONCEPTUAL FRAMEWORK

To realize an increased outflow of S & T students in higher education, an increased intake is needed. Choices in education are made at a young age in the Netherlands. To increase the number of students that choose programs with a stronger emphasis on S & T, already in primary education a positive attitude of students should be fostered. Unfortunately, however, only 12% of the primary schools provide structured technology education (Inspection of Education, 2005-3). To enhance this percentage there are three known domains of teacher knowledge for S & T that can be attended to: subject matter knowledge, pedagogical content knowledge and attitude (Rohaan, Taconis and Jochems, 2008; Walma van der Molen, de Lange, & Kok, 2009). Primary schools cannot improve these domains by only purchasing S & T support materials: professional development is needed (Hagunama, 2008).

There are different ways to achieve professional development. Prebble, Hargraves, Leach, Naidoo, Suddaby, and Zepke (2004) distinguish short courses (for institutional information or training skills and techniques); professional development within working groups (to enhance more complex knowledge and skills); peer assessment and coaching; use of student evaluation of teaching (to improve teaching); and intensive study programs (to change teaching beliefs and practice). This implies that program developers should be aware of the goals they want to achieve and use the approach that is best suited for these goals.

Effective learning environments stimulate learners to retrieve previous knowledge on the basis of which new knowledge can be constructed. Feedback on learning is important and should match the goals that learners want to achieve. Learning environments are more effective when participants feel part of the learning community and are motivated to learn. Successful in-service teacher training courses can be realised in learning communities in which teachers can share their experiences, in programs that extend over a longer period of time, and by means of practical activities that teachers can also use while teaching their students (Brown, Bransford & Cocking, 1999).

When focussing more on the subjects of S & T some exemplary programs and models can be found. Stein, Ginns & McDonald (2007) developed a professional development model for technology education, which showed that three elements (institutional knowledge, pedagogical knowledge, and field/ discipline knowledge) can contribute to the development of personal constructs via enhancement, development, and reflection on experiences that are theoretical, practical, and reflective of nature. By developing personal constructs teachers' knowledge and beliefs regarding technology can improve and practices can be changed. This implies that professional development programs for teachers regarding technology should focus on the development of personal constructs (using previous knowledge to construct new knowledge and skills). Teacher educators should focus on subject matter knowledge, pedagogical knowledge and institutional knowledge.

To enhance teachers' confidence and improve student learning outcomes in science and literacies of science in Australia the professional learning program 'Primary Connections' was developed. Teacher professional learning consisted of: professional learning workshops, exemplary curriculum resources, opportunity to practice science teaching supported with resources, reflections on practice, and principles of learning and teaching. The teaching and learning model (consisting of the phases: engage, explore, explain, elaborate, and evaluate) is closely linked to the constructivist learning theory, according to which each student constructs and elaborates his own knowledge. The project contributed to enhanced teachers' confidence in teaching science and more time spent on the subject. Most teachers appreciated the curriculum resources, found them compatible with the rest of the curriculum and preferred to receive more units. Students were enthusiastic about the program and their learning outcomes increased (Hackling & Prain, 2005). Program developers can use elements of this successful professional learning program when developing a course for primary school teachers.

The professional learning programs investigated in this chapter are based on the goals of VTB-Pro, discerned in three so-called pillars:

> 1. Knowledge of scientific and technological concepts and skill in scientific and technologic reasoning; 2. attitude towards S & T; 3. pedagogical content knowledge, especially in the field of inquiry and design based learning (Walma van der Molen, et al., 2009, p. 30).

These three pillars are taken as a framework of reference for analysing the S & T courses. To gain insight in the working components of the investigated courses and the assumptions at the foundation of these courses, a qualitative evaluation of the learning process outcomes was undertaken. The central question of this chapter is: which components in professionalization programs for primary education teachers are perceived by the teachers themselves as contributing to changes in domains of teacher knowledge regarding S & T?

RESEARCH DESIGN AND METHODS

Participants

Region A: Thirteen individual participants of different elementary schools (some of them in pairs of the same school).
Region B: Fourteen participants of a larger team of elementary school teachers.
Region C: Nineteen participants (a complete elementary school team of teachers).

Teacher Educators And Their Goals

To gain insight in the rationales for the courses the teacher educators were interviewed about their goals for the course and their instruction theories. The answers are summarised in the following.

Region A: In this region the course was carried out by two teacher educators. The goals of the first teacher educator are displayed. The teacher educator wants to show that teaching S & T is not difficult and aims to enhance participants' attitude and enthusiasm, because they have to inspire colleagues in their own schools. Participants learn by doing. The most effects can be achieved with practical ideas and activities that participants can use directly in their practice.

Region B: The teacher educator especially aims at improving the participants' attitude and pedagogical content knowledge. Elementary school teachers should be aware that S & T is very important in today's society, be flexible with their classroom methods, and engage children in inquiry and design based activities. Participants learn by: collecting information themselves; engaging in activities; and explaining and presenting what they have learned to fellow participants. The easier the activities during the course, the more participants appreciate them and the sooner they practice them in their teaching.

Region C: Participants should become aware of what they already know and do regarding S & T in their classrooms; work more from teaching goals; combine S & T with other subjects; and engage children in inquiry and design based activities. The course contents can be used directly in the teaching practice. Teacher educators should respond to participants' signals and adjust the course when needed. The teacher educator is content when participants structurally implement S & T in their curriculum.

Content Description

For teacher educators of the VTB-Pro professionalization programs in the southern part of the Netherlands a compilation of course contents is available. Teacher educators can decide themselves which content is best suited for the course at hand. The content of the course meetings of the investigated courses is described.

Region A: S & T in general, and domains and goals were discussed. A number of activities were performed by participants to inspire them and to show that S & T education does not have to be difficult (for example by making a paper tower). Talents of children were discussed after seeing short videos and multiple intelligence was presented in an activity with choice possibilities. Inquiry and design based learning and the story line approach were addressed. Participants investigated how to: improve an existing product; combine products to design new ones (only thought process); and learn how to use stories in S & T education. Integration possibilities of S & T with other subjects were explored by investigating example lessons. A fellow participants' school was visited to learn about implementing S & T in the curriculum. At the regional VTB support centre S & T learning materials were displayed and recommendations regarding purchasing materials were given.

Region B: Information about S & T, and inquiry and design based learning was presented and participants designed a flying object. Ways of observation, asking the right questions and coaching children at their own level of development were discussed after watching short films, and were practiced while participants designed a S & T lesson for their own class. Knowledge on design based learning was increased and possibilities to integrate different curriculum subjects were explored by designing a house situated in an assigned location and time period. Knowledge on inquiry based learning (also the importance of stimulating curiosity) was increased by watching a number of experiments guided by the teacher educator and composing a lesson in which children could learn certain science concepts. In a final activity the learned knowledge and skills were applied by constructing a chain reaction.

Region C: S & T were discussed and technological artifacts were investigated. Different products were constructed, including a propeller boat, a doormat alarm and flying objects. News articles about S & T were investigated. Talents of children and ways of asking questions were discussed after watching short videos and practiced in some experiments (such as floating and sinking). To focus on learning goals participants combined the purchased school technology materials with a lesson from a method they used in their class. Integration of S & T with language was addressed by choosing between: again combining the school materials with a class lesson, or focusing on the storyline approach to compose a S & T lesson. The integration with mathematics was addressed by focusing on context and constructing a weight balance. Other school materials were investigated and weather instruments and periscopes were made. Finally participants discussed several issues to achieve more structural implementation of the innovation regarding S & T.

Methods

To obtain valid and reliable data regarding perceived positive changes in domains of teacher knowledge and components of the course meetings that contribute to these changes, a variation of methods and collected data were used (multi-method and data triangulation) (see for example Miles & Huberman, 1994; Meijer, Verloop & Beijaard, 2002).

Semi-structured interviews were conducted with the teacher educators after each course meeting and at the end of the course regarding the whole course, to collect data about the components of the meetings and the teacher educators' perceptions on positive changes of participants while executing the different components.

At the end of each meeting and after the whole course participants filled out questionnaires (open questions) to indicate most inspiring and informative components and reasons why this was the case. After the course they also filled out a questionnaire (Likert-scale) regarding these topics.

During the course unstructured observations were conducted by the researcher and audiotapes were made to collect utterances of the participants regarding changes in knowledge, attitude or pedagogy.

Data collected by means of open questions and the recorded data were analysed in accordance with the grounded theory approach of Glaser and Strauss (1967). Starting from the data themselves and by means of open coding and labelling, concepts, categories and relations were obtained. The different steps in this process are accurately documented and can be repeated.

RESULTS

Perceived Positive Changes In Domains Of Teacher Knowledge

To measure participants' perceived positive changes in domains of teacher knowledge a questionnaire (with open questions) was filled out by nine participants from course A, fourteen from course B, and fourteen from course C. Explanations for changes given by two or more participants are presented.

Participants in courses A and C all perceived positive changes in subject matter knowledge. Four participants in course B did not experience positive changes, two of which already had a technology background. When analyzing the three development courses it turned out that participants hardly mentioned changes in their own knowledge. Instead, they often made remarks regarding teaching S & T in their classrooms and knowing more about content possibilities, teaching goals, and strategies. They appreciated the practical examples during the course meetings and the discussion of experiences.

In all three courses most participants perceived positive changes in attitude: being more enthusiastic and less reserved about teaching S & T, and seeing more possibilities to teach these subjects, for example when combining them with other subjects. Also they mentioned being more aware of what they already were practicing in their classrooms regarding these subjects. One participants' attitude was not very positive at the start of the course: 'Another subject to attend to', but afterwards she wrote: 'This is fun, when can we start?!' Some participants did not perceive positive changes in attitude, because their attitude was already positive. Only a few did not perceive positive changes in attitude.

In pedagogical content knowledge almost all participants perceived positive changes. Only one participant did not perceive changes, because she already had pedagogical content knowledge at the start of the course. Participants gained more insights into coaching pupils and ways of stimulating them to investigate, were more aware of S & T in the classroom, and knew more about contents, ideas and strategies to use in their teaching practice.

Contribution Of Different Parts Of The MeetingsTo Changes In The VTB-Pro Pillars

The three courses consisted of similar parts: homework activities (to be executed in the participants own classes); discussion about the executed homework activities; the presented theory; the practical activities; and presentations of practical activities. Participants assessed on a 5-point Likert scale to what degree the different parts of the meetings contributed to changes in the VTB-Pro pillars (1 = no contribution to changes, 5 = total contribution to changes). Here two components have been added: the possibility to discuss with fellow participants (which was often mentioned in the course reflection forms) and the instruction and coaching by the teacher educator (on basis of the literature; see for example ESoE, 2009). Because the results of the three courses were mostly similar, they have been combined and are displayed in table 1.

*Table 1. Assessment of perceived changes of each part of the course meetings**

		ATT	*SMK*	*PCK*	*T*
Part of course meeting**					
Homework activities (executed in participants' own class)	M	4,0	3,7	4,0	3,9
	ST	0,9	0,9	0,7	
Discussion about the executed homework activities	M	3,9	3,7	3,9	3,8
	ST	0,9	0,9	0,8	
The presented theory (before executing practical activities)	M	3,9	3,9	3,9	3,9
	ST	0,9	0,7	0,8	
The practical activities	M	4,3	4,0	4,1	4,1
	ST	0,6	0,7	0,6	
Presentations of practical activities	M	4,0	3,7	3,8	3,8
	ST	0,8	0,7	0,7	
The possibility to discuss with fellow participants	M	4,1	3,8	4,0	4,0
	ST	0,7	0,8	0,7	
Instruction and coaching by the teacher educator	M	4,4	4,2	4,3	4,3
	ST	0,8	0,8	0,8	

*The number of participants that filled out the questionnaire is 36. There are some missing variables, because a few participants didn't fill out these sections (reason unknown). Homework activities: attitude (1 missing), subject matter knowledge (2 missing), pedagogical content knowledge (2 missing). Discussion about the executed homework activities (each pillar 1 missing).

**M = mean, ST = standard deviation, ATT = attitude, SMK = subject matter knowledge, PCK = pedagogical content knowledge, T = total.

Table 1 displays that the categories 'practical activities', 'the possibility to discuss with fellow participants', and 'instruction and coaching by the teacher educator' showed high results. To determine if significant differences could be found some tests were executed. A Kolmokorov Smirnov test showed that all variables were normally distributed except the three mentioned before with highest results. The non-parametrical Independent Samples Kruskal Wallis test showed a significant difference between variables (p=0,004). Some Mann-Whitney post-hoc tests were executed to investigate if significant differences between each of the three variables with highest scores and other variables could be found. The variable 'instruction and coaching by the teacher educator' showed significant differences with all other variables except the variable 'practical activities' (homework activities: p=0,005; discussion of homework activities: p=0,02; theory: p=0,006; presentations of practical activities: p=0,000; discussion with fellow participants:

p=0,018). The variable 'practical activities' showed a significant difference with the variable 'presentations of practical activities' (p=0,021). The variable 'discussion with fellow participants' showed a significant difference with 'instruction and coaching by the teacher educator' (p=0,018). These results suggest that positive changes are mainly caused by 'instruction and coaching by the teacher educator' and 'practical activities'.

Appreciation Of Most Inspiring And Informative Parts Of The Course Using Open Questions

In a second questionnaire at the end of the development project participants were asked to indicate most inspiring and informative parts of the course meetings. Because the answers to both questions were similar, they were combined in the presentation of results. In all investigated courses participants were most inspired and had learned most from activities that they executed during course meetings. As reasons they indicated: getting ideas to use in their classroom; the appreciation of being active themselves; and experiencing what pupils go through when executing an assignment. In regions B and C also appreciation for combining subjects was mentioned and in regions A and C the possibility to communicate with colleagues.

Appreciation Of Parts Of The Course Meetings By Means Of Reflections

At the end of each course meeting participants were asked to fill out a reflection form. In a series of open questions they wrote down what parts of the meeting they found most inspiring and informative. Course teachers were interviewed after each meeting about the same topics.

The written reflections show that participants in all three regions were most inspired and learned most from the practical activities during the meetings. The teacher educators also all stressed the importance of practical activities when asked about most inspiring and informative meeting components.

A Closer Look At Practical Activities

In each course meeting participants were given the opportunity to engage in practical activities. During these activities they for example performed inquiry regarding scientific subjects, designed a technical product, or created a lesson for their pupils in which science or technology were combined with another subject.

To give more insight into practical activities, one activity is discussed more extensively: first by presenting some of participants' quotes, and second by concluding what can be learned from this activity.

Insights into the activity Participants (of region B) were divided into groups and were asked to design a flying object. In one of the groups the participants started to design a paper airplane. After making a prototype the participants were disappointed about the originality of their design.

Participant 1: *'It's just becoming a traditional airplane. With rope.'* Participant 2: *'But do you think that that's fun?'* Participant 1: *'Children are already doing that in the classroom.'* A moment later participant 3 explained to the course teacher: *'We think he's flying all right but we don't find him very original.'*
The teacher educator started to ask questions: *'Let's see?'* (a test follows): *'Yes, he's going down. Can you also make him go up? (...) What makes a plain go up?'* Participant 3: *'The pilot behind the steering wheel.'* Teacher educator: *'And what is this pilot doing?'* Participant 3: *'He's pulling that steering wheel.'* Teacher educator: *'And what happens with that steering wheel?'* Participant 3: *'Then the point is going up.'* (…). Teacher educator: *'How does the steering wheel make the point go up?'* Participant 2: *'With the wings.'* Participant 3: *'Yes, on the wings there are flaps.'* Teacher educator: *'Ok, so these flaps probably are useful when flying.'* Participant 3: *'So they need to be on here too'* (the paper airplane). The teacher educator emphasized that he was able to coach the participants by only asking questions.

The participants decided to make a little flap at the rear of the airplane. But they did not know how to shape the flap. Participant 1: *'Does the flap need to be like this or...?'* Teacher educator: *'Straight. What would happen when it's like this?' (...)* Participant 3: *'I think now he becomes instable, (...) so down.*

The participants tried out different possibilities and tried to come up with explanations for what they were seeing. Participant 1: *'He's going up.'* Participant 2: *'He's not braking anymore.'* Participant 3: *'So he needs to be less up,* (…), *more aerodynamic.'*

Because the teacher educator stimulated the participants to think about the functions of an actual airplane, they kept referring back to it. Participant 1: *'It's only a little flap that is going up on the wings, on an actual airplane.'*

After further tests with going up and down, also movements to left and right were tried out. The teacher educator stimulated the participants to first predict what was going to happen and then to try out what actually took place.

There were again comments about the design. Participant 3: *'I think it's a spectacular thing we designed.'* Participant 1: *'Yes, I think so too.'* And a moment later participant 2: *'Ok, you can do a lot with it already. I think it's more of an experiment airplane.'*

The participants now made predictions themselves and then tried out what actually happened. Participant 3: *'So now the flap is left and this goes flat for a moment. Because now it's not about up and down, but that he's able to go left or right. So now he should go left.'* Another test follows. The responses: *'Yeah, yes!'* Participant 2: *'This is really fun!'*

The participants were asked to present their designed product to fellow participants. The presenting participant showed he learned by example when asking: *'Now the flaps are down, what would happen?'* A response: *'Then he's going to turn.'* Participant: *'We discovered so much.'*

Conclusion Participants progressed at all three VTB-Pro pillars. Their attitude first was rather negative, but turned positive after being asked the right questions by the teacher educator to deepen the learning process and after having experienced successes. They also experienced that investigating in, and understanding S & T can be fun and informative. By experiencing what children go through when executing a task like this, and by modeling the way of asking questions by the teacher educator, participants enhanced their pedagogical-didactical knowledge and skills. Because the teacher educator made sure participants retrieved knowledge regarding actual airplanes, and by improving the design, participants could improve their own subject matter knowledge.

Taking into account the unstructured observations from other activities it turned out that in all three regions teacher educators and participants focused mostly on pedagogical content knowledge and attitude. By participating in the activities participants could experience what children go through when executing an activity, and in addition it made it possible for them to use the activity directly in the classroom.

Interviews With Teacher Educators After Finishing The Courses

After the last course meeting each teacher educator was interviewed and asked to provide his thoughts and views regarding the course at hand. Each interview consisted of questions concerning the degree in which the teacher educator perceived to achieve his teaching goals, opinions regarding the content of the course, and views about the experiences of participants.

The interviews were analysed by splitting the text in fragments, each about a single topic. Each fragment was summarised. Then each summarised fragment was analysed to find topics regarding changes in participants and components of the courses that may contribute to these changes (because this is the main focus of the investigation). The following topics came to the fore: learning of course participants, changes regarding the three pillars of VTB-pro, components of the course that were perceived to contribute to changes, comments about implementation of S & T in participants' schools, comments about improving or maintaining the course at hand. For each region a summary of the interviews regarding these topics is presented.

Region A: Teacher educators need to be flexible and adjust courses when needed. Participants learn most by executing simple activities (learning by doing) that produce successful experiences and that they can use in their classrooms and schools, and by discussing with fellow participants. According to the teacher educator participants already possessed subject matter knowledge and enthusiasm, but at the end of the course they also knew how to start the implementation at their own schools (especially after the fifth meeting). This is important because in a course with individual participants each participant needs to transfer his or her positive attitude to other colleagues.

Region B: Participants have learned about inquiry and design based learning and ways to combine subjects. Because of these new knowledge and skills, a positive attitude was achieved. Changes in subject matter knowledge are more difficult to achieve. Participants learn by executing assignments, and by realising they lack some knowledge that they need to acquire. Possibly participants (and others) could benefit from methods or guidelines for lesson preparation and how to stimulate inquiry based processes. In a follow-up program, the teacher educator would (for example) focus more on training/coaching in the classroom.

Region C: Participants positively changed their attitude and pedagogical content knowledge. The changes in subject matter knowledge are difficult to estimate. A start has been made with the formation of a core S & T team and setting up more structural implementation. The teacher educator believes that especially the possibility to make choices between activities during the meetings has contributed to positive changes. Each course is adapted to meet the wishes and possibilities of the participating schools. With more course meetings, the teacher educator could for example provide an example lesson in the participants' school observed by the participants.

CONCLUSIONS AND RECOMMENDATIONS FOR PROFESSIONAL DEVELOPMENT

Perceived Changes In Domains Of Teacher Knowledge

The results of this research show that the investigated in-service courses for primary school teachers contributed to perceived positive changes, in particularly in pedagogical content knowledge and attitude. When answering open questions most participants indicated to perceive changes in their teaching style and being more aware of elements of S & T that they were already teaching without being aware of it.

In subject matter knowledge the participants in regions A and C experienced more positive changes than those in region B. However when their answers were analysed it turned out that participants in all three courses discussed pedagogical and didactical skills. Changes in their own subject matter knowledge were hardly taken into consideration, although they did mention the importance of enough background information and basic knowledge when asked about the qualities of a good S & T teacher.

Positive changes in the different domains of teacher knowledge for S & T are important because they are assumed to influence pupils' attitude. Rohaan, et al. (2008) for example found that teacher knowledge, and especially an enhanced pedagogical content knowledge, contributed to pupils' learning and interest in technology.

Practical Assignments And Learning From Experience

To take into consideration which components of the investigated courses contributed to the changes in the domains of teacher knowledge, perceived positive changes of participants regarding these components were analyzed (the Likert-items). It turned out that performing practical activities during the course meetings and coaching by the teacher educator contributed to positive changes. This finding was partially confirmed when participants were asked about the most inspiring and informative course components by means of open questions. Participants primarily mentioned the practical activities carried out during course meetings (both in the questionnaire at the end of the course and in the written reflections after each course meeting), because they could experience what children go through and obtain ideas for their own class. The importance of learning from experience also emerged from the interviews conducted with course teachers.

After investigating one activity it was found that course participants perceived a positive change in attitude and pedagogical content knowledge after being coached by the teacher educator (by asking the right questions and focusing on predicting before testing a hypothesis), by executing their own investigations and by engaging in successful experiences. Some subject matter knowledge was enhanced by retrieving previous knowledge and using this to improve a product design. After investigating the unstructured observations regarding other practical activities it turned out that positive changes were perceived by participants in all three pillars, but mostly in pedagogical content knowledge and also in attitude.

In school reform and professional learning it is important to stay close to the teaching practice of participants (Fullan, 2006; ESoE, 2009). In two of the investigated courses primary school teachers participated in courses that took place in their own school, using their own materials when possible. It is advised that teacher educators also provide on-site individual support, for example during teaching practice or by executing an exemplary lesson (Akerson & Hanuscin, 2007). This is seen as a possibility by the teacher educators B and C in the investigated courses. The teacher educators mentioned that it also depends on the requests and wishes of the participating schools: for example, when wishing to improve organisational skills, observation of an exemplary lesson executed by the teacher educator can be informative. This statement is in line with Akerson and Hanuscin (2007), who advise course developers to make room for adaptation of the course to meet both the goals of teachers educators, and participating teachers and their schools.

When implementing an innovation, one of the conditions to take into account is shared meaning by all the participants (Ely, 1999; Fullan, 2007). In the investigated course with participation of a team of teachers an entire course meeting was dedicated to decision making about structural implementation. After completing the courses the teacher educators were interviewed and all emphasized the importance of structural implementation of the innovation. By means of the courses a first step in the implementation was taken: participants gained more knowledge and skills and now see possibilities to integrate S & T into their teaching practice.

DISCUSSION AND RECOMMENDATIONS

We may conclude that in-service courses for primary school teachers should contain many possibilities for participants to engage in practical activities in which they can: get familiar with the domain of S & T, can experience inquiry and design based learning, discuss their experiences with fellow participants, obtain ideas and see possibilities regarding their own classroom, and increase their enthusiasm regarding S & T. Guidance by the teacher educator during the practical activities is essential to create a learning experience for all participants.

In this research project participants have reflected upon each course meeting and also upon the total course after the last meeting. Participants perceived changes in subject matter knowledge, attitude, and pedagogical content knowledge. In future research it can be investigated to what extent these changes have long term effects (for example Akerson, Morrison & Roth McDuffie, 2006, found that 'one course is not enough' when teaching pre-service teachers about the nature of science).

To realise a better inflow in higher S & T education it is important that primary school teachers not only experience changes in their thinking, but also change their teaching practice accordingly. It is possible that the participants saw more possibilities to practice S & T, but have not effectively changed their practice yet (see for example: Lee, Hart, Cuevas & Enders, 2004, who found only perceived positive changes, but no significant changes in teacher classroom practices in the first year of a professional development course). Future research will have to establish to what extent a significant quantitative increase and qualitative improvement has taken place as a consequence of these courses.

Martina van Uum
Martina.van.Uum@esoe.nl

Koeno Gravemeijer
Koeno.Gravemeijer@esoe.nl
Eindhoven School of Education, The Netherlands

ELLEN SJOER AND JACOBINE MEIRINK

19. TRANSPARENCY IN STORYTELLING: AN ANALYSIS OF PRIMARY SCHOOL TEACHERS' EXCHANGES OF THEIR EXPERIENCES WITH SCIENCE AND TECHNOLOGY EDUCATION

INTRODUCTION

The development of science and technology teaching competencies among primary school teachers and bachelor of education students plays an important role in the implementation of science and technology (S & T) curricula in primary schools. In the Netherlands, a program was launched that allows 5000 teachers and 5000 students to take a course to acquire these competencies. The course consisted of 12 parts (including homework) and was aimed at "stimulating, challenging and inciting curiosity" (VTB-Pro, 2008). Such courses are primarily designed to develop new knowledge and skills and focus less often on how to relate the knowledge and skills to the specific context and vision of the schools in which the teachers are working. As a result, the implementation of the newly acquired knowledge and skills in the workplace is often a problem (Briscoe & Peters, 1997; Pierce & Hunsaker, 1996). Cochran-Smith & Lytle (1999) classify this type of professional development with 'knowledge for practice'. With this label they refer to types of professional development in which researchers develop knowledge and theories which can be used by teachers to improve their practice. More contemporary views on teacher professional development underscore a more active role for teachers (Little, 2006; Smith & Gillespie, 2007; Verloop & Kessels, 2006). These views are derived from a situative perspective on learning, in which it is argued that teachers, in addition to knowledge from others, also need to develop new knowledge and skills by examining their own practice. Consequently, it is important to connect professional development activities undertaken outside the job to teachers' daily work contexts (Borko, 2004; Little, 2006). Ideally, this connection is made in close collaboration with experts and researchers from other institutes.

In the context of the VTB program, teachers after completing the VTB-pro course, were puzzling with questions like How do I get my colleagues to participate? How do I ensure that inquiry-based learning and design-based learning become part of the curriculum, linked with other subjects? The implementation of science and technology teaching competencies cannot be achieved by an individual teacher alone, close collaboration with colleagues from both in- and outside the

Marc J. de Vries, Hanno van Keulen, Sylvia Peters, and Juliette Walma van der Molen (Eds.), Professional development for primary teachers in science and technology. The Dutch VTB-Pro project in an international perspective.

school is required. Teacher collaboration is also important to performing well in the long term, as teachers should be able to develop themselves continuously in the area of science and technology. More and more research confirms that teacher collaboration in the workplace provides a powerful learning context for teachers and has a positive impact on teaching and on pupil learning (Levine & Marcus, 2010; Vescio, Ross & Adams, 2008, Meirink, Imants, Meijer & Verloop, 2010; Platteel, 2009). Collaboration allows teachers to exchange ideas, experiences and methods and ask colleagues for advice. Recent studies on teacher collaboration demonstrate that the exchange of experiences is most powerful when teachers make their experiences as transparent as possible for their colleagues (Little, 2002; Little, 2003; Levine & Marcus, 2010).

This chapter presents a study on transparency in exchanging teaching experiences with science and technology education. It deals with the question of how primary school teachers collaborate on developing a science and technology curriculum for their own school practice, and more specifically with how storytelling can contribute to this. The following research question was formulated:

How do primary school teachers exchange their experiences with science and technology education and how transparent are these experiences?

THEORETICAL FRAMEWORK

Teacher collaboration and interdependency

Teacher collaboration is assumed to be a powerful learning context for teachers. Collaboration allows teachers to exchange experiences, to ask for advice from colleagues, and to create new ideas. Furthermore, it is assumed that teacher collaboration is most powerful if the level of interdependence in interaction between teachers is high. For example, if teachers share responsibility for designing a new teaching method they need each other's knowledge and skills to ensure a successful final product. Therefore, a high level of interdependence is assumed to foster teacher learning. Little (1990) distinguishes four levels of interdependency in teachers' informal collegial interaction. The first type of collegial interaction, with a low level of interdependency, is labelled "storytelling and scanning." This type of interaction is characterized by moment-by-moment exchanges and provides empathy and the support of colleagues. The second type of collegial interaction, with an intermediate level of interdependence, is labelled "aid and assistance." In this type of interaction a critical look at one's teaching practice is encouraged. In the next type of interaction, "sharing," teachers routinely share ideas, methods, and experiences. This can make teachers' daily teaching routines accessible to their colleagues which in turn can result in productive discussions. The type of collegial interaction with the highest level of interdependency is labelled "joint work." A collective responsibility for teaching is specific to this type of interaction.

Transparency

These four levels of interdependency show that the exchange of experiences plays a key role in teacher learning. As a result, more recent studies on collaborative teacher learning (Little, 2003; Levine, 2010) focus specifically on what makes exchanging teaching experiences a powerful learning resource. These researchers argue that observing colleagues in their teaching practices and team teaching are important learning activities but that these activities are often difficult to organize because of scheduling conflicts. As a result, "collaborating teachers must represent their work verbally or through artefacts of practice, such as student work samples or lesson plans" (Levine & Marcus, 2010, p. 390). Little (2003) argues that "classroom accounts that surface in interactions tend to rely heavily on a certain shorthand terminology, and on condensed narratives that convey something of the press of classroom life without fully elaborating its circumstances or dynamics" (p. 936). Therefore, teachers need to present their practices as transparently as possible. Transparency refers to the specificity and completeness with which teachers share what they do in their teaching practices. The study by Levine and Marcus (2010) confirms the importance of transparency in presenting teaching practices. Horn (in press) distinguishes two ways in which teachers can present their teaching practice: by providing a replay or a rehearsal. In a replay teachers talk about specific and actual actions in their practices, whereas in a rehearsal they talk about what they or their students generally do or might do. It is assumed that a replay would be more specific than a rehearsal. Hence, teachers would learn more if they were to be as specific as possible about their own teaching experiences.

In this study we use the classification of levels of interdependency and the concept of transparency to analyze primary school teachers' exchanges of their experiences with teaching science and technology in team meetings. Since science and technology education is a not yet well-established subject for many teachers, they need to develop their knowledge of how to teach students this subject. Teachers can develop this knowledge through their teaching experiences and through reflection on those experiences. Professional development activities can play an important role in the development of teachers' knowledge. They can help teachers analyze their experiences, for example in collaboration with colleagues. In this study teachers exchanged their experiences in organized team meetings. Little's classification is based on teachers' collegial interaction in informal settings. In a study by Meirink (2010) however, these four levels of interdependency in collegial interaction were also used to characterize formal types of collegial interaction in secondary education. It was concluded that the four types of interdependency in collegial interaction should be viewed as points on a continuum rather than as fixed points.

METHOD

Respondents

For this study, two teams were observed for a period of one school year. Team A consisted of 10 primary school teachers, the principal and assistant principal of a public school in the western part of the Netherlands. All grades were represented in the team. Team B consisted of 6 teachers and the principal of a Montessori school, also in the western part of the Netherlands. This team also represented all grades.

Two teachers from each team participated in the offsite course to deepen their knowledge of science and technology and the scientific methodology corresponding to relevant disciplines such as physics and technical design. During the course teachers were supplied with a variety of ideas and materials to design science and technology lessons for their own teaching practice. The other teachers varied in their experiences and expertise with respect to science and technology education.

During the year that both teams were observed for this study, they worked, in close collaboration with an external coach, on a shared vision of and coherence in science and technology lessons for their own curriculum. The coach encouraged the teachers to share their ideas and experiences as transparently as possible. For this purpose, he asked questions like: How did students react? Why do you think they reacted the way they did? Why do you think this assignment was a success? What would you change if you are planning to use this assignment next year? In addition the coach provided concepts from Science and Technology education whenever relevant to deepen and extend teachers' knowledge and understanding of Science and Technology. The role of coach should therefore be considered as providing teachers assistance in both the content of and procedure for their professional development.

All meetings of both teams were videotaped.

Analysis

The team meetings were analyzed both in terms of content and the way in which experiences were shared. As a first step, 38 fragments were selected in which teachers shared experiences with science and technology lessons from their own teaching practices. Six fragments were selected from these for a detailed analysis (see Table 1). These six fragments met the following criteria: 1) each fragment lasted more than four minutes; and 2) in each fragment a different teacher shared his/her experience.

Transparency

To characterize how teachers exchange experiences we first looked at the transparency (specificity and completeness) of their stories. Specificity was determined by using the distinction between replays and rehearsals (Horn, in press).

A fragment was labelled as a replay if the teacher talked about specific actions in his/her teaching practice. The label "rehearsal" was used when teachers talked about what they generally do in their practice. To determine the completeness of the teaching experience the six fragments were labelled with sensitizing concepts by two independent researchers. These sensitizing concepts were derived from the concept of Pedagogical Content Knowledge (PCK) (Shulman, 1986; Magnusson, Krajcik & Borko, 1999; Grossman, 1990; Van Driel &Verloop, 1998). Magnusson et al. (1999) distinguish four categories in PCK: (1) knowledge and beliefs about curriculum; (2) knowledge and beliefs about students' understanding of specific topics; (3) knowledge and beliefs about instructional strategies for specific topics; (4) knowledge of student assessment. These four categories were used for an initial analysis of the fragments. Next, they were further specified or adjusted to the specific context of this study. These labels were discussed by two independent researchers until agreement was reached, resulting in four codes that were used to determine the completeness of a teaching experience (see Table 2).

Interdependency

Little's classification of the levels of interdependency was used to better understand how teachers' exchanges might be improved. For example, teams in which teachers merely exchanged their experiences without critically examining them have a lower level of interdependency compared to teams in which teachers make their teaching practices public and expect critical feedback from their colleagues (Meirink, 2010).

RESULTS

Table 1 provides a brief overview of the six fragments that were selected. Among the six fragments was one lesson on technical design (housing), and one with an experiment (in which the empirical cycle could be addressed). Further, there was one real-life lesson where the children went outside the classroom (to a mill). Finally, there were two lessons in which technology education was combined with visual arts education, including handwork and practical arts (fragments 4 and 6).

Table 1. Selected fragments

Fragment	*Team*	*School level*	*Short description of experience*
1	A	Primary grade	Pupils experiment with ice cubes
2	A	Primary grade	School visit to a mill to understand how gear wheels work
3	A	Intermediate grade	An introduction to the subject of "energy"
4	B	Secondary grade	Making penguins
5	B	Primary grade	Constructing a house
6	B	Primary grade	Making ladybugs

Specificity of teaching experiences

The analysis of the six fragments showed that the teachers were encouraged to tell stories and share experiences (Level 1 Little, 1990). During the team meetings, the coach asked for concrete examples of science and technology lessons. In response the teachers began to describe the science and technology lessons they recently taught (see Table 3). Sometimes the teachers brought children's' results to the team meeting. In all fragments the teachers shared their experience with a specific lesson. They talked about their instructional strategies, how students reacted, etc., without referring to what they usually do or how students generally react (with exception of a few short comments, for instance in fragment 5, see Table 3). Therefore, all six fragments were characterized as replays rather than rehearsals.

Completeness of teaching experiences

With respect to the completeness of teachers' exchanges of teaching experiences it appeared that teachers differ in how completely they describe their experience. Table 2 shows the codes which were used to determine the completeness of a story.

Table 2. Codes for completeness

	Code	*Short description*	*Labels*
1	Context	– Who, what, when, where?	– C
2	Execution of a lesson	– How was it organized? – How did the lesson turn out (results)? – How did pupils react (classroom interaction)? – Which materials were used? – Which pedagogy was used?	– Eo – Er – Ei – Em – Ep
3	Goal and assessment	– What was the goal of a lesson or series of lessons and was this goal achieved? What did pupils learn?	– G
4	Addressing Science and Technology goals and methodology	– Are the Tule goals or the map of technical concepts addressed? To what extent was the empirical cycle or the technical design cycle used? (Where appropriate).	– S & T

1. Context

All teachers indicated what their story was about: a theme (lower grades) or the subject they are engaged with in the method. For example: "We have had a theme on baking and so we visited a mill". When they carried out the lesson was scarcely mentioned. Also, the group of pupils that was involved was not specified. Only when the group was differentiated, such as "oldest children," did the teachers mention it. None of the teachers explained how the story fits into the curriculum. With exception of one teacher, who talked about what he would do next in response to a question from the coach, the teachers did not say what took place before the lesson or what they taught before this experience and what comes afterward. In sum, the teachers in this study provided only a limited amount of information about the context of their story.

2. Execution of a lesson

A majority of the six selected fragments consist of an elaboration of the topic of a lesson. Teachers talk about how a certain lesson turned out (code Er). They elaborate on the pedagogy (Ep) and materials (Em) they used, on how they organized the lesson (Eo) and on students' reactions (often an indication of whether they liked it or not)(Ei).

All six teachers elaborate on the pedagogy they used. They often start by giving students the assignment to try out a certain task themselves without any specific instruction. For example, in fragment 3 the teacher says:

> *'One afternoon I said, "When you go to your house look around at all the things that require energy to use." "Yes, but..." they said, and I replied, "That's all, please find out for yourself, if you go home and ask your parents you'll probably come up with something." So the next day, they had written down all kind of things, such as my alarm clock, my toothbrush, so now they did have some awareness that certain things take energy.'*

The introductory lesson on energy had two follow-up lessons which were also briefly introduced by the teacher. The other teachers limited their story to one lesson; half of them also said something about the debriefing (fragment 2, 3 and 6). That could consist of an evaluation of the process ("Which approach did you choose to make ladybugs of papier-mâché"), or a summary of the matter ("I did the debriefing using some pictures, so I showed gear wheels in the classroom also [...]").

A rationale for why a particular instructional strategy was chosen by the teachers was indicated briefly in most of the stories. In the case of the experience with an introductory lesson on energy, it appeared that a colleague conducted the same lesson, but using a different approach. However, the two approaches were not compared and discussed.

In addition, another teacher stated (fragment 4):

> *I do not like examples; so I thought I'd try to explain it [the penguins] the best I can. Well, I regret that I did it that way. [Because?] Because it proved to be more difficult than I thought.*

In the remainder of this teacher's story the complications of this assignment were elaborated, but not the pedagogy.

In addition, teachers mention a lesson's organization and the materials used when it is relevant: like dividing the group in two when visiting the mill and being present at the group table because of using hot water while experimenting with ice cubes. In these six fragments, teachers did not elaborate much on the organization or the materials. Their choices for materials or organization are also not discussed.

Table 3. Analysis of the six selected fragments

	Description of the story	*Labels*		*Description of the story*	*Labels*
1	*Experiment with ice cubes*		*2*	*Visit to mill*	
	The teacher tells:			The teacher tells:	
	1. what she did when (experiment with ice cubes, this week)	1. C		1. where she went (to a mill)	1. C
	2. how she did it (group table), what materials she used (salt, rope, hot & cold water, ice cubes)	2. Eo & Em		2. how it fits into the theme baker	2. C
	3. what the reaction of the children was ('It floats! In cold and warm water, how can that be?')	3. Ei		3. what was demonstrated at the mill (operation of gears and grinding grain into flour)	3. Ep
	4. what it does to her (children see things she is not aware of)	4. Er		4. what the reaction of the children was ('great')	4. Ei
	5. what the experiment showed (melting slowly and quickly)	5. Er		5. what the children have learned (operation of gear wheels)	5. G
	6. what her next question was (asked for explanation)	6. Ep		6. what technical aspects have been addressed: wind energy, transmission of motion	6. S & T
	7. what the explanations of the children were (' cascade of cold water')	7. Ei		7. what was difficult in the organization	7. Eo
	8. what the children wanted to find out more (touch the glass to feel if the water cools)	8. Ei		8. how the lesson was debriefed (images in class)	8. Ep
	9. where she had to focus on (safety: hot tea glass)	9. Eo		9. what she would do next time (theme was too long, however, visit of the mill remains in the program)	9. Eo
3	***Energy***		***4***	*Making penguins*	
	The teacher tells:			The teacher tells:	
	1. how the energy lessons matches the method	1. C		1. what she has done and what the connection was with the theme (winter and snow)	1. C
	2. what the purpose of the first lesson was	2. G		2. what the assignment was	2. Ep
	3. the assignment he gave (look for things at home that costs energy)	3. Ep		3. how she look back on that ("I regret it")	3. Er
	4. what the initial reaction of the children was and how he reacted	4. Ei		4. why it was more difficult than she expected	4. Er
	5. what the results were (alarm clock, my toothbrush)	5. Er		5. how she reverted on the lesson on math	5. Ep
	6. what next assignment was (sheet from the method)	6. Ep		6. what children should do and what went wrong	6. Er
	7. how the children worked on that assignment	7. Er		7. what the next step was (how do you attach a ball and how do you do that nicely)	7. Em & Ep
	8. what the topic of the second lesson was (electricity chain)	8. G		8. what a good solution of one of the children was	8. Er
	9. what the first reaction of the children was (one child came with a picture of the 'wrong' drill, a drill for a foundation instead of a derrick)	9. Ei		9. she realized that you are often helped by the children	9. Er
	10. what additional instructions he gave (drawing the electricity chain, making word fields)	10. Ep		10. what technical aspects have been addressed (fixed link or loosely connected)	10. S & T
	11. what he did in the third lesson (examples in class)	11. Ep			

Table 3 (Continuation). Analysis of the six selected fragments

	Description of the story	*Labels*		*Description of the story*	*Labels*
	Remainder of the fragment: The teacher tells how the preparation of the lessons was shared with a colleague. A colleague tells that she started ' the other way around' with examples in the class and with an experiment with hot and cold bottles and the problem: how to move this coin? Reaction of the children: ' well, then press the button' [...] Further, thematic typing of the next lessons (environmental pollution, fossil fuels with the aim of awareness raising, food chain) and the conclusion that it is possible to make S & T lessons, in line with the method, also at the intermediate level.)				
5	*Building a house* The teacher tells: 1. what the theme was 2. what assignment she has given (build a house with many different materials) 3. what materials she had offered 4. how the assignment could be characterized (open, three-dimensional, both S & T and visual arts) 5. how she instructed the children 6. how it differed from what she normally did 7. what requirements the design had to meet (people had to walk in and out) 8. about the problem that everyone imitate one child's solution 9. what that solution was and how young children usually proceed 10. what other technical problems she encountered 11. what two children made, the reactions of the children and herself and the (interim) results of the children	1. C 2. Ep & Em 3. Em 4. Ep & G 5. Ep 6. Ep 7. Ep 8. Er 9. Er 10. S & T 11. Er & Ei	*6*	*Making ladybugs* The teacher tells: 1. what the assignment was (making papier-mâché ladybugs) 2. what children had to do sequentially 3. what the reaction of the children was (nice, but also disappointment when the balloon was already completely deflated at night) 4. that the number of dots was addressed 5. what the children had to do next 6. what the result was 7. why she had not assessed the ladybugs 8. how she debriefed the lesson (approach children, best solutions were named) 9. what aspects of the visual arts were raised unconsciously: forms and colors 10. that she assumes too much prior knowledge (for example, explaining a cylinder using a kitchen roll, but children did not know what that was. Vocabulary is particularly low)	1. Ep 2. Ep 3. Ei 4. Ei 5. Ep 6. Er 7. G 8. Ep 9. S & T 10. Ep

3. Goals and assessment

When looking at what teachers tell about their goals and assessment procedures in the six selected fragments it appears that the goal is often not explicitly mentioned, or is described only in general terms, e.g. "students will learn about the operation of gear wheels". The teacher in fragment 3, for example, states:

> *We use the Da Vinci method. We are working on "humans and earth" and within that we started on energy. And actually the first lesson was an introduction to energy [...] in the meantime we have also done the second lesson. The first lesson was really about becoming conscious of what energy is, where it comes from, what it costs.*

Only one teacher made a remark about assessment, but it did not come up spontaneously (fragment 6).

As a result of the limited information on goals and assessment procedures, it is often not clear what results were achieved with students. The teachers keep each other informed about students' reactions and response to an assignment. Most of the teachers gave an indication of whether the students were enthusiastic about a lesson or not. However, from this information it is not possible to infer whether the teachers are satisfied with the result and with the progress pupils make. Furthermore, colleagues do not ask questions about or discuss these reactions.

4. Addressing science and technology goals and methodology

In both teams the coach explained and elaborated on science and technology goals and concepts at the request of the teachers. He explained for example the "Tule content and activities" guideline where the main goals of "Nature and Technology" are described in the chapter "Orientation to the world and to yourself" (http://tule.slo.nl). These core goals provide a legal framework for the core of educational content; schools have the freedom to make specific choices themselves and to use their own pedagogical interpretations. Further, the coach provided a "map" containing technical concepts, such as "conveying" (for example, a lever or a gear wheel) and "operating"(for example, a light sensor) for teachers to evaluate their lessons in technical terms. Stories in which teachers address the Tule goals or the technical concept map, or those that make use of the empirical cycle or the technical design cycle are considered to be more complete.

None of the teachers, however, used these concepts in their stories spontaneously. Therefore, the coach asked questions or assigned technical terms to the teachers' answers. From analyzing the stories it became clear that teachers differ in their ability to use technical concepts. For example, one of the teachers told about a lesson in which her pupils had to build a house (fragment 5). The coach then assigned a science and technology concept to her experience.

> *One girl had a box which had a strip of paper folded like this [makes a triangle with her hands] and she had it pinched [demonstrates with a piece of paper] and then she made a loft right away, because that was where she*

slept, very nice. And there was a boy who had a slightly taller box, and that did not work as nicely. [Coach: Why not?] Well he did not like it, because he had a very thin box which made a very strange little roof, and he did not like it at all. And then I said you should try something else.

The coach then reformulates:

This example of the girl and the boy with that sort of roof is clearly a triangular construction. [...] And the boy did not like it, so it's a combination of visual, aesthetic and technical aspects.

Also, most teachers did not think in terms of the design cycle (see example below). In the VTB-pro course teachers learn that a design process (which should be regarded as a cycle) consists of six major phases: 1. formulate a problem; 2. make a list of requirements; 3. generate alternatives and select a few alternatives; 4. draft a design proposal; 5. make the design; 6. test and evaluate the design.

Did the house have to meet certain requirements? (coach)

No, they could make their own house, one pupil had a flat roof, yes that is possible, I also live in a house with a flat roof.

Should people be able to come in and go out? (coach)

Yes, you should have windows, and doors, so, yes.

And what kind of solution did they invent for that? (coach)

Well, one pupil cut out a rectangle and stuck that to the house and of course, everyone imitated that.

[...]

Stuck to the house, so it could not be opened? (coach)

No, it was not possible to open the door.

That is also a good question: how can you open the door? (other teacher)

[...]

That is a nice example, because [...] you just introduced a technical problem: you should be able to open and shut the door. So there must be something in the design that you can use to open it and close it again [...] for example, a hinge would be an appropriate solution. (coach)

From this example it can be concluded that the teacher did not use the phases of a design process explicitly in her lesson.

To sum up, teachers did not address the Tule goals or use technical concepts that were previously introduced. When asked about this, they differed in their ability to answer the question. They used neither the design cycle nor the empirical cycle when appropriate as a connecting thread for their story.

CONCLUSION

The aim of this study was to obtain a more comprehensive understanding of how teacher collaboration and more specifically the exchange of teaching experiences can be a powerful learning context for primary school teachers. For this purpose team meetings of two teams were observed during their first year of collaboration. The meetings were organized at school and can be viewed as a follow-up to a training course in science and technology education which was organized offsite. The following research question was addressed: How do primary school teachers exchange experiences with science and technology education and how transparent are these exchanges of experiences? Transparency was divided according to how *specifically* and how *completely* teachers exchanged their experiences.

With respect to how specifically teachers represented their practice, in this context all the teachers succeeded in talking about their actual actions in a lesson, how students reacted, etc. (cf. replay). Teachers rarely talked about what they or their students generally do in their lessons (cf. rehearsal).

To assess the completeness of teachers' exchanges of experiences four subcategories were distinguished: (1) context; (2) execution of a lesson; (3) goals and assessment; (4) addressing science and technology goals and methodology. These categories were used to determine how completely teachers talk about their experiences.

From the fragments it could be concluded that teachers' information about the context of their experience was limited and can be regarded as incomplete. Teachers were more or less complete in their information about the execution of a lesson in that they tell a lot about what they did, the materials they used and the instructional strategy they chose. However, they failed to provide a rationale for their choices and their colleagues did not ask for this information either. It can be argued that discussions about underlying reasons for certain choices in methods and materials may be productive in developing teachers' knowledge and skills. Also, collaborative analysis of student work or teaching methods can result in productive discussions.

With respect to goals and assessment, teachers were incomplete in their exchanges. Goals remained implicit and as a result it was often unclear whether a teacher reached his/her goals. The teachers did provide information on student reactions to tasks and questions. Some teachers indicated whether students were enthusiastic or not. In this category too, colleagues did not ask questions that could have resulted in more insight into teachers' goals and ideas about assessment.

The final category, addressing science and technology goals and methodology, never occurred spontaneously. Teachers needed help from the coach to identify which science and technology concepts they used, often unconsciously.

From the four categories used to determine completeness it can be concluded that the selected exchanges of experiences might not be transparent enough for teachers to learn from. Teachers indicate that they were inspired by the stories of their colleagues and appreciated the help of the coach. The experiences, however, might have been more complete if the colleagues and/or the coach had asked more critical questions. Levine et al. (2010) among others indicates that taking a critical look at each other's work is a challenge for most teachers. Teacher collaboration is often confined to safe styles of encouragement (Schwarz McCotter, 2001). Positive critical dialogue supports cooperation, but it must be learned and practiced (Platteel, 2009). Critically examining teaching experiences can result in a type of collaboration which can be labeled as "sharing" (cf. Little, 1990). Since "sharing" supposes a higher level of interdependency compared to "storytelling," it may also better foster teacher learning.

Moreover, it appeared that for some teachers it was difficult to relate their teaching experiences to science and technology concepts. Since the collaboration in both teams was aimed at developing a science and technology curriculum it is necessary to develop a shared language for this subject. The coach played an important role in accomplishing this goal.

Practical implications

Some practical implications for future professional development activities can be derived from the results of this study. In general, it can be argued that this study offers new insights in how to optimize professional development trajectories. Storytelling or exchanging experiences is a frequently used activity in such trajectories. Many teachers indicate that they find this activity useful as, for example, they provide new ideas or inspiration for experimenting with different teaching methods. In this study we analyzed teachers' stories on a more detailed level and showed that teachers can learn 'more' from this activity if they exchange their experience as specific and complete as possible. A coach can take the initiative by providing exercises and methods during the meetings that direct teachers in this way and in becoming more critical. Although this study was conducted in an on-site professional development course, the same advice also applies to off-site courses. In these courses coaches also often encourage teachers to exchange their experiences which can be characterized as "storytelling" rather than "sharing." Coaching teachers to be more critical and to deepen a subject is also a necessary step in this context. Once they have become familiar with elaborating on concrete examples and asking critical questions, the presence of a coach may no longer be needed.

A second implication which can be derived from this study is that it takes time for teachers to become familiar with science and technology concepts. Also, it can be difficult for teachers to apply these concepts to their teaching practices. A coach can facilitate this process by providing examples and by helping teachers to designate the scientific and technological concepts they use in their teaching practices, often unconsciously. If school teams want to develop a common curriculum, it is important to develop a common language. Familiarizing teachers with how to translate science and technology concepts into challenging science and technology lessons for pupils is an important task for professional development activities in this context. This may stimulate teachers to experiment more with science and technology lessons which in turn can yield interesting stories for team meetings. In the end, these stories can provide important information for colleagues.

Future research

In this study we examined the transparency with which teachers exchange their experiences, as this is assumed to foster teacher learning. For future research it would be interesting to study more teacher teams and over a longer period of time. Did they develop new knowledge or skills? Are their stories exchanged on a more detailed level, in other words are they more complete? Analyzing team meetings of more teams over a longer period of time should result in a more comprehensive understanding of the 'power' of storytelling. In addition, it should be noted that storytelling should never be used as an isolated professional development activity. A variety in professional development activities is needed for 'effective' teacher learning (Timperley, Wilson, Barrar & Fung, 2007).

In the end all efforts should contribute to the development of professional teachers who can take a critical look at their own teaching practices and those of their colleagues and who are willing to continuously improve their practices. One of the worthwhile professional activities would be telling stories about beta talents of children and their contributions, how to recognize and describe the development of children's talents, how to address these talents and how to recognize whether the teachers succeeded in guiding these learning processes, using stories from their own practice. That would be a next step in teachers' collaborative lifelong professional development with science and technology education in the Netherlands.

Ellen Sjoer
The Hague University and Delft University of Technology, the Netherlands
E.Sjoer@tudelft.nl

Jacobiene Meirink
The Hague University and Leiden University, the Netherlands

ELIANE SEGERS, MARIEKE PEETERS, HELEEN STRATING, CINDY TEUNISSEN, AND LUDO VERHOEVEN

20. EFFECTS OF A DIGITAL LEARNING ENVIRONMENT ON KNOWLEDGE AND ATTITUDE OF STUDENT TEACHERS

INTRODUCTION

To develop a positive attitude towards Science, Technology, Engineering and Mathematics (STEM), the actual experience of *doing* science is an important contributor. Just telling children that science is fun, will not do the trick, and neither will reading in books about it. However, enthusiasm of the teacher, an interesting explanation, discussion in class and designing and conducting experiments themselves, can lead to emotional involvedness, lasting memories and cognitive understanding (Van Graft & Kemmers, 2007). The teacher is the role model, and the attitude they have towards STEM and towards teaching STEM can have a lasting impact on the children's attitude (Pell & Jarvis, 2003; Schroeder, Scott, Tolson, Huang, & Lee, 2007).

Unfortunately, student teachers often show a negative attitude towards STEM, thinking it is difficult subject matter that is hard to teach and integrate into their lessons (Wilkins, 2008); in other words they lack functional scientific and technological literacy (cf. Jenkins, 1990; Van Keulen, 2009). It is important to try to change this attitude of teachers, because if they have a more positive attitude, they are better able to integrate STEM into their lessons (Jarvis, 2004), and will gain a more positive attitude towards teaching STEM.

Often, this less positive attitude towards STEM is attributed to a lack of knowledge of the students. In the Netherlands, most Dutch student teachers studying to become teacher in primary education are female and are specialized in what is called 'culture and society', and not 'science and technology' in high school. They often have misconceptions about the way things work in science and technology (see also Bleicher, 2007). The importance of having knowledgeable teachers with a positive attitude towards STEM becomes clear from a study by Pell and Jarvis (2003) mentioned before, which showed that children form their attitude between eight and thirteen years of age, and that the attitude and knowledge of the teacher are important factors in this process.

Fortunately, teacher attitudes, as well as their knowledge, can change as a result of successful interventions, and these are necessary to effect changes in practice (Kinzer, Cammack, Labbo, Teale, & Sanny, 2006; NRP, 2000). The present

Marc J. de Vries, Hanno van Keulen, Sylvia Peters, and Juliette Walma van der Molen (Eds.), Professional development for primary teachers in science and technology. The Dutch VTB-Pro project in an international perspective.

chapter has the goal to give an example of the implementation of such an intervention in the Netherlands, and to investigate its effectiveness on the attitudes of student teachers in primary education towards teaching STEM. We describe the theoretical framework of a digital learning environment for Dutch student teachers for primary education with a focus on teaching STEM while integrating language education, with as the ultimate goal to further improve teacher education and interactive science and technology education in primary schools in the Netherlands. We will also describe a pilot-study on the effects of this environment on attitudes towards teaching STEM and knowledge of student teachers for primary education.

THEORETICAL FRAMEWORK OF THE LEARNING ENVIRONMENT

In designing any technology for instructional purposes, a theoretical framework with regard to pedagogy and content is necessary (Kinzer, et al., 2006). In our instructional design, two lines of theory were integrated: one on the development of the environment for student teachers, and one on the contents of science and technology education in primary education.

In the design of the *environment for teacher students*, we adopted the successful approach of case-based learning (Kinzer, et al., 2006). In case-based learning, video-taped lessons can be presented to the students in which they see a teacher giving e.g. a lesson on DNA. The lesson is presented in the presence of a classroom context, which is much richer than reading about how a lesson should be taught (i.e. a print-based case). The video-cases form a bridge from theory to practice. The videos can be used in anchored instruction in teacher education (CTGV, 1997), in which the video is used as a common knowledge base on which follow-up tasks, discussion and other activities can be build. This is a powerful means of teaching, as students can reflect on lessons they have all seen. This is very different from talking about their own experiences, as these experiences are more abstract to their peers. Next to videos, other materials can be made available online, such as assignments children have made, lesson plans, etc. The videos do not intend to be prefab examples that the student teachers should exactly copy in their own teaching, but serve as a starting point for inspiration, discussion, and learning.

Kinzer et al. implemented their ideas in the CTELL (Case Technologies to Enhance Literacy Learning) project, which can be viewed online at http://ctell.uconn.edu/home.htm. Previous research has shown that an online implementation of successful multimedia cases of science learning activities of children in interaction with their teacher can positively influence the attitudes of teacher students (Lampert, 1988; Ball, 1996), and can enhance their knowledge and skills that are necessary to implement such activities in their own lessons (Wilber, Kinzer, & Lohnes, 2008).

With regards to the *contents of science and technology education*, we adapted the LOOL-model from Van Graft and Kemmers (2007). LOOL is a Dutch acronym for Learning to Learn by Investigating and Designing. This model introduces seven steps in the teaching of science and technology, building on previous frameworks of e.g., Suchman (1963) and Llewellyn (2002) (see Van Graft &Kemmers, 2007). The Suchman model introduces five steps: confrontation, exploration, hypothesize & test, organize & explain, and reflect on process (see also de Vaan & Marrell, 2006)

Van Graft and Kemmer broke down step three into two: designing an experiment and conducting an experiment. Step four, organize and explain, is replaced by 'concluding' and a final step is added: 'going further', in which the teacher continues working with the children on the concepts just learned. In these five to seven steps, the children start in an open, exploring way, and – with help of the teacher – focus on the problem at hand. This focusing needs to be guided so children can think of and conduct an experiment. Van Graft and Kemmer (2007) refer to guided inquiry in this case. Schroeder et al. (2007) found an important role of guided inquiry in teaching science to students, as well as for enhancing the context of learning, similar to the suggestions in the LOOL model.

Language can be seen as an important factor in lessons on science and technology. All models on inquiry based learning put an emphasis on the interaction between teacher and child, and also on the interactions between the children. The rationale for integrating science with literacy is to generate a synergistic learning benefit caused by teaching of two domains simultaneously, (i.e. literacy and science). Combining both strengthens both, and helps developing academic language and academic thinking. Academic language refers to a more formal language used in texts in science. These texts are more difficult to understand than fictional texts or discursive essays if children do not have this academic vocabulary. Children need help in learning academic vocabulary to become independent learners of science (Snow, 2010). In our project, there is also a focus on the teaching of academic vocabulary (cf. Snow, 2010), drawing on the ideas of the Word Generation project (www.wordgeneration.org).

The importance of integrating language and science has, for example, been shown in a study of Barber, Catz, and Arya (2006). They investigated the effects of a combined science-literacy approach to learning science compared to a science–only and a literacy-only approach. Their research showed that the combined literacy science approach provided advantages in literacy skills, as well as the acquisition of science content knowledge. In their meta-analysis, Schroeder et al. (2007) showed how such strategies as questioning (asking comprehension questions, discussing, etc) and inquiry (student-centered instruction, guided discovery learning) have larger effect sizes in enhancing student achievement in science.

This interaction between teacher and child is presented in the theory of dialogic teaching (Alexander, 2004), and also in the anchored instruction framework (CTGV, 1997), and has been successfully implemented in our own previous work on integrating literacy into the subject areas (Van Elsäcker, Damhuis, Droop, & Segers, 2010). The role of the teacher is, for example, to verbalize what children see while doing experiments. The teacher introduces new vocabulary, can help children in their understanding of what they experience, and can give tools to verbalize their thoughts.

THE DIGITAL LEARNING ENVIRONMENT

In the design of the learning environment (www.interactiefonderzoeken.nl), we made the seven steps from the LOOL-model visible in cases with several movie-clips representing each of the steps for different areas of science and technology. These areas were adapted from the Dutch VTB-program, which is based on the Science Standards in the USA. The Science Standards divide science (and technology) into four different categories: Earth and Space Science, Life Science, Physical Science, Engineering and Technology. A fifth category was added in the Netherlands: Mathematical Systems.

For each series of lessons, the student is provided with several video-clips representing a step from the LOOL-model, goals of the lessons, main vocabulary, lesson materials, results from children (e.g. pictures of filled-out work sheets), and background sources. Each video-fragment is accompanied by questions to enhance higher-order thinking of student teachers, a description of what is going on, the step within the LOOL-model, a transcript of the video, and a description of the interactive aspects of the scene(s) to show how the teacher facilitates the children in developing their (academic) language skills.

We asked student teachers to reflect on an early version of the environment. One wrote: "This is definitely something to remember. The children are clearly fascinated by the subject, and I love the fact that they can go on to investigate things in their own way." And another: "This gives me ideas to start working on teaching interactive science education myself." And: "I realized that I was wondering whether I ask those open questions myself during science and technology lessons." These statements suggest student teachers to become more enthusiastic about teaching science and technology, and also that the environment helps them to rethink their own teaching behavior.

An example of the implementation of the environment in practice is a project on the growth of the garden cherry which comprised a series of lessons (see Figure 1). The central concept in the first lesson was growth. Specific attention is paid to experiences using a central concept. Furthermore, in these experiences the role of communication, literacy and language are integrated. This concept is in accordance with the Preschool Pathways to Science (cf. Gelman & Brenneman, 2004).

In the first lesson, the teacher read a story about a little piglet who wanted to grow, and so ate a lot, but unfortunatelydid not grow. Then, one day, his sweater did not fit anymore and he became angry, because it was his favorite sweater. In the end, the piglet finds out that he has grown, and that's why the sweater did not fit anymore. The story proved to be a perfect starting point for group discussions on growing. The teacher, for example, asked children to bring clothes from when they were a baby. A group discussion included the following questions: Why do those not fit anymore? Why do you grow? Does the teacher also still grow? Why? Then the garden cherry is introduced. Why do plants grow? What do they need to grow? Soon, children came up with light, warmth, water and air. One child suggested that the garden cherry would optimally grow when you would provide it with lemonade. And so the experiment was born. Guided by the teacher, children made different conditions: with and without water, with and without light, etc. The growth under each condition was carefully monitored and depicted in drawings, but also in histograms depicting the height of the garden cherry every day on the classroom board, which made it easy to compare the different conditions. Then, after a week, there were small group discussions in which the teacher talked with the different "experimental groups" about the results of the experiment. In the end, the children presented their results using a puppet theatre. The teacher then went on elaborating on the (academic) vocabulary the children learned about growth.

Although the experiment about growth of the garden cherry is mainstream, integrating it in the LOOL model is not. The teacher from the garden cherry example told us that she had never conducted the experiment in such an interactive way, stretched over several lessons. She specifically noticed that the kindergartners began to use more sophisticated reasoning about the topic, and that their understanding was deeper.

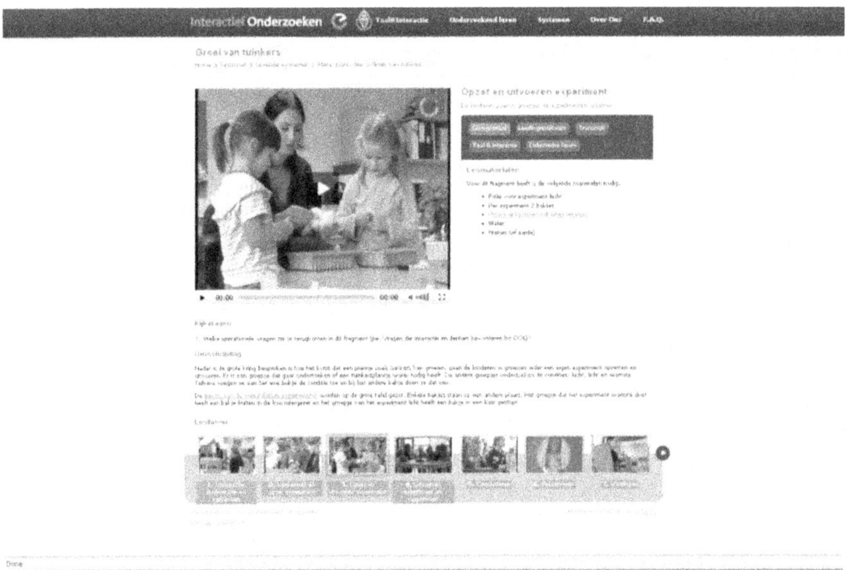

Figure 1: Screendump of the learning environment (www.interactiefonderzoeken.nl).

Another example we put on-line is a heredity and DNA-project. In a series of lessons the children learned about heredity and DNA. This is quite a complex topic, and we therefore also used the idea of "bringing the scientist to the classroom", in close collaboration with the Nijmegen University Science Hub (www.wkru.nl). In the case of the DNA lessons, a PhD student doing research on genetic causes of schizophrenia came to school and gave an introduction about DNA, after which the children could ask questions. This connection of scientists and schools is what science hubs are all about; but more research into the effects of these connections to the attitudes of children towards science is necessary.

PILOT STUDY

From previously cited literature, we know that teachers often have little knowledge about STEM and that their attitudes towards teaching STEM can be low (Wilkins, 2008), and that this can influence their way of teaching STEM and also, on the long term, the attitudes of the children towards STEM (Pell & Jarvis, 2003). A first step in research after constructing a learning environment for teachers is to find out if indeed working with this learning environment leads to knowledge gains and more positive attitudes towards teaching STEM.

We conducted a pilot study on a group of teacher students for primary education doing a minor in science and technology. The research questions of this pilot study were:

1. How does the knowledge of student-teachers about science topics change as a result of working with an interactive learning environment?
2. In what way do explicit (enjoyment, relevance, competence, motivation) and implicit attitude towards teaching science and technology of the student-teachers change as a result of working with an interactive learning environment?

Participants

Three groups of student teachers for primary education participated. Two of the smaller groups were the control group and the one larger group was the experimental group. The control group consisted of 30 students (4 male, 26 female), and the experimental group consisted of 20 students (1 male, 19 female). Students were in their second year of teacher education and could choose for a minor focusing on younger or older children. The students working with the younger children studied a series of lessons on growth of the garden cherry, and the students working with older children studied lessons on DNA. In the control group, 12 focused on garden cherry and 18 on DNA, in the experimental group, the division was 12 garden cherry, 8 DNA.

MATERIALS AND PROCEDURE

A questionnaire was presented to the students before and after the intervention, including questions related to knowledge about either the garden cherry or DNA, and explicit attitude towards science and technology. Also an implicit association test (IAT) to measure implicit attitude towards the teaching of science and technology was administered individually, using a laptop computer.

Knowledge. The knowledge test consisted of 20 yes/no questions, either on garden cherry or on DNA.

Explicit attitude. To measure the explicit attitude towards science and technology, the questionnaire from Denessen et al. (2011, this volume) was used. This questionnaire consists of 20 items on a four-point Likert scale, and is divided in four scales: (1) Teachers' enjoyment for teaching STEM (5 items, for example 'I feel comfortable when teaching STEM', Cronbach's alpha = .84), (2) their perceived relevance of teaching STEM (5 items, for example 'I think it is important to teach STEM', Cronbach's alpha = .80), (3) their perceived levels of competence and efficacy related to teaching STEM (5 items, for example 'For me teaching STEM is very difficult', Cronbach's alpha = .70), and (4) their motivation to invest in their STEM teaching competences (5 items, for example 'I am willing to invest in the development of my abilities to teach STEM', Cronbach's alpha = .85).

Implicit attitude. Implicit attitude was measured with an IAT test (see Denessen et al., this volume for an elaborate description), aimed at tapping teachers' general attitudes towards science and technology, by assessing positive and negative associations with science and technology.

Procedure. The intervention was set up in the following way. First, the second author of this chapter visited the school and introduced the digital environment. Students were asked to do an assignment using the environment and answering the questions that were online (either on DNA or garden cherry). Next, these questions were discussed in small group discussions under the supervision of research assistants conducting the experiment. Then, students designed and taught a lesson about science and technology during an internship. The Posttest was administered after this lesson. The control group followed the same minor, but did not have access to the online environment, or the group discussions.

RESULTS

For *knowledge*, we first checked whether the two topics were equally difficult at Pretest. At Pretest, the DNA group had lower scores than the Garden Cherry group (which makes sense, since DNA is a more complicated topic, and designed for the upper grades), $t(47) = 2.041$, $p = .047$, $d = .591$. We therefore transformed the scores on these tests to z-scores, in order to be able to compare these groups.

To measure effects of the intervention on knowledge, we conducted a univariate analyses of variance with Posttest score as the dependent measure, and Group (experimental, control) as fixed factor, while controlling for prior knowledge using Pretest score as a covariate. We found a main effect of Group, $F(1, 35) = 4{,}796$, $p = .035$, $\eta^2_p = .121$, indicating that the experimental group had higher scores at Posttest. Descriptives can be found in Table 1.

Table 1: Means and standard deviations (z-scores) of experimental and control group at Pretest and Posttest on the knowledge test.

		Mean	*SD*
Pretest	Control group (n=30)	-.119	.994
	Exp. group (n=19)	.188	.979
Posttest	Control group (n=22)	-.336	1.022
	Exp. group (n=16)	.462	.738

GLM repeated measures analyses with Time (Pretest, Posttest) as within subjects variable, and Group (experimental, control) and Topic (garden cherry, DNA) as between-subjects variables were next conducted on explicit attitude and implicit attitude.

For *explicit attitude*, we analyzed the four scales of the questionnaire. For *enjoyment*, there was an interaction between Group * Topic, indicating that in the experimental group, the Garden Cherry students had higher scores than the DNA students at both Pretest ($p = .015$) and Posttest ($p = .036$). effect sizes.

For *competence,* there was an interaction between Time and Group, which can be explained by differences between the experimental and control group at Posttest but not at Pretest; the experimental group had a more positive attitude towards competence at Posttest, $t(41) = 2.538$, $p = .019$, $d = .758$ but there was no significant difference at Pretest, $t(47) = .899$, $p = .379$, $d= .260$.

For *relevance* and *motivation*, there were no effects of the intervention. Table 2 gives an overview of the statistical analyses. Descriptives on the four scales of the explicit attitude questionnaire can be found in Table 3.

Table 2: F, p and partial eta square values of the four scales of the explicit attitude questionnaire in a repeated measures analysis.

	Enjoyment	*Relevance*	*Competence*	*Motivation*
Time	$F(1, 38) = 2.206$, $p = .146$, $\eta^2_p = .055$	$F < 1$	$F(1, 38) = 3.002$, $p = .091$ $\eta^2_p = .073$	$F(1, 37) = 2.314$, $p = .137$, $\eta^2_p = .059$
Group	$F(1, 38) = 3.047$ $p = .089$, $\eta^2_p = .074$	$F < 1$	$F(1, 38) = 2.474$, $p = .124$, $\eta^2_p = .061$	$F < 1$
Topic	$F(1, 38) = 2.297$ $p = .138$, $\eta^2_p = .057$	$F < 1$	$F < 1$	$F < 1$
Time*Topic	$F(1, 38) = .103$, $p = .750$, $\eta^2_p = .003$	$F(1, 39) = 1.270$, $p = .267$, $\eta^2_p = .032$	$F < 1$	$F(1, 37) = 2.267$, $p = .141$, $\eta^2_p = .058$
Group* Topic	$F(1, 38) = 5.842$, $p = .021$, $\eta^2_p = .133$	$F < 1$	$F(1, 38) = 2.278$, $p = .140$, $\eta^2_p = .057$	$F(1, 37) = 2.641$ $p = .113$, $\eta^2_p = .067$
Time*Group	$F(1, 38) = 2.640$, $p = .112$, $\eta^2_p = .055$	$F(1, 39) = 1.355$ $p = .251$, $\eta^2_p = .034$	$F(1, 38) = 6.667$, $p = .014$, $\eta^2_p = .149$	$F < 1$
Time*Group*Topic	$F(1, 38) = .221$, $p = .641$, $\eta^2_p = .006$	$F(1, 39) = 2.998$, $p = .091$, $\eta^2_p = .071$	$F(1, 38) = 1.161$, $p = .288$, $\eta^2_p = .030$	$F < 1$

Table 3: Mean scores and standard deviations of experimental and control group with n's between brackets) on the four scales of the explicit attitude questionnaire.

Scale	*Intervention*	*Mean*	*SD*
Pretest enjoyment	Contr. group (24)	12.333	2.548
	Exp. Group (18)	13.389	2.547
Posttest enjoyment	Contr. group (25)	12.250	2.658
	Exp. Group (17)	14.333	2.910
Pretest relevance	Contr. group (25)	15.200	1.414
	Exp. Group (18)	15.056	1.305
Posttest relevance	Contr. group (25)	15.040	.841
	Exp. Group (18)	15.444	1.097
Pretest competence	Contr. group (24)	12.080	2.548
	Exp. Group (18)	12.765	2.463
Posttest competence *	Contr. group (25)	11.920	2.613
	Exp. Group (17)	13.941	2.461
Pretest motivation	Contr. group (25)	12.280	2.492
	Exp. Group (16)	12.500	2.394
Posttest motivation	Contr. group (25)	12.400	2.483
	Exp. Group (16)	13.312	2.056

For *implicit attitude* there were no effects of the intervention; the Time * Group interaction was not significant, $F(1, 36) = 1.573$, $p = .218$, $\eta^2_p = .042$. There was also no main effect of Time, $F(1, 36) = 2.229$, $p = .144$, $\eta^2_p = .058$, no interaction between Time and Topic, $F < 1$, and no three-way interaction between Time, Group and Topic, $F < 1$. There was no main effect of Group, $F(1, 36) = 1.446$, $p = .237$, $\eta^2_p = .039$, but, in general, the students who studied DNA had a higher IAT score than the students who studied the Garden Cherry, $F(1, 36) = 4.252$, $p = .046$, $\eta^2_p = .106$, suggesting these students to have a less positive implicit attitude to science and technology in general. Table 4 shows the descriptives.

As in the Denessen et al. chapter in this volume, the correlations between implicit attitude and explicit attitude did not reach significance. The IAT scores at Pretest and Posttest did not show a significant correlation with any of the scales of the explicit attitude questionnaire, or with the score on the knowledge test (either Pretest or Posttest). There was one exception: the IAT at Posttest showed a positive correlation with the relevance scale at Pretest, $r(43)=.346$, $p =.023$; suggesting that those who had a more negative attitude at Posttest were the ones with the more positive explicit attitude towards relevance at Pretest. Because of the amount of correlations calculated, this significance should be interpreted with caution. IAT scores at Pretest and Posttest were not correlated, although there was a trend, $r(40) =.272$, $p = .089$.

Table 4: Means and standard deviations of implicit attitude measure (IAT) at Pretest and Posttest for the different groups of internships. A higher score reflects a more negative attitude.

	Group topic	*Mean*	*SD*
IAT Pretest	Garden Cherry	.184	.312
	DNA	.296	.265
IAT Posttest	Garden Cherry	.232	.367
	DNA	.407	.247

DISCUSSION

In this pilot study, we tested the effects of an intervention with a digital learning environment on knowledge and attitude of teacher students. We found positive effects of the intervention for knowledge and explicit attitude towards perceived competence, but not for implicit attitude.

The fact that the intervention had a positive effect on the explicit attitude of competence of the students is as expected, since previous studies already showed that well-designed learning environments can cause attitude changes (Lampert, 1988; Ball, 1996), however, the present is the first to show this on teaching STEM. The effect was not visible in the IAT measure. However, it is interesting to note that the garden cherry group had on average a more positive explicit enjoyment attitude as well as more positive implicit attitude towards science and technology. The correlations between this explicit scale and the implicit measurement did not reach significance at either Pretest or Posttest. This is in line with other social studies, studying relation between implicit and explicit measurements. They are not necessarily intended to be highly correlated, as a high positive explicit attitude can go with a negative implicit attitude (cf. Van Goethem, Scholte&Wiers, 2010). These results are certainly interesting for future research. IAT is a fairly new measurement to be used in educational research, and, as yet, we cannot know exactly what it adds to the explicit measurement or how it predicts classroom behavior.

There were no differences between the other measures of explicit attitude. An explanation could be that the relevance scale was already high in both groups, since they both chose a minor in science and technology and were therefore already convinced of the importance of the topic. For enjoyment and motivation, also no effects were evidenced; perhaps these scales are already shaped in previous years of teacher education, whereas the competence to teach is what needs shaping during a minor in science and technology.

The results have to be interpreted with caution though. The pilot study was set up in a quasi-experimental design, and the effects might be accounted for by differences between the teachers of the two groups of students, as the control group had a different teacher than the experimental group. Also, the intervention was not yet fully developed, and so this pilot study should be seen as a formative study.

Unfortunately, we could not measure the effects of the environment on the knowledge gain and attitude change of the children that were taught by the student teachers. For future research, it would be very interesting to monitor the behavior of both teacher and children in an experimental design in which one group of (student) teachers works with the environment and one group does not. Measures of knowledge and attitude of both teachers and the children in their class can then be taken, as well as observations of the interactions between teachers and children, and between children and their peers. Also, we now measured effects on attitude towards teaching STEM, but not on general attitudes towards STEM. It would be highly interesting to see whether the attitude in general changes, and whether this has long-lasting effects on teaching STEM and whether this transfers to the attitudes of the children.

CONCLUSION

In the present chapter, we have shown the design of a digital learning environment to enhance knowledge and attitude of teachers in teaching science and technology with an additional focus on (academic) language skills. This chapter mainly had a practical focus. We have shown that the perceived competence of student teachers can be improved by teaching them with a digital learning environment. This means that showing student teachers movies of experiences teachers giving science education can function as a good role model for these student teachers. The fact that their perceived competence has been improved can be a consequence of the fact that by observing the experienced role models, they felt that they would also be able to give the science lessons in a comparable way. The fact that a digital learning environment stimulated the attitude/ perceived competence would call for implementing such an environment in the student-teacher curriculum. This may have long-lasting effects on the teachers of the future, as well as on the children (cf. Pell & Jarvis, 2003).

The use of the *environment* has direct implications for the field. It is our believe that the environment can have a larger impact on the attitude of students when it is implemented well into the curriculum, as in the Kinzer et al. (2006) example. This environment is a next step into the improvement of teacher education regarding science and technology, and as such a contribution to primary education. It may help teachers to learn how they may stimulate children's scientific inquiry processes by helping them to organize and structure information, by teaching them how to ask themselves questions or correct themselves, or by helping them to focus their attention, maintain their motivation, and minimize their fear of failure. The meta-analysis by Schroeder et al. (2007) has shown the effects this can have on the achievement of children in science.

Eliane Segers
e.segers@pwo.ru.nl

Marieke Peeters
Heleen Strating
Cindy Teunissen
Ludo Verhoeven
Radboud University Nijmegen, The Netherlands

REFERENCES

Preface

Walma van der Molen, J., Lange, J. de, & Kok, J. (2007). *Professionalisation in science & technology in The Netherlands: Primary education.* Theoretical Framework. Platform Beta Techniek: The Hague, The Netherlands.

Chapter 1

Alberts, B. (2009). Refining science education, *Science*, 323, 437.

Allende, J. (2008). Academies active in education, *Science*, 321, 1133.

Artigue, M. (2010). *Les défis de l'enseignement des mathématiques dans l'éducation de base* (2010), UNESCO, Paris.

Berthélémy, J-C. (2007). Globalization and challenges for education in least developed countries. In Malinvaud, E., Léna, P., Sanchez Sorondo, M. (Eds) (2007) *Globalisation and Education*, Pontifical Academies of Sciences & of Social Sciences, Berlin: de Gruyter.

Black, P. & Wiliam, D. (1998). Assessment and classroom learning, *Assessment in Education 5*(1): 7–74.

Bransford, J., Brown, A. & Cocking, R. (Eds) *How People Learn.* Washington, D.C.: National Academy Press.

Centre Jean Lagarde (2006). *Expérimentation sur la démarche d'investigation*, DVD, CNDP-Scéren, Paris.

Charpak, G., Léna, P. & Quéré, Y. (2005). *L'enfant et la Science*, O. Jacob.

Dehaene-Lambertz, G., Hertz-Pannier, L., Dubois, J., & Dehaene, S. (2008). How does early brain organization promote language acquisition in humans? *European Review, 16*(4), 399–411.

Driver, R. (1983). *The pupil as scientist?* Milton Keynes: Open University Press

Driver, R. (1973). The representation of conceptual frameworks in young adolescent science students. PhD thesis, University of Illinois, Urbana, Illinios.

Duschl, R. A., Schweingruber, H.A. Shouse, A.W. (Eds) (2007). *Taking Science to School: Learning and Teaching Science in Grades K-8* Washington DC: The National Academies Press.

European Commission (2007). *Science Education Now: A renewed pedagogy for the future of Europe*, (The Rocard report), Brussels: EC.

Fleer, M. (2007). *Young children: Thinking about the scientific world.* Watson, ACT; Early Childhood Australia.

Gopnik, A., Meltzoff, A. N. & Kuhl, P.K. (1999). *The Scientist in the Crib.* New York: William Morrow.

Goshwami, U. & Bryant, P. (2007). *Children's Cognitive Development and Learning.* (Primary Review Research Survey 2/1a), Cambridge: University of Cambridge Faculty of Education.

Guesne, E. (1978). Lumière et vision des objets: un example de représentation des phénomènes physiques pré-existant à l'enseignement. In G. Delacôte (Ed) *Physics teaching in schools,* London: Taylor and Francis.

Guichard, F. (2007). *Comment devient-on scientifique? Enquête sur la naissance d'une vocation*, Paris: EDP Sciences.

Harlen, W. & Holroyd, C. (1997). Primary teachers' understanding of concepts in science: impact on confidence and teaching. *International Journal of Science Education 19*(1) 93-105.

Harlen, W. & Allende, J. (Eds) (2009). *Teacher Professional Development in Pre-Secondary School Inquiry-Based Science Education (IBSE)*, InterAcademyPanel Report, http://www.iap.org.

Harlen, W. & Allende, J. (Eds) (2006). *International Colaboration in the Evaluation of Inquiry-Based Science Education (IBSE)*, InterAcademyPanel Report, http://www.iap.org.

Haworth, C., Dale, P. & Plomin, R. (2008). A twin study into the genetic and environmental influences on academic performance in science in nine-year-old boys and girls, *International Journal of Science Education, 30*(8): 1003-1025.

Howard-Jones, P., Pollard, A., Blakemore, S-J., Rogers, P. Goshwami, U., Butterworth, B., Taylor, E., Williamon, A., Morton, J. & Kaufmann, L. (2007). *Neuroscience and Education: Issues and Opportunities*, London: TLRP/ESRC.

James, M. & Pedder, D. (2006). Professional learning as a condition for assessment for learning, in J. Gardner (ed) *Assessment and Learning*. London: Sage.

La main à la pâte (2010). http://www.lamap.fr

Léna, P. (2009). Europe rethinks education, *Science*, 324, 501.

Martin, A (2010). *Building Classroom Success: Eliminating Academic Fear and Failure*. London: Continuum.

Michaels, S., Shouse, A. and Schweingruber, H. (2008) *Ready, Set, Science Putting Research to Work in K-8 Science classrooms*. Washington DC: National Academies Press.

Murphy, C., Neil, P. and Beggs, J. (2007). Primary science teacher confidence revisited: ten years on, *Educational Research,* 49 (4): 415-430.

National Science Foundation (NSF) (1997). *The Challenge and Promise of K-8 Science Education Reform. Foundations 1*. Arlington:VA: NSF.

National Science Foundation (NSF) (1999). *Inquiry Foundations 2*. Arlington: VA: NSF.

OECD (2003) *The Pisa 2003 Assessment Framework* Paris: OECD.

OECD/CERI (2005). *Students with Disabilities. Difficulties and disadvantages: Statistics and Indicators*, Paris: OECD.

OECD/CERI (2007). *Understanding the Brain: the Birth of a Learning Science*. Paris: OECD.

OECD (2008a). *Encouraging Student Interest in science and technology Studies*, Global Science Forum, Paris: OECD.

OECD (2008b). *Policies and Practices Supporting the Educational Achievement and Social Integration of First and Second generation Migrants: A systemic Review*. Paris: OECD.

OECD (2008c). *Trends Shaping Education*, Paris: OECD.

Ormerod, M.B. & Duckworth, D. (1975). *Pupils' Attitudes to Science,* Windsor: NFER.

Osborne, R.J & Freyberg, P. (1985). *Learning in Science: the Implications of 'Children's Science'*, Auckland: Heinemann.

Peer Learning Activity Report (2006). Schools as learning communities for their teachers http://www.cilo.europeesplatform.nl/files/habermehl_teachers.doc

Pollen project (2009). *Seed Cities for Science*. www.pollen-europa.net

Royal Society (2006). *Taking a leading role*. London: The Royal Society.

Royal Society (2010). *Science and mathematics education* 5–14. London: The Royal Society.

Russell, T., Black, P., Harlen, W., Johnson, S. & Palacio, D. (1988). *Science at Age 11. A review of APU Survey Findings 1980–84*. London: HMSO.

Saltiel, E. and Duclaux, M. (2010). submitted.

Sarmant, J.-P., Saltiel, E. Léna, P. (2010). La main à la pâte. Implementing a plan for science education reform in France, in G.E. de Boer (Ed) *The role of public policy in K-12 science education*, Charlotte, USA: Information Age Publishing.

SPACE (Science Processes and Concepts Exploration) Research Reports. *Evaporation and Condensation* (1990), *Growth* (1990), *Light* (1990), *Sound* (1990), *Electricity* (1991), *Materials* (1991), *Processes of Life* (1992), Earth in Space (1992) *Rocks, Soil and Weather* (1992), Forces 1998). Liverpool University Press.

Tiberghien, A., and Delacôte, G. (1978). Conception de la chaleur chez les enfants de 1 à 12 ans. In G. Delacote (Ed) *Physics teaching in schools,* London: Taylor and Francis, 275–82.

Wei Yu (2010) Key Laboratory of Child Development and Learning Science, http://rcls.seu.edu.cn/en/rd_1.html

Zull, J.E. (2004). The art of changing the brain. *Educational Leadership 62*(1): 68–72.

Chapter 2

Börner, K. (2010). *Atlas of science - Visualizing what we know*. Cambridge (MA): MIT Press.
Duschl, R. A., Schweingruber, H. A., & Shouse, A. W. (Eds.). (2007). *Taking Science to School: Learning and Teaching Science in Grades K-8*. Washington: The National Academies Press.
Goswami, U. (2008). Conceptual development and the biological world. In: *Cognitive development - The learning brain* (pp. 109–145). Hove: Psychology Press.
Hanson, J. R. (2007). *Chemistry in the garden*. Cambridge: Royal Society of Chemistry.
IFTF. (2006). *Science and Technology outlook: 2005-2055*. Palo Alto: Institute for the Future.
IFTF. (2010). *The future is a high-resolution game*. Palo Alto: Institute for the Future.
Kleinhans, M. G., Bierkens, M. F. P. and Perk M. van der (2010). On the use of laboratory experimentation: "Hydrologists, bring out shovels and garden hoses and hit the dirt", *Hydrology and Earth Systems Science 14*, 369–382.
McGee, H. (2004). *On food and cooking - The science and lore of the kitchen*. New York: Scribner.
Motivaction and Young Works. (2010). *Bèta Mentality 2011–2016 - Attracting young people to science and technology*. The Hague: Platform BètaTechniek.
Murphy, C., Neil, P., & Beggs, J. (2007). Primary science teacher confidence revisited: ten years on. *Educational Research, 49*(4), 415–430.
Santen, R. van, Khoe, D., & Vermeer, B. (2010). *2030 - Technology that will change the world*. Oxford: Oxford University Press.
Siegal, M. (2008). *Marvelous minds - The discovery of what children know*. Oxford: Oxford University Press.
Traianou, A. (2007). *Understanding teacher expertise in primary science*. Rotterdam: Sense.
Wolfe, J. M., Kluender, K. R., and Levi, D. M. (2009). *Sensation and perception*. Sunderland (MA): Sinauer.

Chapter 3

American Association for the Advancement of Science (1993). *Benchmarks for Science Literacy*. New York: Oxford University Press.
Baartman, L. K. J., & De Bruijn, E. (submitted). Integrating knowledge, skills and attitudes: Conceptualising learning processes towards vocational competence.
Bakker, A., Hoyles, C., Kent, P., & Noss, R. (2006). Improving work processes by making the invisible visible. *Journal of Education and Work, 19*, 343-361.
Bakker, A., Wijers, M., Akkerman, S., & Gravemeijer, K. (2010). *The influence of technology on what vocational students need to learn about statistics: The case of laboratory analysists*. Paper presented at the ICOTS 2010.
Bakker, A., Kent, P., Noss, R., & Hoyles, C. (2009). Alternative representations of statistical measures in computer tools to promote communication between employees in automotive manufacturing. *Technology Innovations in Statistics Education, 3*, 1–31.
Barnett, M. (1995). Literacy, technology and technological literacy. *International Journal of Technology and Design Education, 5*, 119–137.
Benenson, G. (2001). The unrealized potential of everyday technology as a context for learning. *Journal of Research in Science Teaching, 38*, 730–745.
Binkley, M., Erstad, O., Herman, J., Raizen, S., Ripley, M., & Rumble, M. (2010). *Draft White Paper 1. Developing 21st century skills*. University of Melbourne: Assessment and Teaching of 21st Century Skills. See: www.atc21s.org.
Burklund, C., Garvin, K., Lawrence, N., & Yoder, J. (1989). *A guide to developing higher order thinking across the curriculum*. Des Moines: Iowa Department of Education.
Costa, A. L. (n.d.). Components of a well developed thinking skills program. Downloaded August 11, 2010 from: http://www.newhorizons.org/strategies/thinking/costa2.htm.
Costa, A. L. (2000). Describing the habits of mind. In A. L. Costa & B. Kallick (Eds.), *Discovering & Exploring Habits of Mind* (Vol. 1, pp. 21-41). Alexandria, Virginia: ASCD.

Dede, C. (2009). Comparing frameworks for 21st century skills. In J. Bellanca & R. Brandt (Eds.) *21st Century Skills*, pp. 51–76. Bloomington, IN: Solution Tree Press.

De Vries, M. J. (2005). The nature of technological knowledge: Philosophical reflections and educational consequences. *International Journal of Technology and Design Education, 15*, 149–154.

Fazio, R. H. (2007). Attitudes as object-evaluation associations of varying strength. *Social Cognition, 25*, 603-637.

Ferguson, E. S. (1977). The Mind's Eye: Nonverbal Thought in Technology. *Science, 197,* 827–836.

Forman, S. L., & Steen, L.A. (1994). Mathematics for work. *Bulletin of the International Commission on Mathematical Instruction (ICMI), 37,* 1–6. Downloaded January, 4, 2010, from: http: www.stolaf.edu/people/steen/Papers/work.html.

Forman, S.L., & Steen, L.A. (2000). Making authentic mathematics work for all students. In A. Bessot& J. Ridgway (Eds.), *Education for Mathematics in the Workplace*. Dordrecht, the Netherlands: Kluwer Academic.

Goos, M., & Manning, A. (2007). Lousy and lovely jobs: The rising of polarization of work in Britain. *The Review of Economics and Statistics, 89*, 118–133.

Gopinathan, S. (1999). Preparing for the next rung: Economic restructuring and educational reform in Singapore. *Journal of Education and Work, 12*, 295–308.

Gravemeijer, K. (2009). *Leren voor later* [Learning for the future]. Inaugural address, Eindhoven University of Technology, the Netherlands.

Gravemeijer, K. (2010). Mathematics Education and the Information Society. Paper presented at the EIMI Conference, October 2010.

Hancock, C., Kaput, J. J., & Goldsmith, L. T. (1992). Authentic inquiry with data: Critical barriers to classroom implementation. Educational Psychologist, 27, 337–364.

Holbrook, J., & Rannikmae, M. (2007). The nature of science education for enhancing scientific literacy. *International Journal of Science Education, 29*, 1347–1362.

Kaput, J., & Schorr, R. (2007). Changing representational infrastructures changes most everything: The case of SimCalc, Algebra, and Calculus. In G. Blume & K. Heid (Eds.) *Research on technology in the learning and teaching of mathematics: Syntheses and perspectives* (pp. 211–253). Mahwah, NJ: Erlbaum.

Krathwohl, D. R. (2002). A revision of Bloom's Taxonomy: An overview. Theory into Practice, 41, 212-218.

Latour, B. (1986). Visualisation and cognition: Drawing things together. In H. Kulick (Ed.) *Knowledge and Society. Studies in the sociology of the culture past and present*, pp. 1–40. Stamford, CT: JAI.

Lesgold, A. (2009). Better schools for the 21st century. *What is needed and what will it take to get improvement.* University of Pittsburgh.

Levy, F., & Murnane, J. (2005). *How computerized work and globalization shape human skill demands.* First International Conference on Globalization and Learning. Stockholm, Sweden.

Millar, R., & Osborne, J. (1998). *Beyond 2000. Science education for the future. The report of a seminar series funded by the Nuffield Foundation.* Retrieved 16 December 2009 from: http://www.kcl.ac.uk/content/1/c6/02/18/24/b2000.pdf

Murphy, C. (2003). Literature review in primary science and ICT. Futurelab Series, report 5. Retrieved October 14, 2009 from: www.futurelab.org.uk.

Noss, R., Bakker, A., Hoyles, C., & Kent, P. (2007). Situating graphs as workplace knowledge. *Educational Studies in Mathematics, 65*, 367–384.

Osborne, J. (2007). Science education for the twenty first century. *Eurasia Journal of Mathematics, Science & Technology Education, 3*, 173–184.

Pozzi, S., Noss, R., & Hoyles, C. (1998). Tools in practice, mathematics in use. *Educational Studies in Mathematics, 36*, 105–122.

Resnick, L.B. (1987). The 1987 presidential address: Learning in school and out. *Educational Researcher, 16* (9), 13–20.

Rodrigues, S., Tytler, R., Darby, L., Hubber, P., Symington, D., & Edwards, J. (2007). The usefulness of a science degree: The 'lost voices' of science trained professionals. *International Journal of Science Education, 29*, 1411–1433.

Roth, W. M. (2001). Learning science through technological design. *Journal of Research in Science Teaching, 38*, 768–790.

Roth, W. M. (2005). Mathematical inscriptions and the reflexive elaboration of understanding: An ethnography of graphing and numeracy in a fish hatchery. *Mathematical Thinking and Learning, 7*, 75–110.

Savelsbergh, E. (2007). Dynamisch modelleren. Aanzet tot een curriculum [Dynamic modelling. The start of a curriculum]. *Nieuw Archief voor Wiskunde, 8*, 207–213.

Slangen, L., Van Keulen, H., & Gravemeijer, K. (2010). What pupils can learn from working with robotic direct manipulation environments. *International Journal of Technology and Design Education. DOI 10.1007/s10798-010-9130-8.*

Van der Kooij, H. (2002). *Wiskundig vaardig. De bijdrage van wiskunde aan (technische) beroepscompetenties* [The contribution of mathematicsto (technical) professional competence]. Exacte vakken en competenties in het beroepsonderwijs. 's Hertogenbosch, the Netherlands: Cinop.

Van Eijck, M., & Claxton, N.X. (2008). Rethinking the notion of technology in education: Techno-epistemology as a feature inherent to human praxis. *ScienceEducation, 93*, 218–232.

Van Keulen. H., & Walma van der Molen, J. (Eds.). *Onderzoek naar wetenschap en techniek in het Nederlandse basisonderwijs* [Research into science and technology in Dutch primary education]. The Hague, The Netherlands: Platform Bèta Techniek.

Voogt, J., & Pareja Roblin, H. (2010). *21st Century Skils. Discussion paper*. Enschede, the Netherlands: University of Twente.

Wynne, B. (1991). Knowledges in context. *Science, Technology, & Human Values, 16*, 111–121.

Chapter 4

Akerson, V. L., Morrison, J. A., & McDuffie, A. R. (2006). One Course Is Not Enough: Preservice Elementary Teachers' Retention of Improved Views of Nature of Science. *Journal of Research in Science Teaching, 43*, 194–213.

Angier, N. (2007). *The Canon. A whirligig tour of the beautiful basics of science*. New York: Houghton Mifflin.

Appleton, K. (2008). Developing science pedagogical content knowledge through mentoring elementary teachers. *Journal of Science Teacher Education, 19 (6)*, 523-545.

Benson, C. (2009). Making the difference - Twenty years of primary design and technology. In C. Benson, P. Bailey, S. Lawson, J. Lunt & W. Till (Eds.), *Making the difference: Proceedings of the Seventh International Primary Design and Technology Conference (pp. 30-32).* Birmingham: University of Birmingham.

Bybee, R. W., & Fuchs, B. (2006). Preparing the 21st Century Workforce: A New Reform in Science and Technology Education. *Journal of Research in Science Teaching, 43*(4), 349–352.

Dijkgraaf, R., Fresco, L., Gualthérie van Weezel, T., & Calmthout, M. van (Eds.). (2008). *De bètacanon: Wat iedereen moet weten van de natuurwetenschappen.* [The Canon of Science: What everyone should know of the sciences]. Amsterdam: Meulenhoff.

Dijkgraaf, R., Fresco, L., Haring, B., & Groothof, F. (2009). *De Junior Bètacanon LuisterCD.* [The junior Canon of Science audio compact disc]. Diemen: Luisterwijs.

Fischer, E. P. (2001). *Die andere Bildung - Was man von den Naturwissenschaften wissen sollte.* Berlin: Ullstein.

Ganten, D., Deichmann, T., & Spahl, T. (2005). *Naturwissenschaften - Alles, was man wissen muss.* München: DTV.

Garmire, E., & Pearson, G. (Eds.). (2006). *Tech Tally - Approaches to assessing technological literacy.* Washington, D.C.: National Academic Press.

National Science Board (2010). *Science and engineering indicators 2010*. Arlington: National Science Foundation.

Jenkins, E. W. (1990). Scientific literacy and school science education. *School Science Review, 71*(256), 43–51.

Laugksch, R. C. (2000). Scientific Literacy: A Conceptual Overview. *Science Education, 84*, 71–94.

Lewis, T. (2006). Design and Inquiry: Bases for an Accommodation between Science and Technology Education in the Curriculum? *Journal of Research in Science Teaching, 43*(3), 255–281.

Ministry of Education, Culture, and the Sciences (2006). *Core objectives of Dutch primary education.* The Hague: Ministry of Education, Culture, and the Sciences.

Motivaction & YoungWorks. (2010). *Bètamentality 2011-2016 - Attracting young people to science and technology.* Den Haag: Platform Bèta Techniek.

Reid, Geleinse & Van Tol. (2008). *De bètacanon van Fokke en Sukke.* [The Canon of Science according to Fokke and Sukke]. Amsterdam: Catullus.

ResearchNed. (2010). *Technomonitor 2010.* Den Haag: Platform Bèta Techniek.

ROA. (2007). *Technomonitor 2007 - De kenniseconomie: liggen we op koers?* [The knowledge economy: are we on schedule?]. Den Haag: Platform Bèta Techniek.

Rocard, M., Csermely, P., Jorde, D., Lenzen, D., Walberg-Henriksson, H., & Hemmo, V. (2007). *Science education now: A renewed pedagogy for the future of Europe.* Brussels: European Commission.

Rohaan, E., & Keulen, H. van (2010). *Lessuggesties bij de junior bètacanon.* [Teaching ideas for the Canon of the Sciences] Den Haag: Platform Bèta Techniek.

Shulman, L. S. (1987). Knowledge and teaching: Foundations of the New Reform. *Harvard Educational Review, 57(1)*, 1-22.

Shrout, PE. & Fleiss, J. L. (1979). Intraclass correlation: uses in assessing rater reliability. *Psychological Bulletin, 86 (2)*, 420–428.

Traianou, A. (2007). *Understanding teacher expertise in primary science.* Rotterdam: Sense Publishers.

Walma van der Molen, J. (2009). Wat vinden leraren basisonderwijs van wetenschap en techniek? [What do primary teachers think of science and technology?]. In H. van Keulen & J. Walma van der Molen (Eds.), *Onderzoek naar wetenschap en techniek in het Nederlandse basisonderwijs (pp. 157-163).* Den Haag: Platform Bèta Techniek.

Yilmaz-Tuzun, O. (2008). Preservice Elementary Teachers' Beliefs About Science Teaching. Journal of *Science Teacher Education, 19(2)*, 183–204.

Chapter 5

Abell, S. K., Rogers, M. A. P., Hanuscin, D. L., Lee, M. H., & Gagnon, M. J. (2009). Preparing the Next Generation of Science Teacher Educators: A Model for Developing PCK for Teaching Science Teachers. *Journal of Science Teacher Education,* 20, 77–93.

APP (2010). *National Curriculum for England and Wales.* Retrieved from the website of APP: http://nationalstrategies.standards.dcsf.gov.uk/node/263985, 19 March 2010.

Appleton, K. (2003). How do beginning primary school teachers cope with science? Toward an understanding of science teaching practice. *Research in Science Education*, 33(1), 1–25.

Appleton, K. (2007). *Elementary science teaching.* In: Abell & Ledermann: Handbook of Research on Science Education. Mahwah (NJ, USA): Lawrence Earlbaum Associates, 493–536.

Appleton, K. (2008). Developing science pedagogical content knowledge through mentoring elementary teachers. *Journal of Science Teacher Education*, 19, 523–545.

Berg, E. van den, Louman, E., & Marell, J. (2010). *Kennisbasis Natuurwetenschappen en Techniek voor Leerkrachten in het Primair Onderwijs.* [Knowledge base science and technology for teacher in primary education]. Draft version. Den Haag: LOBO.

Boersma, K., Graft, M. van, Harteveld, A., Hullu, E. de, De Knecht-van Eekelen, A. de, Mazereeuw, M., et al. (2007). *Leerlijn biologie van 4 tot 18 jaar. Uitwerking van de concept-contextbenadering tot doelstellingen voor het biologieonderwijs.* [Teaching plan biology 4-18. Elaboration of the concept-context approach into objective for biology education]. Utrecht: NIBI.

Bybee, R. (2010). *The teaching of science: 21st century perspectives.* Arlington: NSTA Press.

Coe, M.A. (2001). http://faculty.mwsu.edu/west/maryann.coe/coe/inquire/inquiry.htm (consulted on February 21, 2010).
Dewey, J. (1910). Science as subject-matter and as method. *Science, 31*, 121–127.
Dewey, J. (1916). Method in science teaching. *The Science Quarterly, 1*, 3–9.
Dijkgraaf, R., Fresco, L., Gualthérie van Weezel, T., & Calmthout, M. van (Eds.). (2008). *De bètacanon*. [The canon of the sciences]. Amsterdam: Meulenhoff.
Duschl, R.A., Schweingruber, H.A., & Shouse, A.W. (2007). *Taking Science to School: Learning and Teaching Science in Grades K-8*. Washington: The National Academies Press. http://www.nap.edu.
ETS (2008). *The Praxis Series Official Guide*. Princeton: Educational Testing Services.
Graft, M. van, & Kemmers, P. (2007). *Onderzoekend en Ontwerpend Leren in Natuur en Techniek*. [Inquiry and design based learning in science and technology]. Den Haag: Platform Bèta Techniek.
Hacker, M., Vries, M. de, & Rossouw, A. (2009). *Concepts and Contexts in Engineering and Technology Education*. New York/Delft: Hofstra University & Delft University of Technology.
Hackling, M. W., & Prain, V. (2005). *Primary connections: Stage 2 Trial: Research Report*. Canberra: Australian Academy of Science.
Harlen, W. (1997). Primary teachers' understanding in science and its impact in the classroom. *Research in Science Education*, 27(3), 323–337.
Harlen, W. (2006). *Teaching, Learning & Assessing Science 5-12* (4th edition). London: SAGE Publications.
Harlen, W. & Qualter, A. (2009). *The Teaching of Science in Primary Schools* (5th edition). London: Routledge.
ITEEA. (2007). *Standards for technological literacy: Content for the study of technology*. Reston: International Technology and Engineering Education Association.
Katehi, L., Pearson, G., & Feder, M. (2009). *Engineering in K-12 Education: Understanding the status and improving the prospects*. Washington: National Academy of Engineering and National Research Council.
Keulen, H. van (2009). *Drijven en zinken, wetenschap en techniek in het primair onderwijs*. [Floating and sinking: science and technology in primary education]. Inaugural address. Eindhoven: Fontys University of Applied Sciences.
Keulen, H. van (2010). *Wetenschap en techniek - IJkpunten voor een domein in ontwikkeling*. [Science and technology – Benchmarks for a developing domain]. Den Haag: Platform Beta Techniek.
La Main à la Pâte (2008). Retrieved on November 28, 2008 from http://lamap.inrp.fr/.
Millar, R. (2010). Analysing practical science activities. College Lane, Hatfield (UK):Association for Science Education, ISBN 9780863574252.
Millar, R., & Driver, R. (1987). *Beyond processes*. Studies in Science Education, 14, 33-62.
OCW (2006). *Core objectives primary education*. Den Haag: Ministerie van Onderwijs, Cultuur en Wetenschappen.
Motivaction, & YoungWorks. (2010). *Bèta Mentality 2011-2016 - Attracting young people to science and technology*. Den Haag: Platform Bèta Techniek.
Murphy, C., Neil, P., & Beggs, J. (2007). Primary science teacher confidence revisited: ten years on. *Educational Research, 49*(4), 415–430.
NSES (1996). National Science Education Standards. Washington: National Academy of Sciences. http://www.nap.edu/catalog/4962.html
National Science Teacher Association (2003). *Standards for science teacher preparation*. Arlington: NSTA.
NSTA. (2010). *Draft Revisions to the 2003 NSTA Standards for Science Teacher Preparation*. Arlington: NSTA.
OECD. (2009). *PISA 2009 Assessment Framework - Key competencies in reading, mathematics and science*: OECD.
QCA (1999) *The National Curriculum for England and Wales*. http://curriculum.qca.org.uk/key-stages-1-and-2/subjects/science/index.aspx

Rocard, M., Csermely, P., Jorde, D., Lenzen, D., Walberg-Henriksson, H., & Hemmo, V. (2007). *Science education now: A renewed pedagogy for the future of Europe*. Brussels: European Commission.

Rohaan, E. (2009). *Testing teacher knowledge for technology teaching in primary education.* PhD Thesis. Eindhoven: Eindhoven University of Technology.

Shulman, L.S. (1987). *Knowledge and teaching: Foundations of the new reform. Harvard Educational Review, 57*(1), 1-21.

Smith, C. L., Maclin, D., Houghton, C., & Hennessey, M. G. (2000). Sixth-grade students' epistemologies of science: The impact of school science experiences on epistemological development. *Cognition and Instruction, 18*(3), 285–316.

Traianou, A. (2007). *Understanding teacher expertise in primary science*. Rotterdam: Sense.

Vaan, E. de, & Marell, J. (2006). *Praktische Didactiek voor Natuuronderwijs.* [Practical pedagogy for nature education]. Bussum: Coutinho.

Wagenschein, M. (1989). *Verstehen Lehren - Genetisch-Sokratisch-Exemplarisch.* Weinheim: Beltz.

Walma van der Molen, J., Lange, J. de, & Kok, J. (2007). *Uitwerking van het Theoretisch Kader voor de Professionalisering van Leerkrachten op het Gebied van Wetenschap en Techniek.* [Elaboration of the theoretical framework for professional development of teachers in the area of science and technology]. The Hague, The Netherlands, Platform Bèta Techniek.

Walma van der Molen, J. (2009). Wat vinden leraren basisonderwijs van wetenschap en techniek? [What do primary teachers think of science and technology?]. In H. van Keulen & J. Walma van der Molen (Eds.), *Onderzoek naar wetenschap en techniek in het Nederlandse basisonderwijs* [Research into science and technology in Dutch primary education] (pp. 157–163). The Hague, The Netherlands: Platform Bèta Techniek.

Worth, K., Duque, M., & Saltiel, E. (2009). *Designing and implementing inquiry-based science units for primary education.* Montrouge: La main à la pâte.

Chapter 6

Alfieri, L., Brooks, P. J., Aldrich, N. J., & Tenenbaum, H. R. (2011). Does discovery-based instruction enhance learning? *Journal of Educational Psychology*, 103(1), 1–18.

Berg, B.L. (1998) *Qualitative research methods for the social sciences*. Allyn& Bacon.

Benthem, van. J., Dijkgraaf R. & Lange, de. J. (2005) *TalentenKracht.* [Curious Minds]. Network search on 23rd jun 2010 at http://www.talentenkracht.nl/content/files/SITE1765/TalentenKracht.pdf.

Bransford, J.D., Brown, A.L., & Cocking, R.R. (Eds.) (2004). *How people learn: brain, mind, experience, and school.* Washington D.C.: National Academy Press.

Cellier, J., Eyrolle, H., & Mariné. (1997). Expertise in dynamic environments. *Ergonomics*, 40(1), 28–50.

Creighton, L.L. (2008) *Becoming an elementary science teacher: Identity development in a science methods course*. Dissertation Abstracts International Section A: Humanities and Social Sciences, 68 (7-A), 2882.

Foo, S. Y., & Hedberg, J. G. (2005). Teacher understandings of technology affordances and their impact on the design of engaging learning experiences.*Educational Media International, 42*(4), 297–316.

Geert, van. P. L. C. & Steenbeek, H. W. (2005). The dynamics of scaffolding. New Ideas in Psychology, 23(3), 115–128.

Geert, van. P.L.C. & Steenbeek, H. W. (2007). *Onderzoeksvoorstelsatelliet Groningen TalentenKracht; periode januari 2008 – december 2010*. [Research proposal Curious Minds Groningen]. Groningen: Rijksuniversiteit Groningen.

Geert, P.L.C. van & Steenbeek, H. W, (2007). *Waar ligt de kracht? Bijlage bij onderzoeksvoorstel satelliet Groningen Talenten- Kracht*. [Where is the power? Appendix to the research proposal of Curious Minds Groningen]. Groningen: Rijksuniversiteit Groningen.

Glaser, B.G., & Strauss, A. (1967). *The Discovery of Grounded Theory: Strategies for Qualitative Research.* Chicago: Aldine

Goovaerts, E. (2005). *Geweld en misbruik in de privé-sfeer: verschillende vormen van psychisch geweld op mannen.* [Violence and abuse in private life: different forms of mental violence on men]. Vrije Universiteit Brussel

Hirsh-Pasek, Golinkoff, Berk, & Singer, E. (2009). *A mandateforplayfullearning in preschool; Presenting the evidence.* Oxford: Oxford University Press.

Linn, M.C., & Eylon, B. (2006). Science Education: Integration Views of Learning and Instruction. In P.A., Alexander, & P.H., Winne (Eds.), *Handbook of Educational Psychology*, second edition (pp. 655–674). New York, Londen: Lawrence Erlbaum Associates.

Martin, Yancey, P., Turner & Barry A.. (1986). Grounded Theory and Organizational Research. The *Journal of Applied Behavioral Science, 22*(2), 141. Retrieved June 21, 2009, from ABI/INFORM Global database. (Document ID: 1155984)

Mortelmans, D. (2001). *Atlas-ti, een inleiding.* Faculteit PSW: Universiteit Antwerpen

Muhr, T. (1997) *Atlasti - The KnowledgeWorkbench. Visual Qualitative Data. Analysis Management Model Building.* Germany: Scientific Software Development.

Osborne, J., Simon, S., & Collins, S. (2003). Attitudes towards science: A review of the literature and its implications. *International Journal of Science Education, 25*(9), 1049–1079.

Schwartz, R.S., Lederman, Norman, G., Abd – El - Khalick, F. (2000). Achieving the reforms vision: The effectiveness of a specialists-led elementary science program. *School Science and Mathematics* 100(4), 181–193.

Seidman, I. (1998). *Interviewing as Qualitative Research. A Guide for Researchers in Education and the Social Sciences.* Teachers College Press: New York.

Steenbeek, H.W. & Geert, van. P.L.C. (2009). *Pilot Expertisecentrum TalentenKracht Groningen – Utrecht Projectplan.* [Pilot Curious Minds Groningen]. Faculteit Gedrags- en Maatschappijwetenschappen: Rijksuniversiteit Groningen

Steenbeek, H.W., Geert, P.L.C. van & Dijk, M.W.G. van (2011). The dynamics of children's science and technology talents: A conceptual framework for early science education. Submitted for publication in *Mind, Brain and Education.*

Steenbeek, H.W., Uittenbogaard, W., Feys, E., Franken, D., Geert, van. P.L.C, de Lange, J., Munk & Post, A. (2009) *Bètatalenten van jonge kinderen in kaart.* [Mapping science talents of young children]. Panama Journaal.

Steenbeek, H.W. & Uittenbogaard, W. (2009). *Bèta-talenten van jonge kinderen in kaart.* [Mapping science talents of young children]. Panama-Post. Reken-wiskundeonderwijs: onderzoek, ontwikkeling, praktijk. *28*(1), pp. 89-100.

Thomas, G. & James, D. (2006). Reinventing grounded theory: some questions about theory, ground and discovery, *British Educational Research Journal*, 32(6) 767–795.

Todman, J.B. & Dugard, P. (2001). *Single case and small-n experimental designs: A practical guide to randomization tests.* Manwah (NJ): Erlbaum.

Uittenbogaard, W., & Feijs, E. (2006). *Marijn en Maurits.* Utrecht: FreudenthalInstitute, Utrecht University. [Video published at www.talentenkracht.nl]

Uittenbogaard, W., & Feijs, E. (2006). *Quinten en de muurtjes.* Utrecht: FreudenthalInstitute, Utrecht University. [Video published at www.talentenkracht.nl]

Uittenbogaard, W., & Feijs, E. (2007). *Jop en de periscoop.* Utrecht: FreudenthalInstitute, Utrecht University. [Video published at www.talentenkracht.nl]

Van Aalderen-Smeets, S.I., Walma van der Molen, J.H., & Asma L.J.F. (2011). Primary teachers' attitudes towards science and technology: A new theoretical framework. *Science Education.* doi: 10.1002/sce.20467.

Chapter 7

Appleton, K., & Kindt, I. (1999). Why teach primary science? Influences on beginning teachers' practices. *International Journal of Science Education, 21*, 155–168.

Bennet, J., Rollnick, M., Green, G., & White, M. (2001). The development and use of an instrument to assess students' attitude to the study of chemistry. *International Journal of Science Education, 23*, 833-845.

Blalock, C. L., Lichtenstein, M. J., Owen, S., Pruski, L., Marshall, C., & Toepperwein, M. (2008). In pursuit of validity : A comprehensive review of science attitude instruments 1935 – 2005. *International Journal of Science Education, 30*, 961–977.

Coulson, R. (1992). Development of an instrument for measuring attitudes of early childhood educators towards science. *Research in Science Education, 22*, 101–105.

Eagly, A., & Chaiken, S. (1993). *The psychology of attitudes.* Belmont, CA: Wadsworth group/ Thomsom Learning.

Gardner, P. (1995). Measuring attitudes to science: unidimensionality and internal consistency revisited. *Research in Science Education, 25*, 283–289.

Goodrum, D., Hackling, M., & Rennie, L. (2001). *The Status and quality of teaching and learning of science in Australian schools* (A research report for the Department of Education, Training and Youth Affairs). Canberra, Australia.

Haney, J., Czerniak, C., & Lumpe, A. (1996). Teacher beliefs and intentions regarding the implementation of science education reform strands. *Journal of Research in Science Teaching, 33*, 971–993.

Harlen, W., & Holroyd, C. (1997). Primary teachers' understanding of concepts of science: Impact on confidence and teaching. *International Journal of Science Education, 19*, 93–105.

Jarrett, O. S. (1999). Science interest and confidence among preservice elementary teachers. *Journal of Elementary Science Education, 11*, 47–57.

Jarvis, T., & Pell, A. (2004). Primary teachers' changing attitudes and cognition during a two-year science in-service programme and their effect on pupils. *International Journal of Science Education, 26*, 1787–1811.

Mulholland, J., & Wallace, J. (1996). Breaking the cycle: Preparing elementary teachers to teach science. *Journal of Elementary Science Education, 8*, 17–38.

Osborne, J., Simon, S., Collins, S. (2003). Attitudes towards science: a review of the literature and its implications. *International Journal of Science Education, 25*, 1049–1079.

Osborne, J., & Dillon, J. (2008). *Science Education in Europe: Critical Reflections.* London: Nuffield Foundation. http://www.nuffieldfoundation.org/fileLibrary/pdf/Sci_Ed_in_Europe_Report_Final.pdf

Palmer, D. (2001). Factors contributing to attitude exchange amongst preservice elementary teachers. *Science Teacher Education, 86*, 122–138.

Palmer, D. (2004). Situational interest and the attitudes towards science of primary teacher education students. *International Journal of Science Education, 26*, 895–908.

Plonczak, I. (2008). Science for all: Empowering elementary school teachers. *Education, Citizenship and Social Justice, 3*, 167–181.

Reid, N. (2006). Thoughts on attitude measurement. *Research in Science & Technological Education, 24*, 3–27.

Skamp, K. (1991). Primary science and technology: How confident are teachers? *Research in Science Education, 21*, 290–299.

Tai, R. H, Liu, C. Q, Maltese, A. V, & Fan, X. (2006). Planning early for careers in science. *Science, 312*, 1143–1145.

Tobin, K., Tippins, D., & Gallard, A. (1994). Research on instructional strategies for teaching science. In D. Gabel, *Handbook of research on science teaching and learning* (pp. 45-93). New York, NY: Macmillan.

Tosun, T. (2000). The beliefs of preservice elementary teachers toward science and science teaching. *School Science and Mathematics, 100*, 374–379.

Trumper, R. (1998). The need for change in elementary school teacher training: The force concept as an example. *Asia-Pacific Journal of Teacher Education, 26*, 7–15.

Van Driel, J. H., Beijaard, D., & Verloop, N. (2001). Professional development and reform in science education: The role of teachers' practical knowledge. *Journal of Research in Science Teaching, 38*, 137–158.

Weinburgh, M. (2007). The effect of Tenebrio obscurus on elementary preservice teachers' content knowledge, attitudes, and self-efficacy. *Journal of Science Teacher Education, 18*, 801–815.

Yates, S., & Goodrum, D. (1990). How confident are primary school teachers in teaching science? *Research in Science Education, 20*, 300–305.

Young, B. J., & Kellogg, T. (1993). Science attitudes and preparation of preservice elementary teachers. *Science Education, 77*, 279–291.

Chapter 8

Ajzen, I. (1991). The theory of planned behavior. *Organizational Behavior and Human Decision Processes*, 50, 179–211.

Ajzen, I. (2001). Nature and operation of attitudes. *Annual Review of Psychology, 52*, 27–58.

Ajzen, I. (2002). Perceived behavioral control, self-efficacy, locus of control, and the Theory of Planned Behavior. *Journal of Applied Social Psychology, 32*, 665–683.

Atwater, M.M., Gardner, C., & Kight, C.R. (1991). Beliefs and attitudes of urban primary teachers toward physical science and teaching physical science. *Journal of Elementary Science Education, 3*, 3–12.

Bandura, A. (1997). *Self-efficacy: The exercise of control.* New York, NY Freeman.

Barmby, P., Kind, P.M., & Jones, K. (2008). Examining changing attitudes in secondary school science. *International Journal of Science Education, 30*, 1075–1093.

Bennett, J., Rollnick, M., Green, G., & White, M. (2001). The development and use of an instrument to assess students' attitude to the study of chemistry. *International Journal of Science Education, 23*, 833- 845.

Blalock, C. L., Lichtenstein, M. J., Owen, S., Pruski, L., Marshall, C., & Toepperwein, M. (2008). In pursuit of validity: A comprehensive review of science attitude instruments 1935 – 2005. *International Journal of Science Education, 30*, 961–977.

Coulson, R. (1992). Development of an instrument for measuring attitudes of early childhood educators towards science. *Research in Science Education, 22*, 101–105.

Eagly, A., & Chaiken, S. (1993). *The psychology of attitudes.* Belmont, CA: Wadsworth group/ Thomsom Learning.

Enochs, L.G., & Riggs, I.M. (1990). Further development of an elementary science teaching efficacy belief instrument: A preservice elementary scale. *School Science and Mathematics, 90*, 694-706.

Gardner, P.L. (1995). Measuring attitudes to science: Unidimensionality and internal consistency revisited. *Research in Science Education, 25*, 283–289.

Hartshorne, R. (2008). Effects of hypermedia-infused professional development on attitudes toward teaching science. *Journal of Educational Computing Research, 38*, 333–351.

Harty, H., Samuel, J., & Anderson, H. (1991). Understanding the nature of science and attitudes towards science and science teaching of preservice elementary teachers in three preparation sequences. *Journal of Elementary Science Education, 3*, 13–22.

Kind, P., Jones, K., & Barmby, P. (2007). Developing attitudes towards science measures. *International Journal of Science, 29*, 871–893.

Koballa Jr., T. R. (1986). Teaching hands-on science activities: Variables that moderate attitude-behavior consistency. *Journal of Research in Science Teaching, 23*, 493–502.

Lumpe, A. T., Haney, J. J., & Czerniak, C. M. (2000). Assessing teachers' beliefs about their science teaching context. *Journal of Research in Science Teaching, 37*, 275–292.

Osborne, J., Simon, S., & Collins, S. (2003). Attitudes towards science: A review of the literature and its implications. *International Journal of Science Education, 25*, 1049–1079.

Pajares, M. F. (1992). Teachers ' beliefs and educational research: Cleaning up a messy construct. *Review of Educational Research, 62*, 307–332.

Palmer, D. (2001). Factors contributing to attitude exchange amongst preservice elementary teachers. *Science Education, 86*, 122–138.

Palmer, D. (2006). Sources of self-efficacy in a science methods course for primary teacher education students. *Research in Science Education, 36*, 337–353.

Pardo, R., & Calvo, F. (2002). Attitudes toward science among the European public: a methodological analysis. *Public Understanding of Science, 11*, 155–195.

Ramey-Gassert, L., Shroyer, M., & Staver, J. (1996). A qualitative study of factors influencing science teaching self-efficacy of elementary level teachers. *Science Education, 80*, 283–315.

Reid, N. (2006). Thoughts on attitude measurement. *Research in Science & Technological Education, 24*, 3–27.

Skamp, K. (1991). Primary science and technology: How confident are teachers? *Research in Science Education, 21*, 290–299.

Tosun, T. (2000). The beliefs of preservice elementary teachers toward science and science teaching. *School Science and Mathematics, 100*, 374–379.

Van Aalderen-Smeets, S.I., Walma van der Molen, J.H., & Asma, L.J.F. (2011). Primary teachers' attitudes towards science: A new theoretical framework. *Science Education*. doi: 10.1002/sce.20467

Wenner, G. (1993). Relationship between science knowledge levels and beliefs toward science instruction held by preservice elementary teachers. *Journal of Science Education and Technology, 2*, 461–468.

Wenner, G. (1995). Science knowledge and efficacy beliefs among preservice elementary teachers: A follow-up study. *Journal of Science Education and Technology, 4*, 307–315.

Yates, S., & Goodrum, D. (1990). How confident are primary school teachers in teaching science? *Research in Science Education, 20*, 300–305.

Chapter 9

Atwater, M.M., Gardner, C., & Kight, C.R. (1991). Beliefs and attitudes of urban primary teachers toward physical science and teaching physical science. *Journal of Elementary Science Education, 3*, 3–12.

Berliner, D.C. (2004). Describing the behavior and documenting the accomplishments of expert teachers. *Technology & Society, 24*, 200–212.

Bitner, B. (1994). *Revised Science Attitude Scale for preservice elementary teachers: re-examined.* Paper presented at the annual meeting of the National Association for Research in Science Teaching, Anaheim, 1994.

Bleicher, R.E. (2007). Nurturing confidence in preservice elementary science teachers. *Journal of Science Teacher Education, 18*, 841–860.

Brandell, G., Nyström, P., Sundqvist, C. (2004). *Mathematics – a Male Domain?* Paper presented at ICME 10, Topic Study Group 26, Copenhagen.

Brophy, J. E., & Good, T. L. (1970). Teachers' communication of differential expectations for children's classroom performance - Some behavioral data. *Journal of Educational Psychology, 61*, 365-374.

Coulson, R. (1992). Development of an instrument for measuring attitudes of early childhood educators towards science. *Research in Science Education, 22*, 101–105.

Donaghue, H. (2003). An instrument to elicit teachers' beliefs and assumptions. *ELT Journal, 57*, 344–351.

Forgasz, H.J., Leder, G.C., and Gardner, P.L. (1999). The Fennema-Sherman Mathematics as a Male Domain scale re-examined. *Journal for Research in Mathematics Education, 30*, 342–348.

Gardner, P. (1995). Measuring attitudes to science: unidimensionality and internal consistency revisited. *Research in Science Education, 25*, 283–289.

Gawronski, B., LeBel, E.P., & Peters, K.R. (2007). What do implicit measures tell us? Scrutinizing the validity of three common assumptions. *Perspectives on Psychological Science, 2*, 181–193.

Greenwald, A. G., McGhee, D. E., Schwartz, J. L. K. (1998). Measuring individual differences in implicit cognition: The implicit association test. *Journal of Personality and Social Psychology, 74*, 1464–1480.

Hofmann, W., Gawronski, B., Gschwendner, T., Le, H., & Schmitt, M. (2005). A meta-analysis on the correlation between the Implicit Association Test and explicit self-report measures. *Personality and Social Psychology Bulletin, 31*, 1369–1385.

Jarvis, T., & Pell, A. (2004). Primary teachers' changing attitudes and cognition during a two-year science in-service programme and their effect on pupils. *International Journal of Science Education, 26*, 1787–1811.

Johnston, J., & Ahtee, M. (2006). Comparing primary student teachers' attitudes, subject knowledge and pedagogical content knowledge needs in a physics activity. *Teaching and Teacher Education, 22*, 503–512.

Jussim, L., & Harber, K. D. (2005). Teacher expectations and self-fulfilling prophecies: Knowns and unknowns, resolved and unresolved controversies. *Personality and Social Psychology Review, 9*, 131–155.

Nosek, B.A. (2007). Implicit-explicit relations. *Current Directions in Psychological Science, 16*, 65–69.

Nosek, B.A., Banaji, M.R., & Greenwald, A.G. (2002). Math = male, me = female. Therefore math ≠ me. *Journal of Personality and Social Psychology, 83*, 44–59.

Nosek, B. A., Greenwald, A. G., & Banaji, M. R. (2005). Understanding and using the Implicit Association Test: II. Method variables and construct validity. *Personality and Social Psychology Bulletin, 31*, 166–180.

Nosek, B. A., Smyth, F. L., Sriram, N., Lindner, N. M., Devos, T., Ayala, A., et al. (2009). National differences in gender-science stereotypes predict national sex differences in science and math achievement. *Proceedings of the National Academy of Sciences of the United States of America, 106*, 10593–10597.

OECD. (2006). *Assessing scientific, reading and mathematical literacy: A framework for PISA 2006.* Paris: OECD.

Osborne, J., Simon, S., & Collins, S. (2003). Attitudes towards science: a review of the literature and its implications. *International Journal of Science Education, 25*, 1049–1079.

Poehlman, T. A., Uhlmann, E., Greenwald, A. G., & Banaji, M. R. (2004). *Understanding and using the Implicit Association Test: III. Meta-analysis of predictive validity* (unpublished manuscript). Yale University: New Haven, CT.

Rosenthal, R. (1994). Interpersonal expectancy effects - a 30-year perspective. *Current Directions in Psychological Science, 3*, 176–179.

Shulman, L. S. (1987). Knowledge and teaching: Foundations of the new reform. *Harvard Education Review, 57*, 1–22.

Schwarz, N. (2008). Attitude measurement. In: W. D. Crano& R. Prislin (Eds.), *Attitudes and attitude change* (pp. 41–60). New York: Psychology Press.

Shrigley, R. L. (1973). *The correlation of science attitude and science knowledge of preservice elementary teachers.* Paper presented at the Annual convention of the national science teachers association.

Tiedemann, J. (2000). Gender-related beliefs of teachers in elementary school mathematics. *Educational Studies in Mathematics, 41*, 191–207.

Tosun, T. (2000). The beliefs of preservice elementary teachers toward science and science teaching. *School Science and Mathematics, 100*, 374–379.

Van Aalderen-Smeets, S.I., Walma van der Molen, J.H., & Asma, L.J.F. (2011). Primary teachers' attitudes towards science: A new theoretical framework. *Science Education.* doi: 10.1002/sce.20467.

Van den Bergh, L., Denessen, E., Hornstra, L., Voeten, M., & Holland, R. W. (2010). The implicit prejudiced attitudes of teachers: Relations to teacher expectations and the ethnic achievement gap. *American Educational Research Journal, 47*, 497–527.

Walma van der Molen, J. (2009). Wat vinden leraren basisonderwijs van wetenschap en techniek? [What do primary teachers think of science and technology?]. In H. van Keulen & J. Walma van der Molen (Eds.), *Onderzoek naar wetenschap en techniek in het Nederlandse basisonderwijs* [Research into science and technology in Dutch primary education] (pp. 157–163). The Hague, The Netherlands: Platform Bèta Techniek.

Walma van der Molen, J., Lange, J. de, & Kok, J. (2009). Theoretische uitgangspunten bij de professionalisering van leraren basisonderwijs op het gebied van wetenschap en techniek [Theoretical considerations for the professionalisation of primary school teachers in science and technology]. In H. van Keulen & J. Walma van der Molen (Eds.), *Onderzoek naar wetenschap en techniek in het Nederlandse basisonderwijs* [Research into science and technology in Dutch primary education] (pp. 29-40). The Hague, The Netherlands: Platform Bèta Techniek

Wigboldus, D. (2003). *A Single Target IAT: Method.* Retrieved 15 september 2010 from http://www.unibielefeld.de/psychologie/ae/AE05/LEHRE/SS2004/Urteilsbildung/WigboldusSTIAT.pdf.

Chapter 10

Abulude, F. (2009). *Students' attitudes towards chemistry in some selected secondary schools in akure south local government area, Ondo State.* Post graduate thesis, the national teachers institute, Kaduna, Nigeria. Retrieved from http://www.scribd.com/doc/34545165/students%E2%80%99-attitudes-towards-chemistry-in-some-selected-secondary-schools-in-akure-south-local-government-area-ondo-state.

Ajzen, I. (2005). *Attitudes, personality and behavior.* Berkshire, United Kingdom: Open University.

Barak, M. & Zadok, Y. (2009). Robotics projects and learning concepts in science, technology and problem solving. *International Journal of Technology and Design Education, 19*, 289–307.

Chen, S. (2006). Development of an instrument to assess views on nature of science and attitudes toward teaching science. *Science Education, 90,* 803–819.

Cheung, D. (2009). Students' attitudes chemistry lessons: The interaction effect between grade level and gender. *Research in Science Education, 39*, 75–91.

De Winter, V., & Van Cleynenbreugel, C. (2010). *Leerkrachtprofielen en onderwijs in wetenschap en techniek in het basisonderwijs: Werkzame bestanddelen voor deskundigheidsbevordering en attitudeverandering (VTB-Pro aanvullend onderzoek).* [Teacher profiles and education in science and technology in primary education: Active ingredients to promote professionalism and a positive change in attitudes (VTB-Pro complementary research).] Internal report (draft), Centre for Experiential Education, Catholic University Leuven, Belgium.

Ho, L., & Kuo, T. (2010). How can one amplify the effect of e-learning? An examination of high-tech employees' computer attitude and flow experience. *Computers in Human Behavior, 26*, 23–31.

Hox, J. (2002). *Multilevel analysis: Techniques and applications.* Mahwah, NJ: Erlbaum.

Jalil, P.A., Sbeih, A.M.Z., Boujettif, M., & Barakat, R. (2009). Autonomy in science education: a practical approach in attitude shifting towards science learning. *Journal of Science Education and Technology, 18*, 476–486.

Jarvis, T. (2006). *Guide for the seed city trainer.* Barcelona, Spain: P.A.U. Education.

Joukes, G. (2010). *Meiden en wetenschap & techniek: Van tegenpolen tot aantrekkingskracht.* [Girls and science & technology: From opposites to attraction.] Maarssen, The Netherlands: Kaldenbach Grafische Producties.

Kemmers, P., Klein Tank, M., & Van Graft, M. (2007). *Onderzoekend en ontwerpend leren bij natuur en techniek: Evalueren van brede ontwikkeling van leerlingen in open onderwijsvormen.*[Inquiry-based learning and learning by design in science and technology: Evaluation of the broad development of pupils in open education forms.] Enschede, The Netherlands: Institute for Curriculum Development (SLO).

Laevers, F. (1993). Deep level learning: an exemplary application on the area of physical knowledge. *European Early Childhood Research Journal, 1*, 53–68.

Laevers, F. (1998). Understanding the world of objects and of people: Intuition as the core element of deep level learning. *International Journal of Educational Research, 29*, 69–85.

Laevers, F. (2005). The curriculum as means to raise the quality of ECE. Implications for policy. *European Early Childhood Education Research Journal, 13*, 17–30.

Laevers, F. (2006a). Onderwijs op een nieuwe leest. Een reconstructie van de actuele paradigma-shift. [A new approach for education. A reconstruction of the actual paradigma-shift]. In: F. Laevers, en E. Bertrands (Eds.), *Draagkracht geven. Uitkomst voor de toekomst.* (pp.5–24). Leuven: CEGO.

Laevers, F. (2006b). *Kijkwijzer voor een Procesgerichte Analyse van Onderwijsleersituaties.* [Monitoring system for a process-oriented analysis of learning situations in education]. Leuven: Centrum voor Ervaringsgericht Onderwijs.

Lewis, T. (2009). Creativity in technology education: providing children with glimpses of their inventive potential. *International Journal of Technology and Design Education, 19*, 255–268.

Oberon (2009). *Een selectie uit de tussenrapportage effectstudie VTB-Pro. Schooljaar 2008/09.* [A selection from the interim rapport effectstudy VTB-Pro. 2008/09]. Utrecht, The Netherlands: Platform Bèta Techniek.

Ogbuehi, P., & Fraser, B. (2007). Learning environment, attitudes and conceptual development associated with innovative strategies in middle-school mathematics. *Learning Environments Research, 10*, 101–114.

Pajares, F. (1996). Self-efficacy beliefs in academic settings. *Review of Educational Research, 66*, 543–578.

Pell, A., & Jarvis, T. (2003). Developing attitude to science education scales for use with primary teachers. *International Journal of Science Education, 25*, 1273–1295.

Shrigley, R., & Johnson, T. (1974). The attitude of in service elementary teachers toward science. *School Science and Mathematics, 74*, 437–446.

Sidawi, M.M. (2009). Teaching science through designing technology. *International Journal of Technology and Design Education, 19*(3), 269–287.

Snijders, T. A. B., & Bosker, R. J. (1999). *Multilevel analysis: An introduction to basic and advanced multilevel modelling.* London: Sage.

Tschannen-Moran, M., Woolfolk Hoy, A., & Hoy, W. K. (1998). Teacher efficacy: Its meaning and measure. *Review of Educational Research, 68*, 202–248.

Van Graft, M., & Kemmers, P. (2007). *Onderzoekend en ontwerpend leren bij Natuur en Techniek.* [Inquiry-based learning and learning by design in science and technology]. The Hague, The Netherlands: Platform Bèta Techniek.

Vazquez-Alonso, A., Manassero-Mas, M. A., & Acevedo-Diaz, J. A. (2006). An analysis of complex multiple choice science technology items: Methodological development and preliminary results. *Science Education, 90*, 681–706.

Walma van der Molen, J. (2007). *Eindrapportage VTB attitude monitor. De ontwikkeling van een attitude-instrument op het gebied van wetenschap en techniek voor leerlingen in het basisonderwijs.*[Final report VTB attitude monitor. The development of an attitude instrument on the domain of science and technology for pupils in primary education.] The Hague, The Netherlands: Platform Bèta Techniek.

Walma van der Molen, J. (2009). Wat vinden leraren basisonderwijs van wetenschap en techniek? [What do primary teachers think of science and technology?]. In H. van Keulen & J. Walma van der Molen (Eds.), *Onderzoek naar wetenschap en techniek in het Nederlandse basisonderwijs* [Research into science and technology in Dutch primary education] (pp. 157–163). The Hague, The Netherlands: Platform Bèta Techniek.

Walma van der Molen, J., Lange, J. de, & Kok, J. (2009). Theoretische uitgangspunten bij de professionalisering van leraren basisonderwijs op het gebied van wetenschap en techniek. [Theoretical considerations for the professionalisation of primary school teachers in science and technology] In H. van Keulen & J. Walma van der Molen (Eds.), *Onderzoek naar wetenschap en techniek in het Nederlandse basisonderwijs* [Research into science and technology in Dutch primary education] (pp. 29-40). The Hague, The Netherlands: Platform Bèta Techniek.

Chapter 11

Hodson, D., (2009). *Teaching and learning about science. Languages, theories, methods, history, traditions and values.* Rotterdam/Taipei: Sense Publishers.

Jarvis, T. (2004). Primary teachers' changing attitudes and cognition during a two-year science in-service programme and their effect on pupils. *International Journal of Science Education, 26*, 1787–1811

Mitcham, C. (1994). *Thinking through technology. The path between philosophy and engineering.* Chicago: Chicago University Press.

Strauss, A. and Corbin, J. (1990). *Basics of qualitative research: Grounded Theory procedures and techniques.* London: Sage.

Vries, M.J. de (2005). *Teaching about technology. An introduction to the philosophy of technology for non-philosophers.* Dordrecht: Springer.

Walma van der Molen, J. (2009). Wat vinden leraren basisonderwijs van wetenschap en techniek? [What do primary teachers think of science and technology?]. In H. van Keulen & J. Walma van der Molen (Eds.), *Onderzoek naar wetenschap en techniek in het Nederlandse basisonderwijs* [Research into science and technology in Dutch primary education] (pp. 157–163). The Hague, The Netherlands: Platform Bèta Techniek.

Chapter 12

Baer, J. (1998). The Case for Domain Specificity of Creativity, *Creativity Research Journal*, 11(2), 173–177

Banks, F. (2009). Research on Teaching and Learning in Technology Education. In A. Jones & M.J. de Vries (Eds.) *International Handbook of Research and Development in Technology Education* (555–568). Rotterdam/Taipei: Sense Publishers.

Brown, A.L. & Kane, M.J. (1988). Pre-School Children can Learn to Transfer: Learning to Learn and Learning from Example. *Cognitive Psychology* 20, 493–523.

Brown, J.S., Collins, A. & Duguid, P. (1989). Situated Cognition and the Culture of Learning. *Educational Researcher* 18(1), 32–42.

Carter, L. (2007). Sociocultural Influences on Science Education: Innovation for Contemporary Times. *Science Education* 92(1), 165–181.

Cross, N. (1984). *Developments in Design Methodology.* Chichester: Wiley.

Jonassen, D.H. (2006). On the Role of Concepts in Learning and Instructional Design. *Educational Technology Research & Development* 54(2), 177–196.

Lave, J., Smith, S & Butler, M.:(1988), Problem Solving as an Everyday Practice. In J. Lave, J. G. Greeno, A. Schoenfeld, S. Smith & M. Butler (Eds.), *Learning Mathematical Problem Solving,* Palo Alto, CA.: Institute for Research on Learning report no. IRL88–0006.

Lee, O. (2005). Science Education With English Language Learners: Synthesis and Research Agenda. *Review of Educational Research* 75(4), 491–530.

Nardi, B.A. (1996), Studying Context: A Comparison of Activity Theory, Situated Cognition Models, and Distributed Cognition. In: Nardi, B.A. (Ed.), Context and Consciousness: Activity Theory and Human-Computer Interaction (69–102). Cambridge, MA: MIT Press.

Newell, A. & Simong, H.A. (1972). *Human Problem Solving.* Englewood-Cliffs: Prentice-Hall.

Novak, J.D. (2005). Results and Implications of a 12-Year Longitudinal Study of Science Concept Learning. *Research in Science Education* 35, 23–40.

Pilot, A. & Bulte, A.M.W. (2006). The Use of "Contexts" as a Challenge for the Chemistry Curriculum: Its successes and the need for further development and understanding. *International Journal of Science Education*, 28(9), 1087–1112.

Porlan, R. & del Pozo, R.M. (2004). The Conceptions of In-service and Prospective Primary School Teachers About the Teaching and Learning Science. *Journal of Science Teacher Education, 15*(1), 39–62.

Vries, M.J. de (2005). *Teaching About Technology. An Introduction to the Philosophy of Technology for Non-Philosophers.* Dordrecht, Netherlands: Springer.

Walma van der Molen, J., Lange, J. de & Kok, J. (2009). *Professionalisation in science & technology in The Netherlands: Primary education. Theoretical framework.* The Hague, The Netherlands: Platform Bèta Techniek.

Chapter 13

Appleton, K. (2003). How Do Beginning Primary School Teachers Cope with Science? Toward an Understanding of Science Teaching Practice. *Research in Science Education 33*(1), 1–25.

Bencze, J. (2001). 'Technoscience' Education: Empowering Citizens Against the Tyranny of School Science. *International Journal of Technology and Design Education 11*(3), 273–298.

Davis, R., S., Ginns, I., S. and McRobbie, C., J. (2002). Elementary School Students' Understandings of Technology Concepts. *Journal of Technology Education 14*(1), 35–50.

Hand, B. & Treagust, D. (1988). Application of a conceptual conflict teaching strategy to enhance student learning of acids and bases. *Research in Science Education 18*(1), 53–63.

Harlen, W. & Holroyd, C. (1997). Primary teachers' understanding of concepts of science: impact on confidence and teaching. *International Journal of Science Education 19*(1), 93–105.

Hestenes, D., Wells, M. & Swackhamer, G. (1992). Force concept inventory. *Physics Teacher 30*(3), 141–158.

Jarvis, T. & Pell, A. (2004). Primary teachers' changing attitudes and cognition during a two-year science in-service programme and their effect on pupils. *International Journal of Science Education 26*(14). 1787–1811.

Knowles, M. (1980). *The Modern Practice of Adult Education, from Pedagogy to Andragogy*. Englewood Cliffs: Prentice Hall..

Kruger, C., J., Summers, M., K. & Palacio, D., J. (1990). An Investigation of Some English Primary School Teachers' Understanding of the Concepts Force and Gravity. *British Educational Research Journal 16*(4), 383–397.

Lewis, T. (1999). Research in Technology Education--Some Areas of Need. *Journal of Technology Education 10*(2), 41–56.

Loverude, M., E., Kautz, C., H. & Heron P., R., L. (2003). Helping students develop an understanding of Archimedes' principle. I. Research on student understanding. *American Journal of Physics 71*(11), 1178–1187.

Posner, G., J., Strike, K., A., Hewson, P., W. & Gertzog, W., A. (1982). Accommodation of a Scientific Conception: Toward a Theory of Conceptual Change. *Science Education 66*(2), 211–227.

Rollnick, M. & Rutherford, M. (1990). African primary school teachers-what ideas do they hold on air and air pressure? *International Journal of Science Education 12*(1), 101–113.

She, H.,-C. (2002). Concepts of a higher hierarchical level require more dual situated learning events for conceptual change: a study of air pressure and buoyancy. *International Journal of Science Education 24*(9), 981–996.

Taylor, A., R., Jones, M., G., Broadwell, B. & Oppewal, T. (2008). Creativity, inquiry, or accountability? Scientists' and teachers' perceptions of science education. *Science Education 92*(6), 1058–1075.

Vosniadou, S. & Ioannides, C. (1998). From conceptual development to science education: a psychological point of view. *International Journal of Science Education* 20(10), 1213–1230.

Yin, Y., Tomita, M., K. & Shavelson R. (2008). Diagnosing and Dealing with Student Misconceptions: Floating and Sinking. *Science Scope* 31(8), 34–39.

Chapter 14

Ackermann, E. (1991). The Agency Model of Transactions: Toward an Understanding of Children's Theory of Control. In J. Montangero & A. Trypho (Eds.), *Psychologie génétique et sciences cognitives* (pp. 63–73). Geneva: Fondation Archives Jean Piaget.

Ackermann, E. (2000). Relating to things that think. *Play of Ideas and Ideas of Play, 13*, 2–4.

Alexander, R. J. (2010). *Towards dialogic teaching: Rethinking classroom talk*. York: Dialogos.

Anderson, V., & Johnson, L. (1997). *Systems thinking basics*. Waltham, MA: Pegasus.

Astolfo, D., Ferrari, M., & Ferrari, G. (2007). *Building robots with LEGO® Mindstorms® NXT: The ultimate tool for Mindstorms™ maniacs*. Burlington, MA: Syngress Publishing.

Barnes, D., & Todd, F. (1995). *Communication and learning revisited: making meaning through talk.* . Portsmouth, NH: Boynton/Cook Publishers Inc.

Bodner, G. M. (1986). Constructivism: a theory of knowledge. *Journal of Chemical Education, 63*(10), 873–878.

Carlsen, W. (1999). Domains of teacher knowledge. In J. Gess-Newsome & N. G. Lederman (Eds.), *Examining pedagogical content knowledge* (Vol. 6, pp. 133–144). Dordrecht: Kluwer Academic Publishers.

Cobb, P., Confrey, J., Disessa, A., Lehrer, R., & Schauble, L. (2003). Design experiments in educational research. *Educational researcher, 32*(1), 9.

Hacker, M., de Vries, M. J., & Rossouw, A. (2009). *CCETE Project, Concepts and Contexts in Engineering and Technology Education* (No. DUE 0314910). New York, Delft: Hofstra University & Delft University of Technology.

Hackling, M. W. (2007). Inquiry and investigation in primary science. In V. Dawson & G. Venville (Eds.), *The art of teaching primary science* (pp. 127–148). Crows Nest, NSW: Crows Nest: Allen & Unwin.

Kafai, Y. B., & Resnick, M. (1996). Introduction. In Y. B. Kafai & M. Resnick (Eds.), *Constructionism in practice. Designing, thinking, and learning in a digital world* (pp. 1–8). Mahwah, NJ: Lea, Publishers.

Kirkpatrick, D. (1996). Great ideas revisited. *Training and Development, 50*(1), 54–59.

Krumholtz, N. (1998). Techno-logic: A Micro-world for Constructivist Science and Technology learning [Electronic Version]. *ICCE98.* Retrieved 2006 6, May from http://www.phantom2.com/papr_tch.html.

Lepper, M. R., Drake, M. F., & O'Donnell-Johnson, T. (1997). Scaffolding Techniques of Expert Human Tutors. In K. Hogan & M. Pressley (Eds.), *Scaffolding Students Learning: Instructional Approaches and Issues* (pp. 108-144). Cambridge, MA: Brookline Books.

Levy, S. T., & Mioduser, D. (2008). Does it "want" or "was it programmed to..."? Kindergarten children's explanations of an autonomous robot's adaptive functioning. *International Journal of Technology and Design Education, 18*(4), 337–359.

Mahner, M., & Bunge, M. (2001). Function and functionalism: A synthetic perspective. *Philosophy of Science, 68*(1), 75–94.

Mioduser, D., Venezky, R. L., & Gong, B. (1996). Students' perceptions and designs of simple control systems. *Computers in Human Behavior, 12*(3), 363–388.

Nersessian, N. J. (2008). *Creating scientific concepts.* Cambridge, MA: The MIT Press.

Norman, D. A. (1993). *Things that make us smart: Defending human attributes in the age of the machine*. Reading, MA: Addison-Wesley.

Nourbakhsh, I., Hamner, E., Lauwers, T., Berstein, D., & Disalvo, C. (2006). *A roadmap for technological literacy and a vehicle for getting there: Educational robotics and the TeRK project.* Paper presented at the The 15th IEEE International Symposium on Robot and Human Interactive Communication (RO-MAN06), Hatfield, UK.

Papert, S. (1993). *The children's machine: Rethinking school in the age of the computer*. New York: Basic Books.

Park, S., & Oliver, J. S. (2008). Revisiting the conceptualisation of pedagogical content knowledge (PCK): PCK as a conceptual tool to understand teachers as professionals. *Research in Science Education, 38*(3), 261–284.

Petre, M., & Price, B. (2004). Using Robotics to Motivate "Back Door" Learning. *Education and Information Technologies, 9*(2), 147–158.

Prebble, T., Hargraves, H., Leach, L., Naidoo, K., Suddaby, G., & Zepke, N. (2004). *Impact of Student Support Services and Academic Development Programmes on Student Outcomes in Undergraduate Tertiary Study: A Synthesis of the Research.*: New Zealand: Ministry of Education.

Resnick, M., & Martin, F. (1991). Children and artificial life. In I. Harel & S. Papert (Eds.), *Constructionism* (Vol. 10, pp. 379–390). Norwood, NJ: Ablex Publishing.

Rohaan, E. J. (2009). *Testing teacher knowledge for technology teaching in primary schools.* Eindhoven: Technische Universiteit Eindhoven.

Shulman, L. S. (1987). Knowledge and teaching. *Harvard Educational Review, 57*(1), 1–22.

Silverman, D. (2006). *Interpreting qualitative data: Methods for analyzing talk, text, and interaction.* London: Sage Publications Ltd.

Slangen, L. A. M. P., Fanchamps, L. J. A., & Kommers, P. A. M. (2008). A case study about supporting the development of thinking by means of ICT and concretization tools. *International Journal of Continuing Engineering Education and Lifelong Learning, 18*(3), 18.

Slangen, L. A. M. P., van Keulen, H., & Gravemeijer, K. (2010). What pupils learn from working with robotic direct manipulation environments. *International Journal of Technology and Design Education* Retrieved 01-08, 2010, from http://www.springerlink.com/content/w7245131xl0088jg/

Smith, D. (1999). Changing our teaching: The role of pedagogical content knowledge in elementary science. In J. Gess-Newsome & N. G. Lederman (Eds.), *Examining pedagogical content knowledge* (Vol. 6, pp. 163–197). Dordrecht: Kluwer Academic Publishers.

Stes, A., Clement, M., & Van Petegem, P. (2007). The effectiveness of a faculty training programme: long-term and institutional impact. *International Journal for Academic Development, 12*(2), 99–110.

van Lith, P. (2006). Masterclass Robotica. Limbricht: Elektuur.

Vosniadou, S., Ioannides, C., Dimitrakopoulou, A., & Papademetriou, E. (2001). Designing learning environments to promote conceptual change in science. *Learning and instruction, 11*(4-5), 381–419.

Wells, G. (2002). Learning and teaching for understanding: The key role of collaborative knowledge building. In J. Brophy (Ed.), *Social constructivist teaching: Affordances and constraints* (Vol. 9, pp. 1–41). London: Elsevier/JAL.

Wisse, M. (2008). Robots, sensoren, algoritmes en motoren (Robots, sensors, algorithms and engines). In R. Dijkgraaf, L. Fresco, T. Gualthérie van Weezel & M. Calmthout van (Eds.), *De bètacanon, wat iedereen moet weten van de natuurwetenschappen (The science canon, what everyone should know about science)* (pp. 185-187). Amsterdam: De Volkskrant en Meulenhoff b.v.

Wyeth, P., Venz, M., & Wyeth, G. (2004). Scaffolding children's robot building and programming activities. *RoboCup 2003: Robot Soccer World Cup VII*, 308–319.

Xun, G. E., & Land, S. M. (2004). A conceptual framework for scaffolding Ill-structured problem-solving processes using question prompts and peer interactions. *Educational Technology Research and Development, 52*(2), 5–22.

Zeidler, D. L. (2002). Dancing with maggots and saints: Visions for subject matter knowledge, pedagogical knowledge, and pedagogical content knowledge in science teacher education reform. *Journal of Science Teacher Education, 13*(1), 27–42.

Chapter 15

Alexander, R. (2004). *Towards Dialogic Teaching: Rethinking Classroom Talk.* Cambridge: Dialogos

Anderson, L.W. & Krathwohl, D.R. (2001). *A taxonomy for learning, reaching, and assessing.* New York: Longman.

Bates, E. (1976). *Language and context: Studies in the acquisition of pragmatics.* New York: Academic Press.

Bloom, B.S., Englehart, M.B., Furst, E.J., Hill, W.H. & Krathwohl, D.R. (1956). *Taxonomy of Educational Objectives: The Classification of Educational Goals. Handbook1: The Cognitive Domain.* New York: Longman.

Boersma, K., Goedhart, M., De Jong, O. & Eijkelhof, H. (Eds.) (2005). *Research and the Quality of Science Education.* Dordrecht: Springer.

Cazden, C.B. (1988), *Classroom discourse. The language of teaching and learning,* Portsmouth, NH: Heinemann.

Damhuis, R. (2008), *Gesprekken zijn de spil van onderwijs. Taalbeleid in uitvoering*, Lectorale rede, *(Conversations, the pivot on which education hinges. Working on language policy. Inaugural address).* Utrecht: Marnix Academie.

Damhuis, R., & de Blauw, A. (2008). High quality interaction in the classroom. A focus for professional learning. *L1 Educational Studies in Language and Literature, 8*(4), p. 107–126.

Damhuis, R., de Blauw, A., & Brandenbarg, N., (2004). *CombiList, een instrument voor taalontwikkeling via interactie. Praktische vaardigheden voor leidsters en leerkrachten.* Nijmegen:

Expertisecentrum Nederlands. (*CombiList, a approach for language development via interaction. Practical skills for pre- and primary school teachers*)

Damhuis, R., de Blauw, A, Tammes, A., & Sytema, S., (2009), *How Dutch primary school teachers improve their classroom interaction strategies.* Paper presented at IAIMTE Conference, Toronto, 23–27 June 2009.

Damhuis, R., Tammes, A., de Blauw, A, Sytema, S., & Mallek, R. (2010). *Intensieve training van interactievaardigheden voor taal- en denkontwikkeling: opbrengsten in vraaggedrag bij PO-leerkrachten en Pabo-studenten (Intensive training of interaction strategies for language and thinking: results in question behaviour with primary school teachers and student teachers).* Paper presented at: ORD (Education Research Conference), Enschede: 23–25 June 2010.

De Blauw, A., & Damhuis, R. (2006a). *CombiList Team Training. Basismodel.* Nijmegen: Expertisecentrum Nederlands (unpublished). *(LIST Team Training. A model)*

De Blauw, A., & Damhuis, R. (2006b). CombiList Team Training voor leerkrachten. In: D.Ebbers (red.), *Retoriek en praktijk van het moedertaalonderwijs. Een selectie uit de 19e Conferentie Het Schoolvak Nederlands* (1–11). Gent: Academia Press. *(LIST Team Training. In: D.Ebbers (red.), Theory and practice of mother tongue education. A selection from the 19th Conference Dutch At School* (1–11).

De Blauw, A, Damhuis, R., Tammes, A, & Sytema, S. (2010). *Language- and thought-provoking interaction : how students teachers improve their strategies in multicultural settings.* Paper presented at SIG 10 and 21 Moving through cultures of learning, Utrecht, September 2-3.

Dickinson, D. K., & Tabors, P.O. (Eds) (2001), *Beginning Literacy with Language. Young Children Learning at Home and School.* Baltimore: Paul H. Brooks Publishing Co.

Driver, R. (1983). *The pupil as scientist?* Milton Keynes, UK: Open University Press. [cf. Wells 1999]

Flick, L. B. & Lederman, N. G. (2006). *Scientific Inquiry and Nature of Science. Implications for Teaching, Learning, and Teacher Education.* Dordrecht, The Netherlands: Springer.

Gass, S. M. & Mackey, A. (2006). Input, Interaction and Output: an Overview. In K. Bardovi-Harlig and Z. Dörmyei (Eds.), Themes in SLA Research. *AILA Review, 19,* 3–17.

Hajer, M. & Meestringa, T. (2009). *Handboek taalgericht vakonderwijs.* Bussum: Coutinho. (2nd edition)

Halliday, M.A.K. (1993). Towards a language-based theory of learning. *Linguistics and Education, 5,* 93–116.

Hammond, J. (1990). Is learning to read and write the same as learning to speak? In F. Christie (ED.), *Literacy for a changing world* (pp. 26–53). Hawthorn: Australian Council for Educational Research. [cf. Halliday 1993].

Harlen, W. & Allende, J. (Eds.) (2009). *Teacher Professional Development in Pre-Secondary School Inquiry-Bsed Science Education (IBSE),* InterAcademyPanel Report, http://www.interacademies.net

Hatch, E. (1978). Discourse analysis and second language acquisition. In E. Hatch (Ed.), *Second Language Acquisition* (pp. 401–435). Rowley, MA: Newbury House.

Krashen, S.D. (1980). The Input Hypothesis. In J.E. Alatis (Ed.), *Current Issues in Bilingual Education* (Gurt 1980) (pp. 168–180), Washington DC.

Kemmers, P., Klein Tank, M. and Van Graft, M. (2007). *Onderzoekend en ontwerpend leren bij natuur en techniek. Evalueren van brede ontwikkeling van leerlingen in open onderwijsvormen.* Den Haag: Verbreding Techniek Basisonderwijs. (Inquiry Learning in Science and Technology. Evaluating Broad Development of Pupils in Open Forms of Education)

Keulen, H. van & Walma van der Molen, J. (2009). (Eds.). *Onderzoek naar wetenschap en techniek in het Nederlands basisonderwijs [Research into science and technology in Dutch primary education].* The Hague, The Netherlands: Platform Bèta Techniek.

Kuijpers, J. & Walma van der Molen, J. (2007). *Wetenschap & Techniek: Een rijke leeromgeving [Science and technology: A rich learning environment.* The Hague, The Netherlands: Platform Beta TechniekLampert, M. (1990). When the Problem Is Not the Question and the Solution Is Not the Answer: Mathematical Knowing and Teaching. *American Educational Research Journal, 27* (1), 29-63.

Lemke , J. (2001). The long and the short of it: comments on multiple timescale studiers of human activity. *The Journal of the Learning Sciences* (10) (1 and 2): 17–26

Leontiev, A. A. (1981). *Psychology and the Language Learning Process.* Oxford: Pergamon,

Linden van der, J. & Renshaw, P. (2004). *Dialogic Learning. Sthifting Perspectives to Learning, Instruction and Teaching.* Dordrecht: Kluwer Academic Publishers.

Mercer, N. & Littleton, K. (2007). *Dialogue and the Development of Children's Thinking. A sociocultural approach.* London: Routledge.

Mercer, N. and Hodgkinson, S. (2008) (Eds) *Exploring Talk in School.* London: Sage.

Mesa, V. & Chang, P. (2010). The language of engagement in two highly interactive undergraduate mathematics classrooms. *Linguistics and Education, 21*, 83–100.

Metz, K. E. (2006). The knowledge building enterprises in science and elementary school science classrooms. In L.B. Flick and N.G. Lederman (eds.). *Scientific Inquiry and Nature of Science,* 105–130. Dordrecht, the Netherlands: Springer.

Nystrand, M. Wu, L, Gamorgan, A. Zeiser, S. & Long, D. (2003). Questions in time: investigating the structure and dynamics of unfolding classroom discourse. *Discourse Processes 35* (2): 135–198.

O'Connor, M. C., & Michaels, S. (1993). Aligning academic task and participation status through revoicing: analysis of a classroom discourse strategy. *Anthropology and Education Quarterly, 24(4),* 318–335.

O'Connor, M.C. & Michaels, S. (1996.) Shifting participant frameworks: Orchestrating thinking practices in group discussion. In D. Hicks (Ed.), *Discourse, learning, and schooling* (pp. 63–103). Cambridge UK: Cambridge University Press.

Renshaw, P.D. (2004). Dialogic teaching, learning and intstruction: Theoretial roots and analytical frameworks. In J. van der Linden & P. Renshaw (Eds.) (2004). *Diaologic Learning. Shifting Perspectives to Learning, Instruction, and Teaching (*1–15). Dordrecht: Kluwer Academic Publishers.

Sharpe, T. (2008). How can teacher talk support learning? *Linguistics and Education, 19,* 132–148.

Snow, C.E., & Kurland, B.F. (1996). Sticking tot the point: Talk about magnets as a context for engaging in scientific discourse. In D.Hicks (Ed.), *Discourse, learning, and schooling* (pp.189–221). New York: Cambridge University Press

Swain, M. (1985). Communicative competence: some roles of comprehensible input and comprehensible output in its development. In S.M. Gass & C.G. Madden (Eds.), *Input in second language acquisition* (pp. 235–253). Rowley, MA: Newbury House.

Swain, M. (1995). Three functions of output in second language learning. In G. Cook, & B. Seidlhofer (Eds.), *Principle & practice in applied linguistics: Studies in honor of H.G. Widdowson* (pp. 125–144). Oxford UK: Oxford University Press.

Swain, M. (2005). The Output Hypothesis: Theory and Research. In E. Hinkel (Ed.) (2005). *Handbook of Research in Second Language Teaching and Learning* (471–483). Mahwah, NJ: Lawrence Earlbaum Associates.

Schwartz, B., Dreyfus, T. & Hershkowitz, R. (Eds.) (2009). T*ransformation of Knowledge Through Classroom Interaction.* London: Routledge / Earli.

Van Graft, M. & Kemmers, P. (2007). *Onderzoekend en Ontwerpend Leren bij Natuur en Techniek. Basisdocument over de didactiek voor onderzoekend en ontwerpend leren in het primair onderwijs.* Enschede: SLO.

Vygotsky, L.S. (1978). *Mind in Society.* Cambridge, MA: Harvard University Press.

Wells, G. (1999). *Dialogic Inquiry. Towards a Sociocultural Practice and Theory of Education.* Cambridge: Cambridge University Press.

Wegerif, R., Boero, P., Andriessen, J. & Forman, E. (2009). A dialogue on dialogue and its place within education. In B. Schwarz, T. Dreyfus & R. Hershkowitz, *Transformation of Knowledge Through Classroom Interaction* pp. 184–199. London: Routledge.

Wells, G. (1999). *Dialogic Inquiry. Towards a Sociocultural Practice and Theory of Education.* Cambridge: Cambridge University Press.

Wertsch, J.V., & Toma, C. (1995). Discourse and learning in the classroom: A sociocultural approach. In L.P. Steffe & J. Gale (Eds.), *Constructivism in education* (pp.159–174). Hillsdale, NJ: Lawrence Erlbaum Associates.

Zion, M & Slezak, M. (2005). It takes two to tango: In dynamic inquiry, the self-directed student acts in association with the facilitating teacher. *Teaching and Teacher Education, 21,* 875–894.

Chapter 16

Aarts, R., Demir, S., & Henrichs, L. F. (2008). *Emergent academic language and literacy development of preschoolers and at-risk adolescents in a multicultural environment.* Paper presented at the American Educational Research Association (AERA). New York, March 26, 2008, New York, March 26, 2008.

Ansari, D. (2009). Effects of development and enculturation on number representation in the brain. *Nature Reviews Neuroscience, 9,* 278–291.

Ansari, D., Donlan, C., Thomas, M. S. C., Ewing, S. A., Peen, T., & Karmiloff-Smith, A. (2003). What makes counting count? Verbal and visuo-spatial contributions to typical and atypical number development. *Journal of Experimental Child Psychology, 85*(1), 50–62. doi:DOI: 10.1016/S0022-0965(03)00026-2

Beck, I. L., Mckeown, M. G., & Kucan, L. (2002). *Bringing words to life: Robust vocabulary instruction.* New York: Guilford.

French, L. (2004). Science as the center of a coherent, integrated early childhood curriculum. *Early Childhood Research Quarterly, 19,* 138–149.

Gelman, R., & Brenneman, K. (2004). Science learning pathways for young children. *Early Childhood Research Quarterly, 19*(1), 150-158. DOI: 10.1016/j.ecresq.2004.01.009

Greenfield, D. B., Jirout, J., Dominguez, X., Greenberg, A., Maier, M., & Fuccillo, J. (2009). Science in the preschool classroom: A programmatic research agenda to improve science readiness. *Early Education and Development, 20,* 238–264.

Henrichs, L. F. (2010). Academic language in early childhood interactions. A longitudinal study of 3- to 6-year-old dutch monolingual children. University of Amsterdam).

Klibanoff, R.S., Levine, S.C., & Huttenlocher, J. (2006). Preschool children's mathematical knowledge: The effect of teacher "math talk". *Developmental Psychology, 42*(1), 59–69.

Lena, P. (2006). Erasmus lecture 2005. from science to education: The need fora revolution. *European Review, 14*(1), 3–21.

MacWhinney, B. (2000). *The CHILDES project: Tools for analyzing talk. 3rd edition.* Mahwah, NJ: Lawrence Erlbaum Associates.

McCrink, K., & Wynn, K. (2004). Large-number addition and subtraction by 9-month-old infants. *Psychological Science, 15*(11), 776-781. doi:10.1111/j.0956-7976.2004.00755.x

Nelson, K., & et. al. (1986). *Event knowledge : Structure and function in development.* Hillsdale, New Jersey: Lawrence Erlbaum.

Peters, S., & DeVries, M. (juli 2010). *Hoe kinderen te interesseren voor wetenschap en techniek.*, juli 2010, from http://www.eco3.nl/eCache/DEF/1/07/945.html

Scheele, A.F., Leseman, P.P.M., Mayo, A.Y., & Elbers, E. (in press). Home language and literacy and three-year-olds' emergent academic language in narrative and instruction genres. *The Elementary School Journal.*

Schleppegrell, M. J. (2004). *The language of schooling: A functional linguistics perspective*. London: Lawrence Erlbaum Associates.

Siegler, R. S., & Ramani, G. B. (2008). Playing linear numerical board games promotes low-income children's numerical development. *Developmental Science, 11,* 655–661.

Snow, C. E., & Uccelli, P. (2009). The challenge of academic language. In D. R. Olson, & N. Torrance (Eds.), *The cambridge handbook of literacy* (pp. 112–133). New York: Cambridge University Press.

Spycher, P. (2009). Learning academic language through science in two linguistically diverse kindergarten classes. *The Elementary School Journal, 109*(4), 359–379. Retrieved from http://dx.doi.org/10.1086/593938

Tenenbaum, H. R., & Leaper, C. (2003). Parent–Child conversations about science: The socialization of gender inequities? *Developmental Psychology, 39*(1), 34–47.

Wong Fillmore, L., & Snow, C. E. (2002). What teachers need to now about language. In C. Temple Adger, C. E. Snow & D. Christian (Eds.), *What teachers need to know about language* (pp. 7–47). McHenry, IL: Delta Systems Co. and the Center of Applied Linguistics.

Yoon, J., & Ariri Onchwari, J. (2006). Teaching young children science: Three key points. *Early Childhood Education Journal, 33*(6), 419–423.

Zur, O., & Gelman, R. (2004). Young children can add and subtract by predicting and checking. *Early Childhood Research Quarterly, 19*(1), 121–137. DOI: 10.1016/j.ecresq.2004.01.003

Xu, F., & Spelke, E.S. (2000). Large number discrimination in 6-month-old infants. *Cognition, 74*, B1-B11.

Chapter 17

Bennett, J., Braund, M., & Lubben, F. (2010). *The impact of targeted continuing professional development (CPD) on teachers' professional practice in science. Main report.* Report commissioned by the DFCS through the National Science Learning Centre. York: University of York, Department of Educational Studies.

Dagevos, J.M., Gijsberts, M. & Praag, van, C. (2003). *Rapportage minderheden 2003. Een evaluatieonderzoek naar de arbeid en sociaal-culturele integratie.* [Report on minorities 2003. An evaluation study on labor and social-cultural integration].The Hague: the Netherlands Institute for Social Research(SCP).

Dede, C., Ketelhut, D. J., Whitehouse, P., McCloskey, E. M. (2009). A research agenda for online teacher professional development. November 26, *Journal of Teacher Education January/February 2009 vol. 60 no. 1 8-1,* doi: 10.1177/0022487108327554.

European Commission (2007). Science education now. A renewed pedagogy for the future of Europe. (The Rocard report). Brussel: EC. Retrieved December 2010 from http://ec.europa.eu/research/science-society/document_library/pdf_06/report-rocard-on-science-education_en.pdf.

Education Council of the Netherlands, (2008). *Een rijk programma voor ieder kind.* [An enriched programme for every child.] The Hague: the Educational Council of the Netherlands. Retrieved August 2010 from http://-www.onderwijsraad.nl/upload/publicaties/276/documenten/een_rijk_programma_voor_ieder_kind.pdf.

Fullan, M. (1993). The professional teacher. Why teachers must become change agents. *Educational Leadership, Volume 50, Number 6, March*, 12-17.

Fullan, M. (2010). Motion Leadership. *The skinny on becoming hange savvy*. Thousand Oaks: Corwin.

Hackling, M. Peers, S. & Prain, V., (2007). Primary Connections: reforming science teaching in *Australian primary schools. Teaching Sciences, 53, 3, September*, 12–17.

Harlen, W. & Holroyd, C. (1997). Primary teachers' understanding of concepts of science: impact on confidence and teaching. *International Journal of Science Education* 19(1), 93–105.

Harlen, W. & Qualter, A. (2009). *The teaching of science in primary schools.* Abingdon: Routledge.

Ingvarson, L., Meiers, M. & Beavis, A. (2005). Factors affecting the impact of professional development programmes on teachers' knowledge, practice, student outcomes & efficacy. *Education Policy Analysis Archives, 13*(10). Retrieved December 2010 from http://research.acer.-edu.au/cgi/viewcontent.cgi?article=1000&context=professional_dev.

Jaffe, R., Moir, E., Swanson, E., & Wheeler, G. (2006). Online Mentoring and Professional Development for New Science Teachers. In C. Dede (Ed.), *Online Teacher Professional Development: Emerging Models and Methods* (pp. 89–116). Cambridge: Harvard Education Publishing Group.

Jarvis, T. and Pell, A. (2004). Primary teachers' changing attitudes and cognition during a two-year science in-service programme and their effect on pupils. *International Journal of Science Education 26*(14)., 1787–1811.

Kat, M. P., Neut, van der, A.C. ,Liebregts, K.H.P., S. Hijkoop, S. Vijfeijken, & Van M.M. (2010). *Wetenschap en techniek op de pabo in 2010. Vervolg op het onderzoek 'Techniek op de pabo' uit 2005.* [Science and technology at teacher education college in 2010. Follow-up of the study: technology at the teacher college in 2005]. Tilburg: IVA.

Kinzer, C. K., Cammack, D. W., Labbo, L. D., Teale, W. H., & Sanny, R. (2006). The need to (re)conceptualize pre-service literacy teacher development: Technology's role and considerations of design, pedagogy and research. In McKenna, M. C., Labbo, L. D., Keiffer, R. E., & Reinking, D. (Eds), *International handbook of literacy and technology* (Vol. 2, pp. 211–233). Mahwah, NJ: Lawrence Erlbaum Associates, Inc.

Lena, P. (2006). Erasmus Lecture 2005 From science to education: the need for a revolution. *European Review*, 14, pp 3–21.

Lena, P. (2009). Europe rethinks science education, *Science*, 324, 501.

Lewis, T. (1999). Research in Technology Education--Some Areas of Need. *Journal of Technology Education* 10(2), 41–56.

Loverude, M., E., Kautz, C., H. and Heron P. R. L. (2003). Helping students develop an understanding of Archimedes' principle. I. Research on student understanding. *American Journal of Physics 71*(11), 1178–1187.

McKinsey & Company (2007). *How the world's-best performing school systems come out on top.* McKinsey & Company. Retrieved August 2010 from http://www.mckinsey.com/ App_Media/ Reports/SSO/Worlds_School_Systems_Final.pdf.

McKinsey & Company, (2010). *Closing the talent gap: attracting and retaining top-third graduates to careers in teaching. An international and market-research-based perspective.* Retrieved December 2010 from http://www.mckinsey.com/clientservice/ Social_Sector/our_practices/Education/ Knowledge_Highlights/Closing_the_talent_gap.aspx

Murphy, C., Neil, P., Beggs, J. (2007). Primary science teacher confidence revisited: ten years on. *Educational Research, 1469–5847, Volume 49, Issue 4, 2007*, Pages 415–430.

Murphy, C. & Scantlebury, K. (2010). Co teaching in international contexts: research and practice. *Series: Cultural studies of science education.* Springer Science + Business Media.

O'Dwyer, L.M., Masters, J., Dash, S., De Kramer, R.M., Humez, A & Russell, M. (2010). *E-learning for educators. Effects of on-line professional development on teachers and their students: findings from four randomized trials.* Technology and Assessment Study Collaborative Lynch School of Education Boston College Chestnut Hill, MA 02467. Retrieved December 2010 from http://www.bc.edu/research/intasc/PDF/EFE_Findings2010_Report.pdf.

OECD, (2005). *Attracting, developing and retaining effective teachers* - Final Report: Teachers Matter. Parijs: OECD. http://www.oecd.org/dataoecd/17/51/43023606.pdf.

OECD, (2009). *Creating effective teaching and learning environments.* First results from Talis. Parijs: OECD.

Osborne, J. F., & Dillon, J. (2008). *Science education in europe: critical reflections.* London: Nuffield Foundation.

Palmer, D. (2004). Situational interest and the attitudes towards science of primary teacher education students. *International Journal of Science Education, 26*, 895–908.

Supovitz J.A., Turner, H. M. (2000). The effects of professional development on science teaching practices and classroom culture. *Journal of Research in Science* Teaching 37(9):963–80.

Verhoeven, L. & Kinzer, Ch. (Eds), (2008). *Interactive literacy education: facilitating literacy environments through technology.* New York: Lawrence Erlbaum Associates, Routledge/Taylor Francis.

Walma van der Molen, J., De Lange, J. & Kok, J. (2008). *Theoretical framework for the professionalisation of teachers in science and technology at primary school.* Den Haag: Platform Bèta Techniek.

Chapter 18

Akerson, V.L. & Hanuscin, D.L. (2007). Teaching nature of science through inquiry: results of a 3-year professional development program. *Journal of research in science teaching, 44*, (5), 653–680.

Akerson, V.L., Morrison, J. A., & Roth McDuffie, A. (2006). One course is not enough: Preservice elementary teachers' retention of improved views of nature of science. *Journal of Research in Science Teaching, 43,* 194–213.

Brown, A., Bransford, J., & Cocking, R. (1999). *How People Learn: Brain, Mind. Experience and School.* Washington DC: National Academy Press.

Council for Education, Youth matters and Culture (2007). *Brief van de minister van Onderwijs, Cultuur en Wetenschap.* [Letter from the minister of Education, Culture and Science.] Retrieved August 20, 2010, from http://rijksbegroting.minfin.nl/2008/kamerstukken,2007/12/12/kst113368.html

Deltaplan Bèta/Techniek (2003a, December). *Actieplan voor de aanpak van tekorten aan bèta's en technici.* [Plan of action to diminish shortages in science and technology workers]. Den Haag: DeltaHage. Retrieved November 24, 2009 from http://www.platformbetatechniek.nl//docs/-Beleidsdocumenten/DeltaplanBetatechniek2003.pdf

Ely, D. P. (1999). Conditions that facilitate the implementation of educational technology innovations. *Educational Technology, 39*, 23–27.

ESoE, 2009. *Professional learning: Research Program of the Eindhoven School of Education (ESoE).* Eindhoven: Eindhoven School of Education. Retrieved August 25, 2010, from http://www.esoe.nl/fileadmin/Esoe/Algemeen/ESoE_research_program_2009.pdf

Fullan, M. (2006). Leading professional learning. *The School administrator,* pp 10-14. Retrieved August 25, 2010, from http://www.michaelfullan.ca/Articles_06/Articles_06b.pdf

Fullan, M. (2007). *The new meaning of educational change.* New York and London: Teacher College Press.

Glaser, B.G. & Strauss, A.L. (1967). *The Discovery of Grounded Theory; Strategies for Qualitative Research.* Chicago: Aldine Publishing Company.

Hackling, M. W. & Prain, V. (2005). *Primary Connections: Stage 2 research report.* Canberra: Australian Academy of Science.

Hagunama, E. (2008). *Papua New Guinea primary school technology teachers: The impacts of support materials on their perceptions and practices.* Unpublished master's thesis, University of Waikato. Retrieved August 20, 2010 from http://waikato.researchgateway.ac.nz/bitstream/10289/2274/1/-thesis.pdf.

Inspection of Education (2005-3). *Techniek in het basisonderwijs.* [Technology in primary education]. Schiedam: Romer.

Kuijpers, J. & Walma van der Molen, J. (2008). *Wetenschap en Techniek: een ontdekkingsreis naar kennis.* [Science and Technology: a journey of discovery to knowledge]. Den Haag: Brochure Platform Bèta Techniek.

Lee, O., Hart, J.E., Cuevas, P., & Enders, C. (2004). Professional development in inquiry-based science for elementary teachers of diverse student groups. *Journal of Research in Science Teaching, 41*(10), 1021–1043.

Meijer, P. C., Verloop, N., & Beijaard, D. (2002). Multi-Method Triangulation in a Qualitative Study on Teachers' Practical Knowledge: An Attempt to Increase Internal Validity. *Quality & Quantity, 36*, 145–167.

Miles, M.B. & Huberman, A.M. (1994). *Qualitative Data Analysis (2nd edition).* Thousand Oaks, CA: Sage Publications.

Prebble, T., Hargreaves, H., Leach, L., Naidoo, K., Suddaby, G., & Zepke, N. (2004). *Impact of student support services and academic development programmes on student outcomes in undergraduate tertiary study: A synthesis of the research.* Wellington: Ministry of Education.

Rohaan, E. J., Taconis, R., & Jochems, W. M. G. (2008). Reviewing the relations between teachers' knowledge and pupils' attitude in the field of primary technology education. *International Journal of Technology and Design Education, 18*(3).

Stein, S. J., Ginns, I. S., & McDonald, C. V. (2007). Teachers learning about technology and technology education: Insights from a professional development experience. *International Journal of Technology and Design Education, 17*(2), 179-195.

Walma van der Molen, J.H., De Lange, J., & Kok, J. (2009). Theoretische uitgangspunten bij de professionalisering van leraren basisonderwijs op het gebied van wetenschap en techniek [Theoretical starting points in the professionalization of elementary school teachers in the field of science and technology] (pp. 29–39). In H. van Keulen & J.H. Walma van der Molen (Eds.), *Onderzoek naar wetenschap en techniek in het Nederlandse basisonderwijs.* Den Haag: Platform Bèta Techniek.

Chapter 19

Borko, H. (2004). Professional development and teacher learning: mapping the terrain. *Educational Researcher, 33*(8), 3–15.

Briscoe, C., & Peters, J. (1997). Teacher collaboration across and within schools: Supporting individual change in elementary science teaching. *Science education, 81,* 51–65.

Grossman, P. L. (1990). *The making of a teacher: Teacher knowledge and teacher education.* New York: Teachers College Press.

Horn, I. S. (in press). Teaching replays, teaching rehearsals, and re-visions of practice: Learning from colleagues in a mathematics teacher community. *Teachers College Record.*

Levine, T. H., & Marcus, A. S. (2010). How the structure and focus of teachers' collaborative activities facilitate and constrain teacher learning. *Teaching and Teacher Education, 26,* 389–398.

Little, J. W. (1990). The persistence of privacy: Autonomy and initiative in teachers' professional relations. *Teachers College Record, 91*(4), 509–536.

Little, J. W. (2002). Locating learning in teachers' communities of practice: opening up problems of analysis in records of everyday work. *Teaching and Teacher Education, 18,* 917–946.

Little, J. W. (2003). Inside teacher community: Representations of classroom practice. *Teachers College Record, 105*(6), 913–945.

Little, J. W. (2006). *Professional community and professional development in the learning-centered school.* Prepared for the National Education Association.

Magnusson, S., Krajcik, J. S., & Borko, H. (1999). Nature, sources, and development of pedagogical content knowledge for science teaching. In J. Gess-Newsome & N. G. Lederman (Eds.), *Examining pedagogical content knowledge* (pp. 95–132). Dordrecht: Kluwer Academic Publishers.

Meirink, J. A., Imants, J., Meijer, P. C., & Verloop, N. (2010). Teacher learning and collaboration in innovative teams. *Cambridge Journal of Education, 40* (2), 161–181.

Pierce, D., & Hunsaker, T. W. (1996). Professional development for the teacher, of the teacher and by the teacher. *Education, 117,* 101–106.

Platteel, T. L. (2009). Knowledge development of secondary school L1 teachers on concept-context rich education in an action-research setting. *Doctoral dissertation*, University of Leiden, Leiden, The Netherlands.

Schwarz McCotter, S. (2001). Collaborative groups as professional development. *Teaching and Teacher Education, 17,* 685–704.

Shulman, L. S. (1986). Those who understand: Knowledge growth in teaching. Educational Researcher, *15*(2), 4–14.

Smith, C. & Gillespie, M. (2007). Research on professional development and teacher change: Implications for Adult Basic Education. *Review of Adult Learning and Literacy, 7,* 205–244.

Timperley, H., Wilson, A., Barrar, H., & Fung, I. (2007). *Teacher professional learning and development. Best evidence synthesis iteration (BES).* Wellington: Ministry of Education.

Van Driel, J. H. & Verloop, N. (1998). 'Pedagogical content knowledge': een verbindend element in de kennisbasis van docenten. [Pedagogical content knowledge: a binding element in the knowledge base of teachers]. *Pedagogische Studiën, 75,* 225–237.

Verloop, N., & Kessels, J. M. W. (2006). Opleidingskunde: ontwikkelingen rond het opleiden en leren van professionals in onderwijs en bedrijfsleven. [Skills training: Developments in the training and learning of professionals in education and business]. *Pedagogische Studiën, 83,* 301–321.

Vescio, V., Ross, D., & Adams, A. (2008). A review of research on the impact of professional learning communities on teaching practices and student learning. *Teaching and Teacher Education, 24* (1), 80–91.

VTB-Pro (2008), *Theoretisch kader voor de Professionalisering van Leerkrachten op het gebied van wetenschap en techniek.* [Framework for the professionalization of teachers on science and technology]. The Hague: Platform Bèta Technique.

Alexander, R. (2004). *Towards dialogic teaching: rethinking classroom talk.* Cambridge: Dialogos.

Ball, D.L. (1996). Teaching learning and the mathematics reforms: What do we think we know and what do we need to learn? *Phi Delta Kappan* 77, 500–508.

Barber, J., Catz, K.N., & Arya, D. (2006). *Improving Science Content Acquisition through a Combined Science/Literacy Approach: A Quasi-Experimental Study.* Paper presented at the American Educational Research Association, Lawrence Hall of Science, University of California Berkeley.

Bleicher, R.E. (2007). Nurturing confidence in preservice elementary science teachers. *Journal of Science Teacher Education, 18*, 841–860.

CTGV (The Cognition and Technology Group at Vanderbilt) (1997). *The Jasper Project: lessons in curriculum, instruction, assessment, and professional development.* Mahwah, NJ: Lawrence Erlbaum Associates.

Denessen, E., Vos, N., Damen, T., Koch, S., Louws, M., & Wigboldus, D. (2011). Explicit and implicit measures of teacher attitudes towards STEM. In M.J. de Vries, H. van Keulen, S. Peters & Juliette W. van der Molen (Eds.). *Professional development for primary teachers in science and technology The Dutch VTB-Pro project in an international perspective.* Rotterdam/Taipei: Sense Publishers.

De Vaan, E., & Marell, J. (2006). Praktische didactiek voor natuuronderwijs [Practical didactics for teaching science]. Bussum: Coutinho.

Elsäcker, W., Droop, M., Damhuis, R., & Segers, E. (2010). Zaakvakken en taal: twee vliegen in één klap. [Case subjects and languages: two birds with one stone]. Nijmegen: National Center for Language Education.

Gelman, R., & Brenneman, K. (2004). Science learning pathways for young children. *Early Childhood Research Quarterly, 19*, 150–158.

Graft, M. van, & Kemmers, P. (2007). Onderzoekend en ontwerpend leren: Onderzoekend en ontwerpende leren bij natuurlijk en techniek: basisdocument over de didactiek voor onderzoekend en ontwerpend leren in het primair onderwijs. [Inquiry-based learning and learning by design: basic document in learning in science and technology in primary education]. The Hague: Platform Bèta Technique.

Jarvis, J. (2004). Primary teachers' changing attitudes and cognition during a two-year science in service programme and their effect on pupils. *International Journal of Science Education, 26,* 1787–1811.

Jenkins, E. W. (1990). Scientific literacy and school science education. *School Science Review, 71*, 43–51.

Kinzer, C. K., Cammack, D. W., Labbo, L. D., Teale, W. H., & Sanny, R. (2006). Using technology to conceptualize preservice literacy teacher education. In M. McKenna, L.D. Labbo, R.D. Kieffer & D. Reinking (Eds.). *International Handbook of Literacy and Technology Volume II* (pp. 211–233). Mahwah, NJ: Lawrence Erlbaum.

Lampert, M. (1988). What can research on teacher education tell us about improving quality in mathematics education? *Teaching and Teacher Education, 4(2),* 157–170.

NRP (National Reading Panel) *Report* (2000). http://www.nichd.nih.gov/publications/nrp/report.cfm

Pell, A., & Jarvis, T. (2003). Developing attitude to science education scales for use with primary teachers. *International Journal of Science Education, 25*, 1273–1295.

Schroeder, C. M., Scott, T. P., Tolson, H., Huang, T., & Lee, Y. (2007). A Meta-Analysis of National Research: Effects of Teaching Strategies on Student Achievement in Science in the United States. *Journal of Research in Science Teaching, 44,* 1436–1460.

Snow, C. E. (2010). Academic Language and the Challenge of Reading for Learning about Science. *Science, 328*, 450–452.

Van Goethem, A. A. J., Scholte, R. H. J. , &Wiers, R. W. (2010). Explicit- and Implicit Bullying Attitudes in Relation to Bullying Behavior. *Journal of Abnormal Child Psychology, 38*, 829–842.

Van Keulen, H. (2009). Drijven en zinken. Wetenschap en techniek in het primair onderwijs. [Floating and sinking. Science and technology in primary education]. Oration: Fontys PABO Limburg.

Wilber, Kinzer & Lohnes (2008). Supporting preservice teacher education through a multimedia case-based online learning environment. In C. Kinzer & L. Verhoeven (Eds.), *Interactive literacy education* (pp. 285–308). Mahwah (CA): Lawrence Erlbaum.

Wilkins, J. L. M. (2008). The relationship among elementary teachers' content knowledge, attitudes, beliefs and practices. *Journal of Mathematics Teacher Education, 11*, 139-164.

Chapter 20

Alexander, R. (2004). *Towards dialogic teaching: rethinking classroom talk.* Cambridge: Dialogos.

Ball, D.L. (1996). Teaching learning and the mathematics reforms: What do we think we know and what do we need to learn? *Phi Delta Kappan* 77, 500–508.

REFERENCES

Barber, J., Catz, K.N., & Arya, D. (2006). *Improving Science Content Acquisition through a Combined Science/Literacy Approach: A Quasi-Experimental Study.* Paper presented at the American Educational Research Association, Lawrence Hall of Science, University of California Berkeley.

Bleicher, R.E. (2007). Nurturing confidence in preservice elementary science teachers. *Journal of Science Teacher Education, 18*, 841–860.

CTGV (The Cognition and Technology Group at Vanderbilt) (1997). *The Jasper Project: lessons in curriculum, instruction, assessment, and professional development.* Mahwah, NJ: Lawrence Erlbaum Associates.

Denessen, E., Vos, N., Damen, T., Koch, S., Louws, M., & Wigboldus, D. (2011). Explicit and implicit measures of teacher attitudes towards STEM. In M.J. de Vries, H. van Keulen, S. Peters & Juliette W. van der Molen (Eds.). *Professional development for primary teachers in science and technology The Dutch VTB-Pro project in an international perspective.* Rotterdam/Taipei: Sense Publishers.

De Vaan, E., & Marell, J. (2006). *Praktische didactiek voor natuuronderwijs.* [Practical didactics for teaching science]. Bussum: Coutinho.

Elsäcker, W., Droop, M., Damhuis, R., & Segers, E. (2010). Zaakvakken en taal: twee vliegen in één klap. [Case subjects and languages: two birds with one stone]. Nijmegen: National Center for Language Education.

Gelman, R., & Brenneman, K. (2004). Science learning pathways for young children. *Early Childhood Research Quarterly, 19*, 150–158.

Graft, M. van, & Kemmers, P. (2007). Onderzoekend en ontwerpend leren: Onderzoekend en ontwerpende leren bij natuurlijk en techniek: basisdocument over de didactiek voor onderzoekend en ontwerpend leren in het primair onderwijs. [Inquiry-based learning and learning by design: basic document in learning in science and technology in primary education]. The Hague: Platform Bèta Technique.

Jarvis, J. (2004). Primary teachers' changing attitudes and cognition during a two-year science in service programme and their effect on pupils. *International Journal of Science Education, 26*, 1787–1811.

Jenkins, E. W. (1990). Scientific literacy and school science education. *School Science Review, 71*, 43–51.

Kinzer, C. K., Cammack, D. W., Labbo, L. D., Teale, W. H., & Sanny, R. (2006). Using technology to conceptualize preservice literacy teacher education. In M. McKenna, L.D. Labbo, R.D. Kieffer & D. Reinking (Eds.). *International Handbook of Literacy and Technology Volume II* (pp. 211–233). Mahwah, NJ: Lawrence Erlbaum.

Lampert, M. (1988). What can research on teacher education tell us about improving quality in mathematics education? *Teaching and Teacher Education, 4 (2)*, 157–170.

NRP (National Reading Panel) *Report* (2000). http://www.nichd.nih.gov/publications/nrp/report.cfm

Pell, A., & Jarvis, T. (2003). Developing attitude to science education scales for use with primary teachers. *International Journal of Science Education, 25*, 1273–1295.

Schroeder, C. M., Scott, T. P., Tolson, H., Huang, T., & Lee, Y. (2007). A Meta-Analysis of National Research: Effects of Teaching Strategies on Student Achievement in Science in the United States. *Journal of Research in Science Teaching, 44*, 1436–1460.

Snow, C. E. (2010). Academic Language and the Challenge of Reading for Learning about Science. *Science, 328*, 450–452.

Van Goethem, A. A. J., Scholte, R. H. J., & Wiers, R. W. (2010). Explicit- and Implicit Bullying Attitudes in Relation to Bullying Behavior. *Journal of Abnormal Child Psychology, 38*, 829–842.

Van Keulen, H. (2009). Drijven en zinken. Wetenschap en techniek in het primair onderwijs. [Floating and sinking. Science and technology in primary education]. Oration: Fontys PABO Limburg.

Wilber, Kinzer & Lohnes (2008). Supporting preservice teacher education through a multimedia case-based online learning environment. In C. Kinzer & L. Verhoeven (Eds.). *Interactive literacy education* (pp. 285–308). Mahwah (CA): Lawrence Erlbaum.

Wilkins, J. L. M. (2008). The relationship among elementary teachers' content knowledge, attitudes, beliefs and practices. *Journal of Mathematics Teacher Education, 11*, 139–164.

INDEX

www.ingramcontent.com/pod-product-compliance
Lightning Source LLC
Chambersburg PA
CBHW072243240926
51846CB00012B/149
* 9 7 8 9 4 6 0 9 1 7 1 0 3 *